D0921046

Water Wells

Implementation, Maintenance and Restoration

Wiley Series in Water Resources Engineering

The 1992 *International Conference on Water and the Environment: Issues for the 21st Century* served as a timely reminder that fresh water is a limited resource which has an economic value. Its conservation and more effective use becomes a prerequisite for sustainable human development.

With this in mind, the aim of this series is to provide technologists engaged in water resources development with modern texts on various key aspects of this very broad discipline.

Professor J.R. Rydzewski

Irrigation Engineering
Civil Engineering Department
University of Southampton
Highfield
SOUTHAMPTON
S09 5NH
UK

Design of Diversion Weirs: Small Scale Irrigation in Hot Climates
Rozgar Baban

Unit Treatment Processes in Water and Wastewater Engineering
T.J. Casey

Water Wells: Implementation, Maintenance and Restoration
M. Detay

Dam Hydraulics
D.L. Vischer, W.H. Hager

Water Wells

Implementation, Maintenance and Restoration

Michel DETAY

Doctor of Sciences,
Director of Applied Research Program on Water Resource Management
at Lyonnaise des Eaux (France)

Translated by :

M.N.S. CARPENTER

Doctor of Geology,
Member of the Société française des traducteurs

JOHN WILEY & SONS
Chichester ▪ New York ▪ Weinheim ▪ Brisbane ▪ Toronto ▪ Singapore

1997

MASSON
Paris ▪ Milan ▪ Barcelona

Original French language edition: *Le forage d'eau. Réalisation, entretien, réhabilitation*
Copyright © 1993, Masson, Paris.

First English language edition, 1997
Copyright © 1997 by Masson

Wiley Editorial Offices

Baffins Lane, Chichester,
West Sussex PO19 1UD, England

National	01243 779777
International	(+44) 1243 779777

e-mail (for orders and customer service enquiries):
cs-books@wiley.co.uk

Visit our Home Page	on	http://www.wiley.co.uk
	or	http://www.wiley.com

John Wiley & Sons, Inc., 605 Third Avenue,
New York, NY 10158-0012, USA

VCH Verlagsgesellschaft mbH, Pappelallee 3,
D-69469 Weinheim, Germany

Jacaranda Wiley Ltd, 33 Park Road, Milton,
Queensland 4064, Australia

John Wiley & Sons (Canada) Ltd, 22 Worcester Road,
Rexdale, Ontario M9W 1L1, Canada

John Wiley & Sons (Asia) Pte Ltd, 2 Clementi Loop #02-01,
Jin Xing Distripark, Singapore 0512

British Library Cataloguing in Publication Data

A catalogue record for this book is available from the British Library

ISBN 0 471 96695 9 (Wiley)
ISBN 2 225 85622 2 (Masson)

Printed in Belgium
Bound in France

Contents

Preface XI

Chapter I BASIC CONCEPTS OF HYDROGEOLOGY 1

1.1 The water cycle 2
1.2 Hydrological systems 4
    1.2.1 Identification and recharge of groundwater bodies 4
    1.2.2 The water balance 5
        a) Balance for a drainage basin 5
        b) Balance for a groundwater basin 6
        c) Balance for an aquifer 6
1.3 Aquifer characteristics 7
    1.3.1 Unconfined aquifers 9
    1.3.2 Confined aquifers 10
    1.3.3 Semi-confined or leaky aquifers 11
1.4 Reservoir and aquifer characteristics 12
    1.4.1 Groundwater flow 13
    1.4.2 Darcy's Law 14
    1.4.3 Permeability 16
    1.4.4 Porosity 17
    1.4.5 Hydrodynamic parameters 18
        a) Transmissivity 18
        b) Storage coefficient 19
        c) Diffusivity 20
1.5 Aquifer exploitation criteria 20
    1.5.1. Evaluation of water reserves 21
    1.5.2 Evaluation of water resources 23
    1.5.3 Extraction strategies 24
1.6 Prospection methods 24
    1.6.1 Preliminary studies 25
    1.6.2 Field studies 26
    1.6.3 Photo-interpretation 26
        a) Aerial photography 26
        b) Remote sensing 27
    1.6.4 Electromagnetic methods 27
    1.6.5 Seismic methods 28
        a) Principles of seismic refraction 29
        b) Seismic reflection 33
    1.6.6 Electric survey methods 33
        a) Potential difference method 34
        b) Resistivity method 34
        c) Electric logging 36
        d) Electric panel arrays 39
    1.6.7. Gravimetry 41
1.7 Water chemistry 42
    1.7.1 Dissolution 42
        a) Salts 43
        b) Gases 43
    1.7.2 Chemical attack 43
    1.7.3 Physico-chemical properties of waters 44

a) Water colour 44
b) Turbidity 45
c) pH 45
d) Conductivity 45
e) Hardness 46
f) Alkalinity 46
g) Chemical substances present in water 47
1.7.4 Isotopic characteristics 49
1.7.5 Qualitative requirements 49
1.8 Conclusion 50

Chapter II WELL DESIGN AND CONSTRUCTION 53

2.1 Drilling techniques 53
2.1.1 Percussion drilling 54
2.1.2 Drilling by cutting 55
2.1.3 Rotary drilling 55
2.1.4 Down-the-hole (DTH) hammer drilling 58
a) Conventional down-the-hole hammer drill 58
b) The Odex system for soft unstable formations 58
2.2 Drilling fluids 59
2.2.1 Properties of drilling fluids 60
a) Density 60
b) Viscosity 63
c) Cake and mud filtrate 63
d) pH 63
e) Sand content 63
f) Thixotropy 65
g) Conditioning the mud 65
2.2.2 Bentonitic muds 65
a) Bentonite 65
b) Special muds for swelling marls 66
c) Muds with emulsified oil 66
2.2.3 Polymer muds 67
a) Natural polymers 67
b) Synthetic polymers 68
c) Biodegradable polymer mud 68
2.2.4 Compressed air 70
a) Rotary drilling 70
b) Drilling with a DTH percussion tool 70
2.2.5 Stabilized foam 70
a) Foam for rotary drilling 71
b) Foam for air drilling 71
2.3 Water well equipment 71
2.3.1 Casings 73
2.3.2 Screens 74
a) Open area and screen slot size 78
b) Positioning of the screens 78
2.3.3 Gravel pack 81
2.4 Cementation 83
2.4.1 Cementation through the outside of the casing 84
2.4.2 Cementation through the inside of the casing 85
a) Float shoe method 85
b) Free plug method 86
c) Cementation with a guided plug 86
d) Partial cementation 87
2.4.3 Well head 87
2.5 Development of water wells in alluvial formations 87
2.5.1 Cleaning 88
2.5.2 Development 88

a) Methods of development...88
b) Principle of development with air-lift pumping..95
c) Monitoring well development 99
2.6 Conclusion 100

Chapter III WELL HYDRAULICS 103

3.1 Basic concepts 103
3.1.1 Flow superposition 103
3.1.2 Influence of starting up water extraction..105
3.2 Metrology 106
3.2.1 Choice and layout of observation wells (piezometers)......................106
3.2.2 Measurement methods 107
a) Measurement of flow rates 107
b) Measurement of water-level 109
3.2.3 Implementation of a pumping test 111
3.3 Pumping tests in a steady-state regime 111
3.3.1 Methodology..111
3.3.2 Interpretation 113
3.3.3 Concept of critical velocity..114
3.3.4 Comparison of several tests on the same catchment structure...........115
3.3.5 Calculation of head losses 116
3.3.6 Determination of the maximum operational yield capacity 118
3.4 Pumping tests in a non-steady state regime 118
3.5 Interpretation of pumping tests in a confined aquifer 120
3.5.1 Theis's bi-logarithmic method..121
3.5.2 Jacob's semi-logarithmic method...126
a) Determination of T 128
b) Determination of S 128
3.5.3 The critical radius 130
3.5.4 Analysis of the Theis equation after cessation of pumping 131
3.5.5 Anomalies encountered with the Theis graphical method..................133
3.5.6 The skin effect and post-production..134
3.6 Interpretation of pumping tests in semi-confined (leaky) aquifers 140
3.6.1 Semi-confined aquifers..141
a) Leakage coefficient (or leakance)..142
b) Leakage factor...142
3.6.2. The Hantush-Walton bilogarithmic method 142
3.6.3 The Hantush-Berkaloff semi-logarithmic method 143
3.6.4 Jacob's bilogarithmic method using depression curves 146
3.7 Interpretation of pumping tests in unconfined aquifers 147
3.7.1 General rules 147
3.7.2 Unconfined aquifer with strong drawdown 148
3.7.3 Boulton's bilogarithmic method 149
3.7.4 Berkaloff-Boulton semi-logarithmic method 153
3.8 Conclusion 155
Notation used 156

Chapter IV SUPERVISION AND FINAL ACCEPTANCE TESTS.....................159

4.1 Preliminary checks before drilling 159
4.1.1 The environment 160
4.1.2 Equipment 161
4.1.3 The well site 161
4.2 The "day-to-day checks" 162
4.2.1 Sample collection 163
4.2.2 Depth drilled..163
4.2.3 Water inflow..163

4.2.4 Conformity of the equipment 164
4.2.5 Particular incidents 164
4.3 Specific checks 165
   4.3.1 Determining the nature of the formations 165
     a) Drilling rate or rate of penetration 166
     b) Pressure on the tool 166
     c) Reflected impact 166
     d) Injection fluid pressure 167
     e) Data processing 167
     f) Torque 168
     g) Other parameters 168
   4.3.2 Measurement of permeability 168
   4.3.3 Downhole logging 170
     a) Electric log 171
     b) Acoustic log 173
     c) Nuclear methods 173
     d) Borehole diameter measurement 173
     e) Thermometric measurements 174
     f) Micro-current meter 174
     g) Interpretation 175
   4.3.4 Test for water-bearing layers 176
     a) Method 176
     b) Relation between drawdown and discharge 176
   4.3.5 Equipment of the well 177
     a) Installation of lining equipment 178
     b) Emplacement of additional gravel 180
     c) Cementation 181
   4.3.6 Well development 182
   4.3.7 Pumping tests 182
   4.3.8 Well data sheet 183
4.4 Acceptance of the water well 185
   4.4.1 Acceptance pumping tests 186
   4.4.2 Video camera inspection 187
   4.4.3 Well logging tests 187
4.5 Conclusion 187

Chapter V WATER WELL PROTECTION 189

5.1 Catchment protection zones 190
   5.1.1 Groundwater regimes 190
   5.1.2 Effects of pumping on aquifers 191
     a) Shape of the cone of depression 191
     b) Growth of the cone of depression 193
     c) Groundwater flow under conditions of pumping 193
   5.1.3 Protection zones 194
     a) Estimating the purifying capacity of terrains 195
     b) Drawdown 197
     c) Transfer time 197
     d) Distance 198
     e) Limits of water flow 198
     f) Current technical regulations in member-states of the European Union ... 198
     g) Summary 199
5.2 Conclusion 200

Chapter VI WATER WELL MANAGEMENT 203

6.1 Well operation 203
     a) Adapting the pump to the hydraulic characteristics of the well 203
     b) Knowledge of historical well data 204
     c) Technical equipment 205

6.2 Well maintenance 206
   6.2.1 Basic principles 206
   6.2.2 Regular maintenance 207
      a) Maintenance of the pumping system 207
      b) Maintenance of the catchment structure 209
   6.2.3 Monitoring methods 210
      a) Quantitative monitoring 210
      b) Qualitative monitoring 212
      c) Checking of constituent materials 213
   6.2.4 Monitoring equipment 213
   6.2.5 Frequency of measurements 214
   6.2.6. Summary 216
6.3 Ageing of well structures...216
   6.3.1. Corrosion phenomena 217
      a) Electrochemical corrosion 219
   6.3.2. Clogging phenomena...235
      a) Mechanical clogging 235
      b) Chemical clogging 238
      c) Biological clogging 240
   6.3.3. Failures linked to the resource 244
      a) Rainfall deficit..244
      b) Hydraulic perturbations..244
   6.3.4. Failures linked to management of the catchment structure 244
      a) Overextraction..245
      b) Unadapted extraction 245
   6.3.5. Ageing diagnosis...246
      a) Warning signs...248
      b) Identifying causes...248
   6.3.6. Summary 250
6.4 Protection of catchment works 251
   6.4.1. Passive protection...251
   6.4.2. Cathodic protection 252
   6.4.3. Choice of materials...253
6.5 Artificial recharge of aquifers 254
   6.5.1 Clogging of recharge basins 256
   6.5.2 Functioning of aquifer recharge 257
6.6 River-bank effects 259
   6.6.1 Nature of the problem...260
   6.6.2 Choice of study site..261
   6.6.3 Geology and hydrodynamics of the site 261
   6.6.4 Characterization of the biogeochemical system..262
   6.6.5 Summary...265
6.7 Conclusion 265

Chapter VII RESTORATION OF WATER WELLS 267

7.1 Treatment of sanding up..269
   7.1.1 System for uniform inflow distribution 272
7.2 Treatment of mechanical clogging 272
   7.2.1 Physical treatment 273
      a) Air-lift pumping 273
      b) Controlled overpumping 273
      c) Compressed-air treatment...273
      d) Water injection 274
   7.2.2 Chemical treatment 275
7.3 Treatment of carbonate clogging 280
   7.3.1 Preventive treatment with polyphosphates 283
   7.3.2 Mechanical treatment...284
      a) Scraping..284

b) Use of explosives 285
7.3.3 Acidization treatment 285
a) Principle of acidization 285
b) Introduction of additives 286
c) Carrying out of acidization 286
d) An example of acidization 288
7.4 Treatment of iron-manganese clogging 290
7.5 Treatment of biological clogging 291
7.5.1 Estimation of contaminated aquifer volume 294
7.5.2 Definition of successive annular volumes for treatment 295
7.5.3 Treatment protocol 297
7.5.4 Sodium hypochlorite treatment 298
7.6 Regeneration after corrosion 302
7.6.1 Preventive measures 302
7.6.2 Inspection 303
7.6.3 Restoration 303
7.6.4 Treatment of chemical corrosion 304
7.6.5 Treatment of electrochemical corrosion 305
7.6.6 Treatment of bacterial corrosion 305
7.7 Cleaning and pumping 305
7.8 Role of well age 306
7.9 General course of action prior to preventive maintenance 307
7.10 Abandonment of wells 308
7.11 Conclusion 309

Chapter VIII MANAGEMENT TOOLS 313

8.1 Electronic data processing 314
8.2 General management tools 316
8.2.1 Engineering research tools 317
8.3 Interdisciplinary tools 318
8.3.1 Management of historical data 320
8.3.2 Management of time-dependent data 321
8.3.3 Strategic management 322
8.4 Trends and perspectives 324
8.4.1 The commitment of hydrogeologists to these new technologies 326
8.4.2 Computer science as a tool in advanced communication 327
8.5 Conclusion 327

Chapter IX CONCLUSION 329

Chapter X BIBLIOGRAPHY 333

General bibliography 333
Basic references 334
References cited - main published articles 336
Main journals 348
Main professional associations 351

APPENDICES 353

GLOSSARY 357

INDEX 375

Preface

Readers of this book do not require a detailed knowledge of hydrogeology. A priority is given to applications and field-based experiments. The book was written with the aim of providing water well operators with the necessary basic information for managing groundwater resources, while also facilitating the understanding of hydrogeological mechanisms in time and space. In fact, the study of underground hydraulics is complex and it is difficult to envisage the 3-D geometry of a groundwater body during abstraction. This also applies, among many other factors, to the nature and evolution of piezometric levels, the direction of flow, the margins of the extracted aquifer, potential pollution hazards, propagation velocities and interferences between different wells within the same well field.

The approach followed is deliberately oriented towards data acquisition methods as well as the processing and synthesis of hydraulic data. It is applied to the solving of real problems encountered in the operation of water wells.

Several main objectives are pursued in this work:

— *To present the basic concepts of water well management*, that is, to guide the reader in acquiring an approach and discipline of work that enables a dynamic overview of the system comprising the aquifer, the water well and it's environment.

— *To help provide the technical knowledge* indispensable for understanding the phenomena involved. With this in mind, the author has deliberately avoided theoretical discussions which do not lead to easily demonstrable practical applications. Otherwise, the reader may consult the bibliography if more detailed information is required on a particular point.

— *To equip the reader with the vocabulary used by groundwater operators*, which is needed to enable a dialogue between the numerous hydroscience specialists and which forms the basis for their collaboration.

— *To bring about an awareness and prediction of potential difficulties*, identifying and ranking problems so as to be able to react purposefully. In fact, starting an operation means being able to foresee the difficulties in advance and, by the same token, seeking to avoid them. Nowadays, the problems which must be confronted are at the same time universal and interdependent.

— *To contribute to an appreciation of the magnitude of the problem*. The aim of hydrogeology is to predict the behaviour of water resources that are developed. These predictions should be made not only for the long term but also over long distances and in three dimensions. Local concepts should be abandoned in order to place the well in its geological, ecological and industrial setting. Ultimately, groundwaters and surface waters can no longer be dissociated.

— *To promote the use of tools for managing water resources,* which are based on the processing of collected information: data acquired from well bores and compiled in databases, time-variation of hydrodynamic characteristics in wells, environmental data, legislative and operational context, modelling, etc.

The present work is composed of nine chapters which follow a sequential pattern:

— The *basic concepts of hydrogeology,* aiming to set out the fundamentals of subsurface hydraulics by describing the interactions between groundwaters and surface waters. This is supplemented by the main exploration methods used in hydrogeology, the criteria for developing the resource and the essentials for understanding water chemistry, this latter being a crucial requirement for the operator.

— *Well design and construction* includes a summary of the main drilling techniques, drilling fluids, the emplacement and scaling of equipment, cementation methods and well development. This chapter aims to provide a good understanding of the different stages in the design and implementation of water wells. A state of the art is specified and the well operator is given the means of checking the satisfactory running of a project.

— The chapter on *well hydraulics* specifies the principal methods for interpreting the pumping tests used to characterize water wells and aquifers. It is essential that the operator should be able to identify correctly the characteristics of the boreholes and groundwater bodies in question, since this is the fundamental principle behind the management of well fields. This chapter is the most "numerate" in the book, and many concrete examples are given in order to assist comprehension. The author has deliberately restricted himself to aquifers of a relatively simple nature. It would be appropriate to consult specialists for more complex types of study.

— The *supervision and final acceptance of work* lists those points which must be monitored during implementation. Some important parameters are presented and discussed that are to be checked before, during and after completion of the well. This a crucial chapter for all persons supervising and subsequently accepting water wells.

— *Water well protection* is summarized in such a way as to incorporate the legal and operational aspects of catchment work implementation and the long-term control of groundwater resources. These considerations are brought together in the context of regulatory frameworks which exist ultimately at the national, European or even worldwide level. The synthesis is drawn up in view of the various adaptations that must be made in the application of water law in different countries.

— The chapter concerned with *groundwater management* draws up a basis for the maintenance and management of catchment basins. The concept of ageing in water wells in introduced along with the problem that this entails for the operator.

— The *restoration of water wells* is a synthesis of the different available techniques for restoring borehole structures. Of particular interest is the panoply of clogging treatments - chemical, physical and biological - that can be applied according to the phenomena observed in the well.

— *Management tools* are briefly described in terms of recent techniques for monitoring, archiving and data management.

— The *bibliography* is divided into operational sections, listing some basic reference works, background reading and the articles cited in the text, in addition to the main international journals and professional associations. Finally, the *appendices* contain conversion tables and functions needed for the interpretation of pumping tests.

This book brings together a very large amount of practical information. It is the fruit of several years of work and synthesis of data. While not always innovative, the methods presented in this work have the merit of being made available in one single volume, described in terms that can be readily understood by a technical audience. The methods described here are not only based on the in-house expertise of Lyonnaise des Eaux (French group working in over 100 countries) but are also derived to a large extent from bibliographic compilations and the practical experience of the author.

Although hydrogeology is not a new science, it is nevertheless continuing to evolve. As a result of previous studies, many of the examples in this book are drawn from the work of the following: G. CASTANY, J. MARGAT, Y. EMSELLEM, G. de MARSILY, H. SCHOELLER, M. CASSAN, J. FORKASIEWICZ, E. BERKALOFF, A. MABILLOT, H. CAMBEFORT, R. BREMONT, J. GOGUEL, H. DARCY, C. LOUIS, J-J FRIED, E. LEDOUX, G. SCHNEEBELI, M. HUG, A. HOUPEURT, J-C ROUX, R. LAUGA, B. GENETIER, C. MEYER DE STADELHOFEN, CH. V. THEIS, H. BOUWER, W. WALTON, J. MOUTON, R. BOWEN, SN. DAVIS, R. DEWIEST, J. BEAR, V.T. CHOW, A. VERRUIJT, M.S. HANTUSH, F.G. DRISCOLL, ...

Many of the ideas in the present work are due to collaborations with members of Lyonnaise des Eaux, "Centre International de Recherche Sur l'Eau et l'Environnement" (CIRSEE), SAFEGE and DEGREMONT. A large number of specialists have also participated in the production of this book through their comments, corrections and support. I would also like to thank various local water authorities, engineering design offices, well borers, academics, researchers and industrial partners.

Techniques for controlling groundwaters are destined to occupy a priviledged place in the hydrogeology of the 21st century. In fact, even though the main hydrogeological regimes are now well characterized, the parameters and relations governing groundwaters are clearly identified and the modelling of aquifers has become commonplace, methods for influencing the behaviour of groundwaters are indispensable for purposeful management of the environment while respecting aquatic ecosystems. It has become urgent for humankind to react and bring some influence to bear on this heritage.

In present-day societies, it can be clearly seen that the formidable stakes involved in the control of water require an increasingly astute response. Groundwaters represent a key element in this undertaking, which serves to underscore the crucial importance of the material presented in this synthesis. This book is the result of considerable know-how which needs to be augmented still further in order to master the tools that will be equal to the tasks of tomorrow.

CHAPTER I

Basic concepts of hydrogeology

"Nature, to be commanded, must be obeyed"

F. Bacon

Over the last decade, the techniques of water research, extraction and protection have greatly progressed, thus leading to a better understanding of groundwaters and their involvement in the hydrological cycle. Actions that affect the behaviour of groundwaters have as their ultimate goal the integrated management of water resources and the environment. It has become commonplace to say that if our management of water resources is not considerably improved, then the future of the human race will be compromised along with that of many other species. In this century of explosive technological change, a failure to recognize subsurface hydraulic phenomena or the leading role played by modern groundwater control methods will rapidly lead to water resources that are poorly developed or underextracted.

In the long term, actions carried out on the aquatic environment should take account of the ecosystems and the hydrogeological arguments, as well as surface waters in the much wider context of the hydrosphere. This approach enables the calibration of integrated models, as well as abstraction models, recharge models, pollution danger alert models, etc. It thus becomes possible to exert some influence on what may be called the hydrogeological setting.

The movement of water within the soil and subsoil is just one stage of a vast circulation process occurring at the surface of the Earth, i.e. the global hydrological cycle.

Infiltration is the main source of supply to groundwaters. It comes from effective precipitation, which corresponds to that fraction not taken up in runoff or evapotranspiration. The effect of infiltration is to maintain a constant flux which recharges the groundwaters. This explains why the subsurface contains water in almost all countries where it rains; the amount of water being a function of various hydrodynamic parameters of the aquifer such as porosity, permeability and storage coefficient. Beneath a certain depth (the water table), the water content ceases to rise with increasing depth, i.e. the ground is saturated. It should be stressed that subsurface waters are subject to gravitational forces in the unsaturated zone. The circulation of groundwaters is governed by subsurface hydraulics and remains subject to a number of different factors including transmissivity, hydraulic gradient and storage coefficient. Other parameters such as aquifer configuration and structure lead to a more detailed definition of this general picture

Hydrogeology is the science of studying the mechanisms of storage and flow in groundwaters. All surveys and assessments need to be carried out in relation to a specific area of land or rock volume (hydrographic basin, groundwater basin or aquifer) with reference to a mean duration. This chapter gives a summary presentation of hydrogeology and its methods.

1.1 The water cycle

In order to arrive at a better understanding of the origin, storage and flow of groundwaters it is necessary to appreciate the functioning of the hydrological cycle, that is to say, the distribution and circulation of waters on Earth. The statistical distribution of water by volume in the different natural compartments is as follows:

— *Oceans:* $1\,320 \times 10^6$ km^3, or 97.20% of the total volume of water.
— *Snow and ice*: 30×10^6 km^3 (2.15%),
— *Groundwaters above 800 m*: 4×10^6 km^3 (0.31%),
— *Groundwaters beneath 800 m*: 4×10^6 km^3 (0.31%),
— *Unsaturated terrains*: 0.07×10^6 km^3 (0.005%),
— *Freshwater lakes*: 0.12×10^6 km^3 (0.009%),
— *Salt-water lakes*: 0.10×10^6 km^3 (0.008%),
— *Rivers*: 0.001×10^6 km^3 (0.0001%),
— *Atmosphere*: 0.013×10^6 km^3 (0.001%).

The global cycle starts with the evaporation of water which, under the influence of solar radiation, is transformed into vapor. Evaporation takes place from the surface of free bodies of water (oceans, seas, lakes and rivers) as well as from vegetation. Both of these phenomena — evaporation and transpiration — are brought together under the term *evapotranspiration*. Subsequently, the water vapour condenses as clouds which then give rise to precipitation (rain and snow). This accounts for almost all the water input to the soil (cf. Figure 1-1).

Three processes intervene in subsequent stages of the cycle:

— Part of the precipitation flows towards drainage systems and into free bodies of water: this constitutes the surface runoff.

— Another fraction percolates into the subsoil and contributes to the replenishment of groundwaters: this process is called infiltration.

— The remaining part evaporates and is re-introduced into the cycle. Sublimation occurs in the case of snow and ice, since there is a direct transition from the solid into the gaseous state

In France, it is estimated that 55% of the precipitation is re-evaporated, 25% goes into runoff and 20% infiltrates into the subsurface.

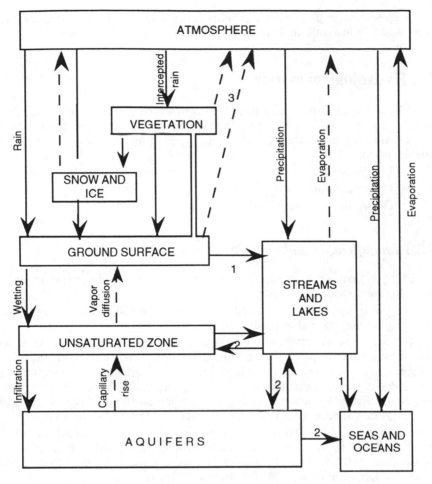

Figure 1-1
*Schematic diagram of the hydrological cycle showing: 1 surface runoff, 2 groundwater
flow, 3 evapotranspiration (after EAGLESON, 1970).*

The hydrological cycle can thus be represented as an equation which takes
account of the water balance:

$$P = E + R + I$$

where:

E : evapotranspiration,
P : precipitation,
R : runoff,
I : infiltration.

Deep-seated groundwaters are deliberately left out of account since their contribution is insignificant in relation to the volume of surface waters.

1.2 Hydrological systems

A hydrological system is a dynamic system which corresponds to a sequence in space and time and which forms part of the water cycle. It is recognizable by its spatial and temporal characteristics. Three types of hydrological system can be distinguished:
— The drainage basin.
— The groundwater basin.
— The aquifer with its groundwater body.

1.2.1 Identification and recharge of groundwater bodies

The physical characteristics and recharge criteria of groundwaters in hydrological systems are as follows:
— A *drainage basin* is bordered by watersheds which delimit the catchment area of a watercourse and its tributaries.

The only input of water into a drainage basin, which is assumed to be closed, comes from effective precipitation. This corresponds to the amount of water from precipitation that remains available at the ground surface after subtraction of losses due to true evapotranspiration. Empirical equations have been established to estimate losses by potential and/or true evapotranspiration. The most widely used equations are due to L. TURC (1954) and C.W. THORNTHWAITE (1948).

— The *groundwater basin* is that part of a drainage basin situated beneath the ground surface. Its volume is occupied by groundwaters, and its boundaries are constrained by the geological structure. Recharge of the groundwater basin takes place through the infiltration of effective precipitation.

— The *aquifer* is identified by its geological setting, and is the spatial unit of investigation of groundwaters. A groundwater basin is made up of one or more aquifers. The aquifer is recharged by effective infiltration, i.e. the amount of water actually reaching the groundwater body.

Furthermore, all the data relative to the system in question should be referred to a given date or a defined mean duration, generally taken as a year. The data treatment should be carried out according to two essential conditions:
  • The longest possible hydrological period, chosen in relation to the time-span covered by records (ten years at least);
  • The shortest possible sampling frequency compatible with the measurement frequency (daily, weekly, monthly or annual.

The exploitation of a water well introduces a certain number of perturbations into the hydrodynamic regime of the groundwater body. It is important to have a good knowledge of the perturbations in order to determine the principal characteristics of the catchment site. But before discussing the effect of these perturbations, it is appropriate to throw as much light as possible on the initial hydrological regime. This means that information concerning the piezometric surface needs to be supplemented by an estimation of flow rates as well as a true mass balance of waters entering and leaving the groundwater system.

1.2.2 The water balance

The calculation of a water balance provides a means of checking the consistency of data with respect to the recharge and flow behaviour of hydrological systems. It enables some insight into the hydrological equilibrium of a system. In natural regimes, the water balance reflects the difference between input and output flow rates. This balance derives from the equilibrium state of the water cycle and schematizes the hydrodynamic behaviour of the system in question. For short observation times, it is sometimes necessary to include a storage increment (ΔW) which may be positive or negative:

$$inflow\ rate = discharge\ rate \pm \Delta W$$

In a non-steady state regime, the water balance becomes:

$$natural\ inflow + imported\ inflow = discharge\ rate + abstraction\ rate \pm \Delta W$$

The different components of the water balance, expressed in terms of flow rates, are brought together in Table I-I.

TABLE I-I — *Components of the water balance (after G. CASTANY, 1982).*

Inputs	Losses and outflow
DRAINAGE BASIN	
Precipitation	Effective precipitation
Effective precipitation	True evapotranspiration
	Total discharge
GROUNDWATER BASIN	
Recharge rate	Groundwater flow
Infiltration	Abstraction
AQUIFER	
Effective infiltration	
Storage increment ΔW	

a) Balance for a drainage basin

As mentioned above, the inflow of water into a drainage basin is supplied by effective precipitation, *EP*, while the outflow is represented by the total discharge rate, *QT*.

The annual balance of the drainage basin of the Hallue (Somme department, France) may be taken as an example. Period 1966-1970. Surface area of the basin: 219 km^2.

$$EP - QT$$
$$50 \ hm^3/yr - 52 \ hm^3/yr$$

It can be seen that the value of PE is very close to QT. The difference of $2 \ hm^3/yr$ is of the same order of magnitude as the precision of the measurements (10 - 15%).

The mean annual balance of the Zorn (eastern slope of the Vosges). Drainage basin area: 682 km². Period 1959-1962.

As explained above, it is necessary to take account of a storage increment, ΔW, in the case of a short period. In this example, $\Delta W = 92 \ hm^3/an$.

$$PE = QT + \Delta W$$
$$373 \ hm^3/yr = 175 \ hm^3/yr + 92 \ hm^3/yr$$

b) Balance for a groundwater basin

The rate of inflow is represented here by infiltration, I, while the outflow is the groundwater discharge rate, QW.

The mean annual balance for the Hallue.

$$I = QW$$
$$48 \ hm^3/yr = 48 \ hm^3/yr$$

c) Balance for an aquifer

The rate of inflow is equal to the effective infiltration, EI. The outflow is represented by the groundwater discharge rate, QW, in some cases augmented by abstraction flow rates, QEX.

TABLE I-II — *Mean annual balance of the unconfined aquifer of the Crau alluvial deposits (southern France). The total surface area of the aquifer is 520 km² and rates of flow are given in m^3/s (after J. BODELLE and J. MARGAT, 1980).*

Rates of inflow	Rates of discharge
Effective infiltration...............1.5	Underground losses......................1
Infiltration of irrigation	(e.g. seepage into the sea)
waters...5.5	Rising springs and ditches............6
Inflow from adjoining aquifers...1	Abstraction...................................1
Total (252 hm³/yr))....................8	Total (252 hm³/yr)........................8

N.B.: This example corresponds to an aquifer strongly affected by irrigation.

Example: mean annual balance for the Beauce limestone. Surface area: 5 966 km². Period 1955-1974:

$$IE = QW + QEX$$
$$465 \ hm^3/yr = 382 \ hm^3/yr + 83 \ hm^3/yr$$

Other examples are presented in Tables I-II and I-III.

TABLE I-III — *Mean annual balance of a leaky aquifer in a multilayered system of Eocene sands and groundwater reservoirs at the base of the Tertiary in the western Aquitaine Basin (after J. BODELLE and J. MARGAT, 1980).*

Inflow rates	Discharge rates
Input from margins...................1.9 Inflow from leakage of overlying or underlying aquifers................2.2	Rising springs, leakage from watercourses..............................1 Concealed discharge to the sea..0.5 Leakage through the roof..........1.45 Abstraction...............................1.15
Total (130 hm^3/yr).....................4.1	Total.......................................4.1

N.B.: Area of basin = 50 000 km^2.

1.3 Aquifer characteristics

An aquifer is a permeable hydrological formation which allows significant discharge of groundwaters as well as the capture of appreciable amounts of water by economical means. Due to this fact, an aquifer may be compared with a mineral deposit in which the ore — being water — is more or less renewable. The aquifer is both a hydrological and a hydrodynamic system; it can be defined in terms of five sets of quantifiable characteristics:

— *The reservoir* is a finite volume of space characterized by its boundary conditions, dimensions and/or geometry as well as its internal organization or structure. It corresponds to a formation or group of formations.

— *Hydrodynamic, hydrochemical and hydrobiological mechanisms*, which give rise to the three functions of a reservoir with respect to groundwater, i.e.: storage, conduit (transfer of water or energy) and medium for geochemical exchanges.

— *A sequence of the global water cycle*, including various interactions with the environment that are reflected in three types of behaviour: hydrodynamic, hydrochemical and hydrobiological. The sequence is characterized by an impulse/response couple that is expressed by a transfer relation or function.

— *The spatial variability of these characteristics.*

— *Time conditions,* since all measurements are referred to given date (initial state) or a mean duration (temporal variability of characteristics). These latter are based on records and can lead to forecasts.

Three main types of terrain may be distinguished on the basis of their capacity to transmit groundwaters: water-bearing terrains, where waters circulate freely, aquiclude or semi-permeable terrains showing very slow circulation and, finally, aquifuge or impermeable terrains. The geological formations which make up the different aquifer systems are mostly sedimentary in nature; they exhibit highly variable geometries (thickness and extent) and hydrodynamic characteristics (storage, permeability).

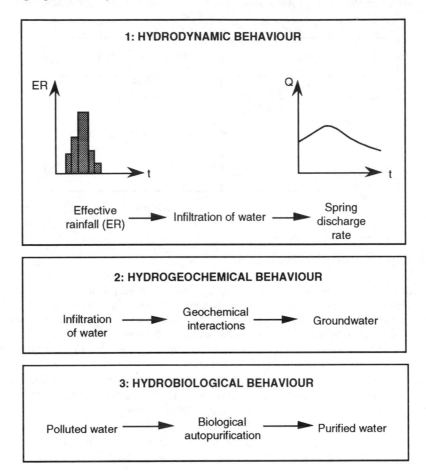

Figure 1-2
Response of a reservoir to a pulse of precipitation (after G. CASTANY, 1982).

There are two main types of groundwater reservoir: homogeneous aquifers and heterogeneous aquifers.

— *Homogeneous aquifers* showing pore-space permeability, are made up of sands, gravels and sandstones. They are associated with alluvial deposits occupying valley floors and account for part of the groundwater of major sedimentary basins (Paris Basin, Aquitaine Basin). The groundwater discharge rates are generally low.

— *Heterogeneous aquifers*, showing fracture permeability, are generally composed of limestones but also comprise volcanic, metamorphic and granitic rocks. Within limestone massifs, the fractures are commonly open, thus providing veritable passage-ways which allow the very rapid circulation of groundwaters.

In certain rock-types (e.g. Chalk of the Paris Basin), both porosity- and fracture-related permeabilities may coexist, although the latter type is generally predominant.

As mentioned previously, an aquifer can be identified by the hydrogeological formation from which it is composed. In the following, it is convenient to take account of the existence and flow of groundwaters as well as the effects of water/rock interactions.

An aquifer is a dynamic system which exhibits three types of behaviour in relation to it's groundwaters. These behaviours result from the mediation of storage characteristics in response to external excitations (or pulses) which are imposed at the boundaries of the system. The behaviour of the aquifer can be seen in terms of impulse, transfer and response, three factors which regulate the discharge rate and which govern the hydrochemical characteristics - even the hydrobiology - of the outflowing water.

As pointed out by G. CASTANY, the aquifer reacts to three types of disturbance (cf. Figure 1-2) that are expressed by recharge at its boundaries:

— *Hydrodynamic pulses*, affecting both the stored waters and the flux. Brings about an intake of water or variations in pressure and/or head.

— *Hydrochemical influence*, with inputs of heat as well as mineral and organic substances.

— *Hydrobiological effects* through the activity of micro-organisms.

The geometry of the aquifer depends on the characteristics of its geological and hydrodynamic boundaries; in this context, we speak of the *boundary conditions*. In simplified terms, the base of the aquifer (substratum) is composed of an impermeable formation. In contrast, its upper boundary can be of three types:

— Hydrodynamic with free fluctuations: *unconfined aquifer*.
— Impermeable and geological: *confined (or artesian) aquifer*.
— Semi-permeable and geological: *leaky (or semi-confined) aquifer*.

1.3.1 Unconfined aquifers

The water-bearing formation is not saturated throughout its entire thickness. Between the groundwater body and the ground surface, or up to the base of the overlying clayey formation where this exists, there is an unsaturated zone containing some air in its pore volume. The top surface of the groundwater body is termed the water table or piezometric (potentiometric) surface.

The water table is defined by the set of piezometric levels measured at different points at a given date; it is represented on a map by groundwater contours (or lines of equal piezometric level). Since the water table corresponds to the upper boundary of the aquifer, it is the hydrodynamic limit of the system.

In the case of an unconfined aquifer, the standing water-level is always found beneath ground level.

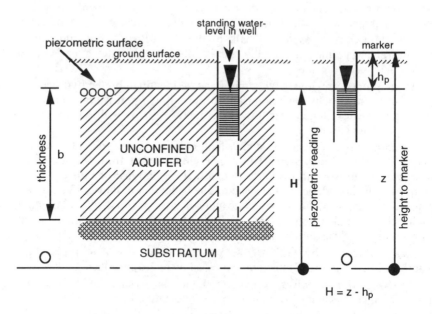

Figure 1-3
Sketch diagram of an unconfined aquifer (after G. CASTANY, 1982).

1.3.2 Confined aquifers

The water-bearing formation is saturated throughout its thickness. At the top, it is bounded by a permeable or semi-permeable layer. The standing water-level remains virtual as long as a borehole or a piezometer has not attained the aquifer, lying always above the base of the overlying impermeable layer. A water well of this type is called *artesian*. When the standing level is above the ground surface, the well is said to be artesian; it flows naturally without pumping.

Artesian wells are used to abstract groundwaters in the Albian sands of the Paris Basin, where the base of the confining bed is at a depth of the order of 600 m. Taking account of the subsurface conditions, the aquifer is subject to a geostatic pressure equal to the weight of the overlying rock column (mean density corresponding to 2.5 bar per 10-m-thick slice) (cf. Figure 1-4). For the Albian sands of the Paris Basin, where the base of the confining bed is at 600 m, this pressure is 150 bars. The groundwater in this case is artesian, since even though the standing level is currently higher than ground level, the conditions for artesian flow are satisfied.

Several distinct groundwater bodies — whether confined or unconfined — may be superposed at any given point.

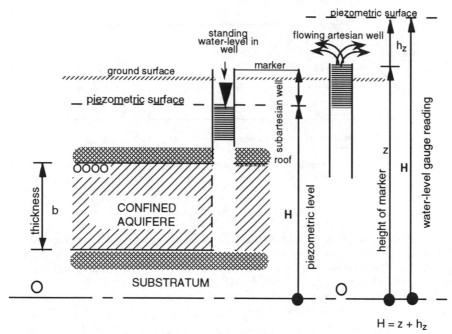

Figure 1-4
Sketch diagram of a confined aquifer (after G. CASTANY, 1982).

1.3.3 Semi-confined or leaky aquifers

The importance of leakage mechanisms relies on the fact that considerable discharge may occur across an impermeable or semi-permeable horizon when it has a large surface area and there is a pressure difference from one side to the other.

The confining bed and/or substratum of an aquifer is often made up of a semi-permeable hydrogeological formation. Under certain favourable hydrogeological conditions (difference of pressure head), the semi-permeable layer allows the exchange of water with the overlying or the underlying aquifer. This phenomenon, which is known as leakage, implies the presence of an aquifer containing semi-confined groundwaters.

The analysis of groundwater behaviour is nearly always undertaken assuming that the hydrodynamic flow is governed by relations applicable to single layers. Exchanges between superposed groundwater bodies subject to different pressures, as well as between the groundwater body and its boundary layer, are very rarely taken into account. However, such exchange phenomena do exist in nature. To be convinced of their reality, it is merely necessary to appreciate the orders of magnitude involved in bringing leakage mechanisms into action.

Let us consider a good-quality aquifer, extending over an area of 100 x 100 km, which is separated from another less aquiferous body by 50 m of impermeable terrain (cf. Figure 1-5).

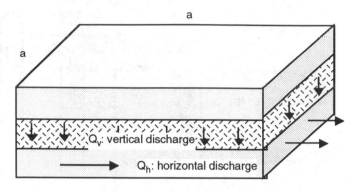

Figure 1-5
Schematic example of leakage

Assuming that there is a pressure difference between the two water-sheets, either due to natural hydrodynamics or arising from exploitation of the better-quality body, then it is straightforward to estimate the horizontal flow rate of this latter using Darcy's Law. Similarly, it is possible to calculate the rate of percolation from one layer to the other. In order for the horizontal flow to become comparable to the vertical flow, it can be shown that — for a pressure difference of 10 m — the vertical permeability should be equivalent to 5×10^{-6} of the main permeability. Now, this pressure difference can easily attain 10 m in certain groundwater bodies of the Aquitaine Basin, while the Albian aquifer of the Paris Basin shows a depression of 100 m.

Moreover, in terms of mass balance, it may be shown that leakage is an extremely powerful mechanism that allows very high rates of percolation despite vanishingly small permeabilities. This is because leakage involves considerable surface areas. It also forces us to reconsider with great prudence the basic concepts of cover and impermeability in hydrogeology.

1.4 Reservoir and aquifer characteristics

The laws and principles governing the circulation of groundwaters can be deduced from the fundamental laws of fluid mechanics. As a preliminary, it is necessary to assume that groundwaters exhibit laminar flow over most of their transport path. Turbulent flow may sometimes occur in the immediate proximity of a well (e.g. screens), where it arises from increased velocity of water circulation. However, this phenomenon remains of only limited importance in space.

1.4.1 Groundwater flow

As mentioned above, groundwaters are under most circumstances subject to laminar flow. The pattern of flow in an aquifer can be schematized by a grid of flow lines and groundwater contours (i.e. lines of equal hydraulic head) making up a *flow net* (cf. Figure 1-6).

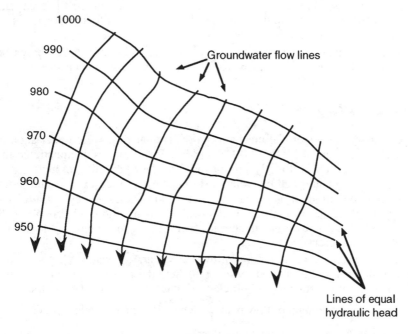

Figure 1-6
Pattern of groundwater flow.

The groundwater flux can be calculated between two flow lines (cf. Figure 1-5) using the equation:

$$Q = L \ b \ K \ \frac{\Delta h}{\Delta x}$$

where:

Q : flux in m³/s,

L : distance in metres between two flow lines,

b : thickness of the saturated zone of the aquifer in metres,

K : hydraulic conductivity in m/s,

$\dfrac{\Delta h}{\Delta x}$: dimensionless hydraulic gradient.

It is shown in a later section that this equation can be derived directly from Darcy's Law.

In practice, two main types of flux can be identified:

— *Lateral flux.* In a subsurface aquifer, water flow takes place from recharge zones (intake) towards discharge zones (drainage and outlet). The driving force for the flow is the difference in elevation giving rise to hydraulic gradients.

— *Leakage flux.* In a deep aquifer, by contrast, the ascending and descending fluxes are predominant due to the draining of subsurface aquifers by rivers. These fluxes are subvertical.

Hydrogeological studies carried out over the last few years have shown the importance of leakage fluxes, which are the main cause of groundwater flow.

1.4.2 Darcy's Law

Preliminary studies on the mechanisms of laminar flow were undertaken by G. HAGEN (1839) and J.M. POISEUILLE (1846). These studies demonstrated that the value of the flux is directly proportional to the hydraulic gradient. Later, H. DARCY (1856) confirmed this relationship and established experimentally the fundamental equation of geohydrology, now known as Darcy's Law. The equation is applicable in the field under well defined conditions, enabling the calculation of groundwater discharge from the coefficient of hydraulic conductivity of a reservoir

Darcy's Law may be stated as follows: a volume of water, Q, percolating vertically down through a sand cylinder of height l and cross-section S, is a function of the proportionality coefficient K, characteristic of the sand, and the loss of head per unit length of flow path, $\frac{h}{l}$, which is a dimensionless ratio.

For which the expression can be written:

$$Q = K S \frac{h}{l}$$

where:

Q : discharge rate in m³/s,
K : coefficient of hydraulic conductivity in m/s,
S : cross-sectional area in m²,
h : head (height of the water column) in metres, proportional to weight of water column,
l : height of sand cylinder in metres.

The $\frac{h}{l}$ ratio, which is known as loss of head or otherwise hydraulic gradient, can be denoted as i.

Then, the expression of Darcy's Law becomes:

$$Q = K S i$$

This formula is theoretically only applicable under precisely defined conditions, notably when dealing with laminar flow. But groundwater flow can become locally turbulent, especially in the vicinity of catchment works.

O. REYNOLDS (1883) defined these two types of flow, and introduced a ratio known as the Reynolds number to characterize them.

$$Re = \frac{v \, d}{V}$$

where:

Re : Reynolds number, a dimensionless parameter,

v : Darcy velocity (LT^{-1}), corresponding to the hypothetical macroscopic discharge rate of water moving uniformly through a saturated aquifer,

d : characteristic length (generally the mean diameter of grains in a porous medium),

Vr : kinematic viscosity.

The dimensionless Reynolds number represents the ratio between forces of inertia and forces of viscosity.

This number serves to calculate the linear loss in head occurring in a conduit under an established steady-state regime. In this classic case of fluid transfer, a laminar flow regime may be characterized by Re < 2 000 and a turbulent regime by Re > 4 000. Between the two, a transitional regime is said to occur where laminar and turbulent flow succeed each other in time and space.

This assumes that the conduit is rectilinear, the discharge rate is relatively stable and the diameter of the conduit is homogeneous. However, although a fluid - in this case water - will flow within an aquifer, this is the about the only similarity with a conduit. Darcy's Law as well as the Hagen-Poiseuille Law both draw a relation between the loss of potential energy (or pressure head) and the velocity of flow; although applicable to the relatively simple geometry of a straight-walled conduit, they are no longer valid in the case of groundwater flow.

In this way, a Reynolds number is defined for an aquifer on the basis of data that are relatively easy to obtain:

— The Darcy velocity, which should not be confused with the actual flow velocity.

— The mean diameter of particles making up the aquifer formation.

It is now acknowledged that the boundaries between the different regimes are no longer as proposed by L.F. MOODY (1944):

— The laminar flow regime comes to an end at Re ≤ 1,

— The turbulent flow regime starts at Re ≥ 10.

The transition between laminar and turbulent flow is a function of a large number of parameters, but it is generally accepted that the laminar flow regime ceases for Reynolds numbers greater than unity.

Theoretically, it can be assumed that the circulation of groundwaters in a porous medium obeys the Darcy Law for Reynolds numbers less than unity, with transitional flow behaviour occurring at Reynolds numbers in the range 1-10. Nevertheless, as shown here, the Darcy Law can be applied in practice to any kind of water circulation regime provided that a certain number of approximations are made.

To characterize the behaviour of water in a rock, it is necessary to define two quantities which must be clearly distinguished: permeability and porosity.

— *Permeability* is a coefficient characterizing the degree to which water is free to circulate through a formation.

— *Porosity*, expressed in percentage, is the volume proportion of a terrain corresponding to voids that could be occupied by water.

1.4.3 Permeability

Permeability is the capacity of a geological formation, whether consolidated or not, to allow the transmission of a fluid under the influence of a hydraulic gradient. It reflects the resistance of a medium to the flow of water traversing a formation. Two measures can be used to quantify permeability: hydraulic conductivity and intrinsic permeability.

TABLE I-IV — *Hydraulic conductivity in relation to grain-size distribution of unconsolidated sediments (sand and gravel formations). After G. CASTANY, 1982*

K (m/s)		10^1	1	10^{-1}	10^{-2}	10^{-3}	10^{-4}	10^{-5}	10^{-6}	10^{-7}	10^{-8}	10^{-9}	10^{-10}	10^{-11}
Granulometry	homogeneous	clean gravel				clean sand		very fine sand			silt		clay	
	inequigranular	coarse and medium gravel		gravel and sand		sand and clay - coarse and fine silts								
Degree of permeability		VERY HIGH TO HIGH						POOR					NIL	
Type of formation		PERMEABLE					SEMI-PERMEABLE						IMPER-MEABLE	

— The *hydraulic conductivity*, denoted as K in Darcy's Law.

This corresponds to the volume of mobile water in m^3 transmitted perpendicularly to the flow direction in unit time *(s)* through a unit cross-section in m^2, under the effect of a unit hydraulic gradient and within the conditions of validity of Darcy's Law (at a temperature of 20° C). Hydraulic conductivity has the dimensions of velocity and is expressed in m/s.

— The *intrinsic (or specific) permeability*, denoted as k, is the volume of liquid (in m^3) having unit kinematic viscosity (centipoise) that passes through a unit cross-section area (m^2) perpendicular to the flow direction under the effect of a unit hydraulic gradient. It is expressed in m^2 or in Darcys.

Various types of permeability may be distinguished:
- *Small-scale permeability* is related to the matrix of a rock; it applies, for example, in the case of a sandy layer where the fluid circulates between the grains.
- *Large-scale permeability* is related to fracturing as seen, for example, in a limestone or a fractured granite.
- *Channel permeability* is related to karstic cavities in limestone terrains.

1.4.4 Porosity

Two types of porosity are recognized: absolute porosity and effective porosity.

— *Absolute porosity* is the property of a geological formation to contain voids or pore-spaces, which may be interconnected or not. It is expressed in percentage, being defined as the ratio between the pore-space volume Vv of a medium and the total volume Vt of a sample.

$$\textit{Porosity: } n = \frac{V_v}{V_t}$$

— *Effective porosity*, denoted as n_e, is the volume of mobile water, Ve, that a saturated reservoir can contain and subsequently release when fully drained, divided by the total volume of the reservoir, Vt.

$$\textit{Effective porosity: } n_e = \frac{V_e}{V_t}$$

However, a reservoir is never completely drained of its water content. This is the reason why effective porosity is more commonly used in hydrogeology than the more theoretical concept of absolute porosity.

A sample of Albian sand from the Paris Basin, a cube with sides of 10 cm having a total volume of 1 000 cm^3, contains a void-space of 280 cm^3 as measured by porosimeter. Its porosity is thus equal to 28%.

Although porosity is independent of grain size, it is a function of the arrangement of the grains with respect to each other as well as the spread of the grain-size distribution.

The average effective porosities for the main types of reservoir are indicated in Table I-V.

It should be pointed out that a porous formation is not necessarily permeable. On the contrary, a permeable formation is by definition porous.

Porosity is strongly influenced by the grain fabric (packing), decreasing from 47.6% for cubic to 25.9% for rhombohedral (or trigonal). In practice, natural media (aquifers) have porosities of a few percent, being generally less than 10%.

A porosity of 15% within an aquifer can already be considered as exceptional.

TABLE I-V — *Mean values of effective porosity in the main types of reservoir formation.*

Reservoir type	*Effective porosity (%)*
Coarse gravel	30
Fine gravel	20
Gravel and sand	15 -25
Alluvium	8 - 10
Coarse sand	20
Fine sand	10
Coarse sand and clay	5
Clay	3
Fractured limestone	2 - 10
Fractured sandstone	2 - 15
Fractured granite	0.1 - 2
Fractured basalt	8 -10
Shales	0.1 - 2
Dune sand	38
Tufa	20

1.4.5 Hydrodynamic parameters

These parameters are characteristic of the reservoir function of an aquifer. They can be determined in the field by means of pumping tests.

a) Transmissivity

The discharge capacity of a well in an aquifer depends on the hydraulic conductivity K and the thickness e of the aquifer. It can also be estimated by means of a parameter T, denoting transmissivity, using the equation:

$$T = K e$$

where:

T : is expressed in m^2/s,

K : in m/s,

e : in metres.

The parameter T represents the discharge of a water-bearing layer over its entire thickness per unit width, subject to a unit hydraulic gradient (cf. Figure 1-7).

Transmissivity should not be confused with hydraulic conductivity, which is calculated over unit thickness.

Long pumping tests are used to calculate the transmissivity, based on measurements of drawdown and recovery both in the control well and also in observation wells.

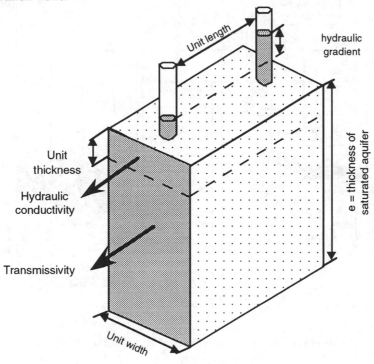

Figure 1-7
Transmissivity and hydraulic conductivity of an aquifer

b) Storage coefficient

The storage coefficient is a dimensionless quantity denoted as S, which represents the ratio between the volume of water released or stored per unit surface of aquifer and the corresponding unit change in total head, Δh.

For an unconfined aquifer, this involves the volume of water released under gravity (draining of the terrain). S varies in the range 1×10^{-2} to 2.5×10^{-1} and is placed in the same category as the effective porosity (n_e) of the aquifer.

For confined aquifer, the volume of water to be considered is expelled by decompression of the aquifer. S varies from 1×10^{-3} to 1×10^{-4}. The storage coefficient is calculated on the basis of long pumping tests, using data just from water-level gauges (observation wells).

A knowledge of the parameters T and S makes it possible to predict the long-term behaviour of an aquifer, notably with respect to different operational configurations (modelling).

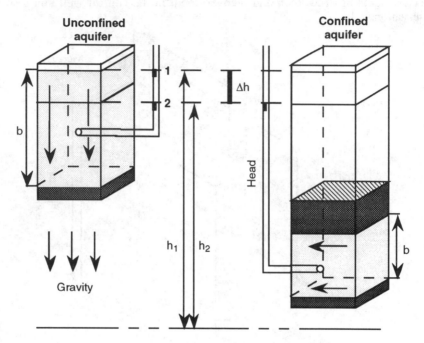

Figure 1-8
Processes of water release from aquifers (after G. CASTANY, 1982)

c) Diffusivity

Diffusivity governs the propagation of disturbances in the aquifer. It is equivalent to transmissivity, T, divided by the storage coefficient, S.

$$Diffusivity = \frac{T}{S}$$

Diffusivity is expressed in m^2/s, since T is also expressed in m^2/s and S is dimensionless.

1.5 Aquifer exploitation criteria

Planning the extraction of groundwaters depends on the evaluation of water reserves and resources. This involves the collation and synthesis of all the data acquired during prospection and field experiments.

The *reserve (or storage)* is the quantity of water contained in a hydrological system at a given date or stored over a period of time. It is expressed in terms of volume (hm^3 or km^3).

The *resource* is the quantity of water that can be extracted from a defined volume over a given period of time. Evaluation of the resource is based on the hydrodynamic and hydrochemical behaviour of the aquifer. The resource is measured in terms of mean discharge rate (m^3/s, hm^3/yr or km^3/yr).

Criteria for the exploitation of groundwater bodies are linked to various constraints that are described below.

— Supply and demand; the water resource should, in fact, satisfy the requirements of user demand. In this framework, the evaluation of a resource should be a compromise between the production potential - defined by the physical and technical constraints of the available resource - and the qualitative, quantitative and economic requirements imposed by the user.

— Planning constraints are liable to change in time and space, and can be placed in various different categories:

- Technical (hydrogeological structure of the aquifer, hydrodynamic parameters).
- Socio-economic (cost of extraction, well discharge capacity.
- Ecological (environmental impact of resource exploitation, permissible drawdown levels).
- Political (planned water and exploitation policy, funding to be agreed).

— Space and time constraints. The evaluation of a reserve or a resource must be carried out with reference to a water resource system (a space delimited at a given moment in time or for a defined duration). Evaluations should be performed within a spatial frame adapted to that imposed by the demand. The period covered by the forecasts is variable, ranging from 5 to 30 years. It is clear that these forecasts will become more random as the period lengthens.

— Variability in time and space. The requirements of water demand may change with time in terms of quality and especially quantity. As a consequence, the resource evaluation will evolve in both space and time. Hence, it is not unalterable and should be periodically updated.

1.5.1. Evaluation of water reserves

As described above, a groundwater reserve is a volume of water stored during a mean duration in a delimited portion of aquifer. Four types of reserve may be distinguished (cf. Figure 1-9).

— *Total groundwater storage.* This corresponds to the quantity of mobile water contained in a volume comprised between the substratum and the upper boundary of the aquifer (water table for an unconfined aquifer or impermeable top wall for a confined aquifer). The mean total groundwater storage is bounded at its top by the mean annual piezometric surface.

— *Regulating storage.* This is the volume of mobile water contained in the zone of fluctuating piezometric level.

— *Permanent storage*. This is the fraction of the total reserve that is not replenished. In the case of unconfined groundwater, it is delimited by the mean minimum water table. For confined bodies, the permanent storage is very similar to the total groundwater storage.

— *Extractable (or mineable) storage*. This is the maximum volume of water that can be extracted from the total storage of an aquifer under economically acceptable conditions. It is related to the water resource represented by the exploitation of a reserve, and is determined by the constraints discussed above.

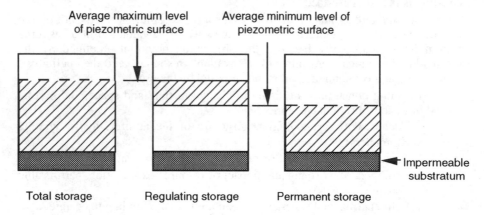

Figure 1-9
Types of storage in an unconfined aquifer (after G. CASTANY, 1982).

The groundwater storage, W, is evaluated using the volume, V, of the slice of aquifer and either the effective porosity n_e (in the case of an unconfined aquifer) or the storage coefficient S (for confined aquifers).

$$W = V \cdot n_e$$

$$W = V \cdot S$$

Replenishment of the groundwater reserve of an aquifer is brought about by the intake of effective infiltration. Under natural regimes, this input compensates for the discharges due to underground flow.

In this way, two parameters may be defined: the turnover rate and the turnover time (cf. Table I-VI).

- *Turnover rate* is the mean annual recharge of the aquifer, IE, expressed in volume, divided by the mean total groundwater storage, WM. This ratio is expressed as a percentage.

$$Turnover\ rate = \frac{IE}{WM} = \frac{QW}{WM}$$

- *Turnover time* is the theoretical duration needed for the cumulative recharge volume of an aquifer to become equal to its mean total storage,

WM, which is equivalent to the discharged volume of groundwaters over the long term, *QW*. The turnover time is expressed in years.

$$Turnover\ time = \frac{WM}{IE} = \frac{WM}{QW}$$

TABLE I-VI — *Turnover parameters for the mean total groundwater storage of certain aquifers in France (after J. BODELLE and J. MARGAT, 1980).*

Aquifer	Turnover rate (%)	Turnover time (years)
Unconfined aquifers		
- Champigny Limestone	7	14
- Beauce Limestone	3	33
- Alsace valley alluvium	2.8	35
- Lorraine karstic limestone	1.2	80
Confined aquifers		
Bunter Sandstone, Lorraine	0.15	6 300
Albian sands, Paris Basin	0.05	200 000

Example of storage evaluation for groundwaters in an unconfined aquifer: alluvial deposits of the Moselle downstream from Metz.

The total storage, WT, is calculated as follow:

$$WT = V \times n_e$$

but since $V = A \times e$

then, $WT = A \times e \times n_e$

where:

V : volume of the aquifer,

A : total surface area of the aquifer (2.10^8 m^2),

e : thickness of the aquifer (4 m),

n_e : effective porosity of alluvium (0.2).

Using these values, we obtain the total storage:

$$WT = 4\ m \times 2.10^8\ m^2 \times 0.2 = 160\ hm^3$$

The regulating storage, *WR*, is estimated with a mean fluctuation in groundwater-level of the order of 2 m.

$$WR = 2\ m \times 2.10^8\ m^2 \cdot x\ 0.2 = 80\ hm^3$$

1.5.2 Evaluation of water resources

The evaluation of extractable groundwater resources makes use of complex methods and means. Hydrogeology should be at the same time quantitative and qualitative. Thus, it is indispensable to utilize mathematical models for hydrodynamic simulations in steady state and non-steady state regimes. The following distinctions can be made:

— A natural renewable water resource, whose unit of evaluation is the drainage basin.

— A natural groundwater resource - whether renewable or not - has the groundwater basin as its unit of evaluation.

— An extractable groundwater resource, whose unit of evaluation is the aquifer, corresponds to the maximum quantity of water available in the aquifer.

The last mentioned category is the most significant to the hydrogeologist. In such cases, evaluation methods rely on taking several parameters into account such as the boundary conditions, hydrodynamic variables, characteristics of the aquifer/catchment system, groundwater storage, etc.

1.5.3 Extraction strategies

The volume of water mobilized by a well, compared with the natural renewable groundwater resource, is determined according to three types of extraction strategy.

— In the case of an unconfined aquifer:
  - More or less continuous extraction of a fraction of this resource. However, this method does not make use of the maximum productive capacity of the aquifer.
  - Full capture at a mean discharge rate equal to the resource intake, using the hydrodynamic behaviour of the aquifer to ensure a year-on-year modulation of groundwater flow rates.

— In the case of a confined aquifer:
  - Excess capture, leading to the progressive exhaustion of the non-renewable groundwater resource (i.e. the permanent storage).

1.6 Prospection methods

In the context of assessing the potential for groundwater extraction in a particular area, water-bearing formations can be geologically characterized by:

— Lithology or rock-type, including unconsolidated (sand, gravel, etc.) as well as consolidated material (sandstone, limestone, granite, etc.).

— Spatial distribution (geometry in space): depth and thickness.

— Structural setting: faulting and folding, leading eventually to a subdivision of the aquifer into compartments.

— Lateral facies variations (e.g. gradations from sand into clayey sand or clay).

— Boundary conditions, viz
  - Possible zones of outcrop.

• Nature of the formation at the foot-wall and top-wall of the aquifer.

• Lateral trapping of the aquifer by a impervious fault.

The geological characteristics of an aquifer are generally schematized and presented as maps, cross-sections, logs (or well profiles) and block diagrams.

The use of prospection methods (photographic interpretation, geophysics) in hydrology enables the determination of some basic parameters essential for assessing the potential of an aquifer, in particular:

— The depth and thickness of a water-bearing formation.

— The lateral extent and thus the volume of the aquifer.

— The location of fractures affecting underground formations.

— The nature of formations at the top-wall of the aquifer.

— The delimitation of zones invaded by brackish or salt water.

— The location of exsurgences (gushing springs and seeps).

This information generally leads to the precise emplacement of mechanically drilled wells and the definition of their targets (final depth, depth to top-wall of the aquifer, thickness of overlying formations). The data obtained during drilling then allows a calibration and refinement of the interpretations.

The conventional hydrogeological approach and the procedure adopted by specialists can be broadly subdivided into four stages which are complementary to each other:

— *Preliminary studies*, which involve the acquisition of available documents covering the area in question, leading to a preliminary approach to the geological, climatological, hydrogeological and geomorphological environment.

— *Field studies*, enabling the hydrogeologist to make the supplementary observations that are indispensable at a particular site. After completion of the field survey, the hydrogeologist takes account of the observed geological setting (thickness of weathered zone, standing water-level, nature of the bedrock, etc.), as well as socio-economic constraints and means of access in order either to implement location of the well if the data are sufficient or to define a perimeter within which supplementary investigations need to be carried out.

— *Photographic interpretation*. This makes it possible to specify, among other features, the land forms (plateaux, drainage basins, low ground, type of drainage pattern, etc.) as well as the geological and structural elements (elongation direction, foliation, stratification, fracturing).

— *Supplementary investigations*. These commonly make use of geophysical exploration methods which are often employed to validate or eliminate hypotheses drawn up by the hydrogeologist. Electromagnetic, seismic and electric techniques are successively described in later sections of this chapter.

1.6.1 Preliminary studies

It is pointless to repeat work that has already been undertaken. This is why it is necessary to consult existing documents (maps, reports, syntheses of information) which allow the hydrogeologist to acquire a rational overview of the subject covering, as it does, various disciplines such as hydrogeology, geomorphology, structural geology and climatology.

In France, a large amount of information is available to the public through the Subsurface Databank of the BRGM.

On the basis of these documents, the hydrologist is able to define zones that are more or less favourable for groundwater recovery in order to target field studies.

1.6.2 Field studies

Field studies carried out by the hydrologist provides the basic technical details, equally as regards the geological as the socio-economic factors, that are essential for the satisfactory siting of water wells.

At this stage of the project, the hydrogeologist can check and study in more detail the data given in preliminary documents, and then direct the investigation according to the anticipated results. This work is frequently supplemented by photographic interpretation that can be performed directly at the site under consideration.

In all cases, the field investigation is an essential stage in the location of a borehole. It represents a direct addition to preliminary studies.

1.6.3 Photo-interpretation

Two types of photographic survey are generally used in hydrogeology: conventional aerial photogrammetry and satellite imagery. The latter is also known as remote sensing.

a) Aerial photography

This type of survey represents a most valuable documentary source. It is an efficient means of supplementing existing maps (topographic, geological and soil surveys) which provide the essential information for choosing the location of a borehole.

Photo-interpretation is a rapid and cheap method for drawing up a structural or even a geological sketch map, but it is above all used for identifying fractures at the local and regional scale. In principle, it is possible to trace all the morphotectonic lineaments which are underscored by the drainage pattern or which are simply picked out as light and dark areas on the photograph. In this way, a regional-scale image of fracturing is obtained which enables a statistical analysis.

b) Remote sensing

This technique has been used over the past twenty years and is based on data obtained from artificial satellites in Earth orbit. It comprises a certain number of advantages:

— The periods of the shots and the repetitivity of the information makes it possible to select the most interesting images.

— Satellite imagery leads to a better integration of major fractures (on the scale of several km) which are usually few in number.

— Even though the contrast of black-and-white images is poor, various existing enhancement treatments can be used with advantage (e.g., digital image processing can simplify the information by emphasizing certain trend directions).

— Image processing also enables the production of documents at larger scales (1:100 000; 1:50 000).

However, there are some disadvantages which place considerable limits on this technique:

— Although the large scale of observation (1:200 000) is an advantage in regional syntheses, it becomes a drawback when it is necessary to transfer to the scale of the terrain.

— The lack of relief on these images hinders satisfactory correlation between the remote sensing data and the ground truth, to such an extent that the user is then obliged to fall back on conventional methods.

— Image processing increases the costs and brings about a certain deterioration in the quality of information.

Nevertheless, where high groundwater discharges are sought, satellite imagery can prove to be a valuable guide to the hydrogeologist in the selection of borehole sites. In particular, the more recent satellites (Spot and Landsat D) provide a better resolution at ground level (30, 20 or 10 m) and offer the possibility of obtaining stereoscopic views (Spot). As a result, such images can reveal just as much detail as aerial photographs at a scale of 1:50 000.

Otherwise, satellite remote sensing in the infrared also leads to promising results concerning the hydrogeology of fractured terrains. In fact, geological discontinuities exhibit a variation in thermal behaviour which is probably a function of several phenomena, some of which can be indirectly detected from space due to their effect on infrared emission. This is notably the case with the hydraulic condition of ground features.

1.6.4 Electromagnetic methods

The same principle is common to all electromagnetic methods. A conducting formation buried beneath the surface is subjected to an alternating magnetic field (the primary field Hp), thus inducing the flow of electric currents which in turn create a secondary magnetic field Hs. Measurements are carried out on the secondary field Hs. The instruments used allow the determination of the apparent resistivity of a slice of terrain situated near the surface.

The depth range of investigation depends notably on the type of apparatus used and the resistivity of cover formations; it varies from about 10 m to 80 m (airborne surveys). While seismic waves are propagated at acoustic velocities, electromagnetic waves travel practically at the speed of light. The presence of a discontinuity or "reflecting horizon" is expressed by its observed effect on electromagnetic impedance.

— Orders of magnitude of resistivity:
  • A few ohm-meters in clayey terrains.
  • Several tens of ohm-meters in marly limestones.
  • Several hundred ohm-meters in massive limestones.

— Some examples of application:
  • In mining geology: detection of veins.
  • In civil engineering and hydrology: location of karstified zones, marly substratum at shallow depth, areas of surficial drainage.

Many electromagnetic methods have found an application in the field of hydrogeology. Among these, we can mention radar, electromagnetic well surveying and Very Low Frequency methods.

1.6.5 Seismic methods

A disturbance originating at ground level will be propagated in the form of waves through the subsurface as shock waves which are progressively attenuated. Seismic surveying is based on a study of the travel-times of these waves. Travel-time depends on the nature and structure of the geological formations.

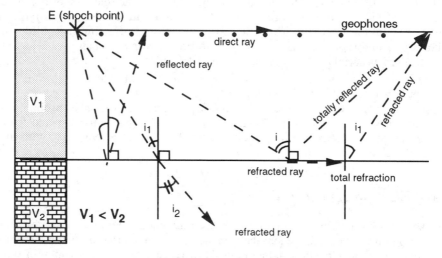

Figure 1-10
Trajectory of seismic waves at the interface of two heterogeneous media.

These waves, which are represented by seismic ray-paths, are of three types (cf. Figure 1-9):
— Compressional waves.
— Shear waves.
— Surface waves.

Since these waves obey the laws of optics, seismic surveying makes use of reflection and refraction to study the nature of the traversed terrains. The study of the elastic properties of rocks enables a distinction to be made between porous and/or fractured zones and compacted rocks (limestones, sandstones, crystalline rocks, etc.) as well as between saturated and unsaturated zones and, finally, between unconsolidated and consolidated formations. The measurements are carried out essentially with regard to the two parameters trajectory (ray-path) and velocity.

Operations in the field comprise the following:
— Setting up the seismic array with a point source (explosive shot, vibrator) and receivers (geophones).
— After signal amplification, recording of the transit-times of waves between the source and receivers.

Since each lithological formation is characterized by a seismic velocity, geophysical seismic surveys are able to provide, even before exploratory drilling, some predictive cross-sections of the terrain showing the thicknesses of beds and the location of certain types of tectonic discontinuity (faults).

Two seismic propection methods are used:
— *Seismic reflection*, which considers solely the reflected rays.
— *Seismic refraction*, which is mainly concerned with fully refracted rays.

In hydrogeology, seismic refraction is preferred to seismic reflection for the following reasons.:
— Better results are obtained in the depth range 0-200 m.
— Better implementation on small profiles.
— Lower costs.

a) Principles of seismic refraction

The distance of the seismographs from the shotpoint is known and the time of first arrival of the shock is read off the recording.

If the velocity of elastic waves in the terrain is uniform, then the travel-time will be proportional to distance. But since the velocity increases with depth, the seismic waves are refracted with the result that they follow concave-upward trajectories. Consequently, the first arrival of the shock corresponds to an acoustic (or seismic) ray-path that penetrates deeper as the geophone is placed farther from the source.

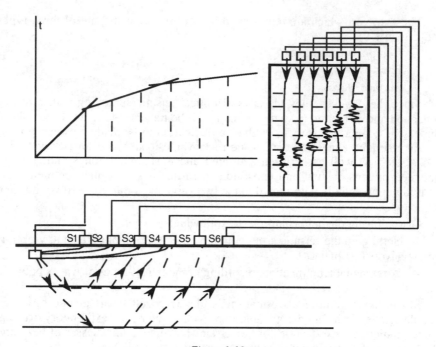

Figure 1-11
Principle of seismic refraction. A cross-section of the terrain is given (at bottom), with
chart recording (top right) and calculated time-distance curves (top left)
(after J. GOGUEL, 1967).

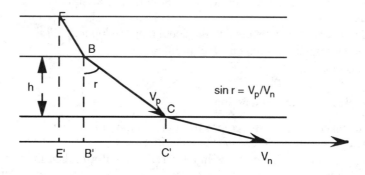

Figure 1-12
Principle used to calculate depth for a horizontal layer.

Seismic ray-paths can be defined as having trajectories perpendicular to the
wave front. Seismic waves emitted from the source are governed by laws
analogous to geometrical optics, with the refractive index being replaced by the
reciprocal of velocity. The travel-time along a ray-path descending to a given

depth depends on all the formation velocities between the ground surface and this depth.

Once the travel-times are recorded as a function of source-geophone distance, the problem is then to calculate the variation of shock-wave velocity with depth (cf. Figure 1-12).

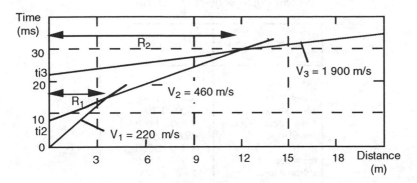

Figure 1-13
Time-distance relationship for three velocities of seismic wave.

— *Interpretation:* the obtained data make it possible to draw time-distance curves with shotpoint-to-geophone distances plotted along the x-axis and first arrival times at the receiver along the y-axis. The slopes of the straight lines on this diagram give the propagation velocities of wavelets for various trajectories that correspond to the first arrivals of shock waves at the receiver (cf. Figure 1-13).

The coordinates of the intersections of these straight lines enable a calculation of the depths at which changes occur in the geological structure. The thickness of the upper layer, or the depth to the first formation boundary, is given by the following equations:

$$h_1 = \frac{R_1}{2} \sqrt{\frac{V_2 - V_1}{V_2 + V_1}}$$

where:

h_1 : depth in metres to the base of the first layer,

V_1 : seismic velocity in the upper layer,

V_2 : seismic velocity in the layer situated beneath the upper layer,

R_1 : horizontal distance in metres, measured off the diagram, between the receiver and the vertical projection of the first velocity change (x-axis coordinate of the crossover point).

V_1 and V_2 are calculated from the reciprocals of the slopes of linear segments on the time-distance diagram.

Table I-VII presents the mean seismic velocities for a variety of different terrains. These velocities are approximate, being only valid for formations that are relatively close to the surface.

In this discussion, only the case of horizontal layers is considered. It is nevertheless straightforward to detect the dip of strata by carrying out two profiles in opposite directions following the same line.

TABLE I-VII — *Approximate seismic (or acoustic) velocities for different types of material.*

Nature of medium	Velocity (m/s)
air	330
water	1 400 - 1 700
alluvium, dry sand	500 - 1 200
waterlogged alluvium	1 600 - 2 000
humid sand	600 - 1 500
sand	1 800 - 2 200
sea water	1 400 - 1 500
volcanic tuff	1 600 - 2 500
lava	2 500 - 5 000
hard sandstone	1 800 - 3 500
hard shale	2 500 - 4 000
chalk	1 500 - 3 000
marl	2 000 - 3 500
limestone	2 000 - 4 500
weathering deposits	1 000 - 2 000
granite	3 000 - 5 000
basalt	3 500 - 6 000
metamorphic rock	3 000 - 5 000
limestone with dolomite	2 500 - 3 500

Nevertheless, seismic refraction suffers from some limitations regarding its use:

— In the case of relatively thin layers, the interpretation can be somewhat difficult and may lead to inaccuracies in estimating the thickness of the upper or underlying layers.

— Another limitation is the impossibility of detecting a seismically slower layer situated beneath a faster layer. It is extremely difficult to pick out the area of no signal reception. This case is relatively rare in routine exploration, but such circumstances render it impossible to investigate the terrain behind the concrete liner of underground workings.

In order to mitigate these disadvantages, seismic refraction may be combined with electric logging. This is notably applied in certain basins to determine the thickness of a high-resistivity aquifer (alluvium, weathering deposits) lying on top of a substratum of equally high resistivity (limestones, granites).

b) Seismic reflection

This method is relatively little used in hydrogeology because of the difficulty of obtaining distinct reflections from depths of less than 200 m. The techniques of data acquisition and processing have been very markedly improved over the past few years, and it is now possible to filter the signals correctly and make use of seismic reflection at shallow depths.

Small-scale seismic reflection surveys provide a valuable adjunct to seismic refraction, covering a range of investigation from a few tens of metres to some hundred metres depth.

1.6.6 Electric survey methods

The methods of electrical prospection are based on the conductivity of subsurface formations, or their capacity to conduct an electric current, whether natural or artificial. They do not involve measuring the magnetic field.

Electric survey methods make use of:

— Natural (telluric) currents:
  • Spontaneous (or self) polarization.
— Artificial currents:
  . Electrical potential (mapping of potential difference, with earthing, etc.).
  . Resistivity method (electric logging, resistivity rectangles).

The most commonly employed method in hydrogeology is resistivity. The electrical conductivity of rocks is more commonly expressed by its reciprocal, the resistivity, which has the units of ohm-metre.

TABLE I-VIII — *Mean values of resistivity for rocks and waters.*

Medium	Porosity (%)	Permeability $(cm.s^{-1})$	Resistivity $(ohm.m)$
Claystone	35	$10^{-8} - 10^{-9}$	70 - 200
Chalk	35	10^{-5}	30 - 300
Volcanic tuff	32	10^{-5}	10 - 100
Marl	27	$10^{-7} - 10^{-9}$	10 - 50
Sandstone	3 - 35	$10^{-3} - 10^{-6}$	50 - 1 000
Dolomite	1 - 12	$10^{-5} - 10^{-7}$	200 - 10 000
Limestone	3	$10^{-10} - 10^{-12}$	50 - 1 000
Gneiss	1,5	10^{-8}	10 - 10 000
Quartzite	< 1	10^{-10}	1 000 - 10 000
Granite	1	$10^{-9} - 10^{-10}$	10 - 1 000
Gabbro	1 - 3	$10^{-4} - 10^{-9}$	6 000 - 10 000
Basalt	1,5	$10^{-6} - 10^{-8}$	300 - 15 000
Clays	45	$10^{-7} - 10^{-9}$	2 - 20
Lavas		$10^{-5} - 10^{-9}$	100 - 15 000
Sea water			0.1 - 0.3
River water			5 - 100
Groundwater			0.2 - 50

Resistivity ρ is the resistance (R) of a cylinder having unit length (L) and unit cross-section (S). It is expressed in ohm-metres in the following relation:

$$\rho = R\frac{S}{L}$$

The resistivity of rocks depends mainly on:
— Porosity and permeability (and thus indirectly the degree of fracturing).
— Content of water of imbibition and its resistivity (salinity).
— Clay content (shaliness) of the formation.

Disregarding the clay content, a lithological formation will have a higher resistivity as its water content decreases. Similarly, the resistivity of a formation becomes lower as the clay content rises.

a) Potential difference method

A direct or alternating current passing along a fixed line AB will induce a potential difference (PD) between a fixed electrode M and a movable electrode N.

The survey consists of drawing up a map of PD by displacing electrode N, bearing in mind that the distribution of equipotential lines in the subsurface is a function of the presence and homogeneity of rock masses of variable conductivity/resistivity.

However, the interpretation of PD maps remains qualitative.

A special type of electrical potential method involves *earthing*. One of the current input electrodes is placed at a point on the conductor (e.g. an ore body or resurgence of an underground river) while the other electrode is placed at infinity. Since the conductor is everywhere at roughly the same potential, the ore body or underground river is delimited by equipotential lines (cf. Figure 1-14). However, some very particular conditions are required to reveal underground flow and, in this context, earthing may be employed as a strategy in support of prospection.

b) Resistivity method

The basic principle of the resistivity method consists of sending an electric current — usually a direct current — of intensity i between two electrodes A and B, and measuring the potential difference between two other electrodes M and N. The set of electrodes makes up a quadripole with any proportions.

The resistivity method may be applied in two different way according to the objective sought:
— Electric logging, leading to a quantitative interpretation.
— With panel-type arrays, leading to a qualitative interpretation.

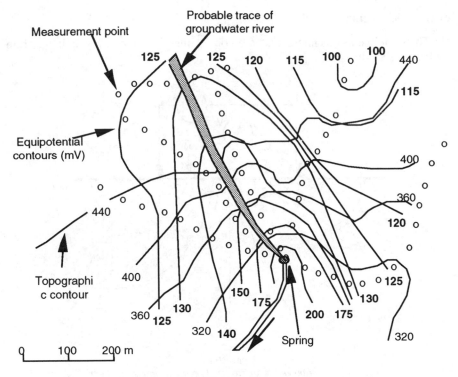

Figure 1-14
Example of the determination of the course of an underground river using the earthing method (after a document, courtesy of SAFEGE).

The potential difference ΔV between M and N can be written as:

$$\Delta V = V_M - V_N$$

$$\text{where } V_M = \frac{\rho\, i}{2\pi} \left(\frac{1}{AM} - \frac{1}{BM} \right)$$

from which we obtain $\Delta V = \frac{1}{2\pi} \left(\frac{1}{AM} - \frac{1}{AN} - \frac{1}{BM} + \frac{1}{BN} \right) \rho \cdot i$

The true resistivity of the terrain is given by the following equation:

$$\rho = K \frac{\Delta V}{i}$$

where ρ is in units of ohm.m, ΔV in mV and i in mA. K is a coefficient characteristic of the geometry of the quadripole AMNB, which is expressed in metres.

c) Electric logging

The principle of this method is based on the identification of layers in a terrain by their resistivity.

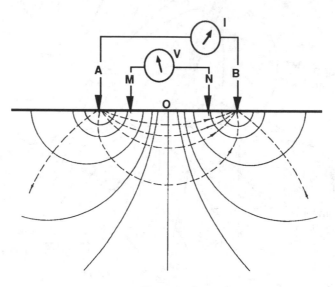

Figure 1-15
Operational arrangement for electric logging (Schlumberger-type)

Resistivity varies as a function of two main parameters:

— *The lithological composition*: the more clayey a terrain, the lower its resistivity. Thus, a clayey sand is a better conductor than a clean sand or a sandstone. By the same token, a compact limestone has a higher resistivity than a fractured or weathered limestone.

— *The water content and the degree of mineralization of the water*: a water-saturated terrain is a far better conductor than a dry terrain; the higher the degree of mineralization of the imbibition water, the higher the conductivity of the terrain. The presence of highly mineralized water (salt water, for example) in a terrain leads to high conductivities, thus masking the electrical properties of the underlying formations.

In this way, depending on the geological setting, resistivity surveys make it possible to determine the lithological composition of the formations encountered, as well as their degree of fracturing and their potential invasion by brackish or salt waters. The measurement of resistivity values is carried out through the interpretation of electric logging data. The implementation of electric logging involves introduction of an electric current (I) into the ground by means of two electrodes (A, B). As a result, a pattern of equipotentials is set up which, in particular, generates two curves emerging at the surface at points M and N (receiver electrodes). This allows measurement of the potential difference. The most commonly used arrangements are the Schlumberger-type array (composed of an in-line quadripole AMNB with MN ≤ 1/5 AB; cf. Figure 1-15) and the Wenner-type array characterized by a linear layout with AM = MN = N.

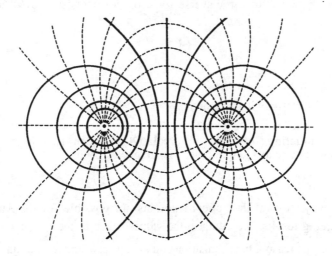

Figure 1-16
Distribution of current flow (dotted lines) and equipotentials (solid lines) in the field and extending beneath ground level.

A knowledge of the potential difference $\Delta V = V_M - V_N$, the current I and the geometry of the AMNB array (otherwise called a quadripole) enables calculation of the mean resistivity of a certain rock volume situated broadly to the right of MN. The depth investigated is a function of the distance AB, as well as the resistivities of the formations and their spatial distribution. By increasing the AB or the MN spacing, it is generally possible to enhance the depth of investigation. A plot of apparent resistivity ρa versus the AB half-distance is drawn up on log-log paper; the curve representing the relation $\rho a = f(\frac{AB}{2})$ is then interpreted in order to obtain true resistivity values for the successive formations.

According to the starting hypothesis assumed (number of layers, resistivity or thickness), application of an algorithm leads to a best fit of the theoretical curve to the field data.

The interpretation of electric logging data is carried out by comparing, either manually or automatically by computer, the field-based curves with theoretical nomograms. Thus, the thickness and true resistivity can be obtained for each defined layer. Only the resistivity can be determined for the deepest formation.

The following principles are notably used as a basis for this interpretation:

— A high-resistivity formation intercalated between two conducting formations gives rise to a transverse (or crosswise) resistance R_t, defined as follows:

$$R_t = e\,\rho$$

where:

e : thickness in metres,

ρ : resistivity in ohm-metres.

— A conducting formation intercalated between two high-resistivity formations gives rise to a longitudinal conductance C_l, with:

$$C_l = \frac{e}{\rho}$$

In this way, several theoretical solutions are possible for the same set of electric logging results, according to the values given to the products $e.\rho$ or the ratios $\frac{e}{\rho}$. For this reason, the acquisition and synthesis of the maximum amount of geological information is necessary for the interpretation of electric survey data in a given area., in order to rule out certain theoretical solutions at the expense of more geologically realistic solutions.

Recent water research carried out in basement terrains show a direct relation between the longitudinal conductance, the failure rate and the discharges obtained in water wells (cf. Table I-IX).

The total longitudinal conductance is derived from electric logging results using the position of the final upgoing branch of the $\rho a = f(AB/2)$ diagram.

It corresponds to the sum of $\frac{\xi i}{\chi i}$ ratios, in which ξi represents the thickness of the different layers making up the complex and χi represents their resistivity.

Table I-IX indicates statistically that high values of total longitudinal conductance are associated with relatively high discharges. Conversely, low values of conductance correspond to low discharges.

As far as possible, it is indispensable to perform electric logging in proximity to wells having a known geological cross-section (calibration well). If this is not the case, it is still preferable to carry out at least one exploratory borehole after a geophysical electric survey in order to calibrate and refine the interpretation of electric logging results.

TABLE I-IX — *Relations between discharge, success rate and total longitudinal conductance.*

Longitudinal conductance (mho)	Number of measurements	Percentage of successes	Discharge obtained (in m^3/h)
< 0.1	48	< 25%	< 1
0.1 - 0.2	61	78%	2
0.2 - 0.5	88	98%	2,8
0.5 - 2.5	76	66%	3.5

In hydrogeology, the use of electric logging is well adapted to the following cases:

— In alluvial environments (river valleys), to determine the thickness and quality of alluvium, locating the standing water-level in certain cases and identifying clayey lenses.

— In areas of granitic or volcanic basement, to establish the importance of weathering zones.

— In sedimentary terrains, to identify clayey zones and determine the degree of fracturing (by its effect on permeability).

— In coastal environments, to delimit the salinity front (freshwater/ saltwater boundary).

Electric logging, which is based on resistivity profiles and their lateral extensions, leads on to the siting of prospect boreholes and the drawing up of predictive lithological profiles (nature and thickness of formations, drilling target).

d) Electric panel arrays

In an electric panel array, the electrodes A, M and N are all situated on the same axis and electrode B is placed at infinity along a direction perpendicular to AMN.

The method consists of measuring apparent resistivity between M and N, for different positions of electrode A. In addition, electrodes M an N are displaced by the same step along the axis of the array while maintaining a constant spacing between themselves. Thus, electrode A is moved from one side to the other of the receiver pair MN.

Six resistivity measurements are performed at each position of the pair MN, covering a depth interval beneath this position which decreases as electrode A comes closer to MN (cf. figure 1-17). Two series of measurements are taken for each position of the receiver pair, with current injection on the left and on the right. The discrepancies between these two series can be explained in terms of heterogeneities occurring from one side to the other of MN.

Electric surveys of this type are generally presented with injection on the right, on the left or in the middle of the array. The interpretation of the results so obtained is only qualitative, being a function of the lateral variation of resistivity in the terrain.

Panel-type arrays are commonly employed to reveal or confirm the presence of major structural features (fault traces), in which case the array is operated at right angles to the inferred structural trace. Such surveys enable the emplacement of exploratory boreholes near faults, in order to confirm their role as drains or impervious boundaries (e.g. in the Chalk or in basalts).

It should be mentioned that so-called square and double-rectangle arrays are frequently used in geoelectric prospecting. The square array enables the measurement of anisotropy corresponding to the fracturing direction, the coefficient of anisotropy depending on the intensity of fracturing and the mean apparent resistivity. The double-rectangle array allows a horizontal investigation in two directions for a practically constant depth; these directions are parallel and perpendicular to the anisotropy trend derived from the square array. The interpretation of double-rectangle data leads to great precision in the location of the borehole.

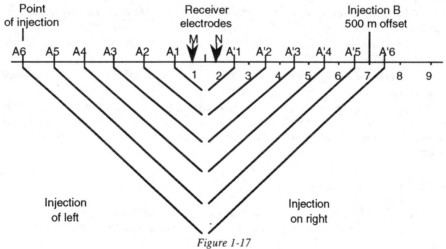

Figure 1-17
Operating principle of the panel-type electric survey (after SAFEGE document).

1.6.7. Gravimetry

Gravimetric methods require competent personnel, while their implementation is relatively cumbersome and costly. This explains why such methods are little used in hydrogeology. On the contrary, microgravimetry is easier to apply for solving a particular problem, and is being increasingly used to locate karstic cavities at shallow depths.

The principle of gravimetry is based on the force of attraction between masses, giving rise to an acceleration g - due to gravity - which affects all bodies placed near the surface of the globe. If the Earth were perfectly spherical and homogeneous, the value of g would be the same at all points on its surface. However, this not the case and g is seen to vary for a number of reasons, some of which are geological (variable density of different rock-types, erosion, tectonics, uprise of magmas). Other reasons are non-geological in origin (effect of elevation, position of the Sun and Moon, instrumental drift, etc.).

The skill in applying this method amounts to identifying the non-geological factors, and then being able to demonstrate and explain the causes of variation in gravity linked to certain anomalies of interest in hydrogeology. These anomalies are said to be positive when they are expressed as an increase in g (excess of mass in the subsurface) and negative when they are linked to a mass deficit. The density of a rock formation varies with its mineralogical composition, its porosity and the degree of fluid saturation in its pores. As an indication, the densities of some potentially water-bearing rocks are reported in Table I-X.

TABLE I-X — *Densities of some potential reservoir rock-types (after C. MEYER DE STADELHOFEN).*

Rock-type	Probable density range (g/cm^3)
Granite	2.5 - 2.7
Basalt	2.7 - 3.3
Quartzite	2.6 - 2.7
Micaschist	2.5 - 2.9
Gneiss	2.7 - 2.8
Amphibolite	2.8 - 3.2
Coal	1 - 1.5
Chalk	1.9 - 2.7
Sandstone	1.8 - 2.7
Limestone	2.6 - 2.7
Dolomite	2.4 - 3
Sand	1.4 - 2
Moraine	1.8 - 2.2
Weathering deposit	1.6 - 2
Clay	1.6 - 2.2
Gravel	1.8 - 2.1

A conventional gravimetric survey involves the determination of the *Bouguer anomaly*. This corresponds to the difference between the measured and theoretical values of g at the point of measurement. According to the particular context, it is evidently appropriate to take account of the various corrections applied to the measured value of g.

$$\text{Bouguer anomaly} = g \text{ measured} - g \text{ theoretical}$$

The theoretical value of g is given by:

$$978.0490 \left(1 + 0.0052884 \sin^2 \lambda + 0.0000059 \sin^2 2\lambda\right)$$

which applies to all points on the Earth's ellipsoid having zero elevation. It is expressed in gal (named after Galileo) and has a value of 978.0490 at zero elevation on the equator, while, at other latitudes l and at zero elevation, the value of g is calculated with the expression given above. In the CGS system, the gal corresponds to a gravitational acceleration of 1 cm/s^2.

Since the Bouguer anomaly is exclusively the result of geological causes, such anomalies can be linked with the study of water-bearing structures or reservoirs.

1.7 Water chemistry

The water in geological formations is in reality an aqueous phase whose composition can be highly complex due to its great power of dissolution.

In principle, given the intimate contact existing between rocks and their impregnated waters, groundwaters should be saturated with respect to all the substances present in a soaked terrain. In fact, two situations can arise:

— The groundwater remains stationary and does not take part in the circulation of surficial (or connate) waters, in which case it can become highly saline.

— The groundwater circulates within the terrain and progressively leaches the rocks, taking up low levels of mineralization that are variable according to geography and climate.

This passage into solution is brought about by dissolution and chemical reactions with substances present in the formation traversed by the water.

1.7.1 Dissolution

The dissolution of a substance involves the loss of its cohesiveness, which itself originates from Coulomb-type electrostatic forces. The hydrating power of water (which is a strongly bipolar molecule) leads to a partial or complete destruction of the various electrostatic bonds between the atoms and the molecules of the substance entering solution, thus forming new bonds and structures in the liquid state. This process is known as solvation. Dissolution is said to occur when solvation is complete.

a) Salts

During their residence in the subsurface, groundwaters take up a certain number of mineral substances into solution, including limestone ($CaCO_3$), dolomite($(Ca, Mg) 2CO_3$), gypsum ($CaSO_4 2H_2O$), sodium chloride (NaCl) and potassium chloride (KCl.)

The amount of dissolution of these substances, which make up most of the soluble matter in sedimentary rocks, is necessarily of considerable importance. Nevertheless, some other constituents of reputably insoluble rocks — such as silica and the silicates — are liable to be dissolved in very small quantities.

b) Gases

Waters can also enter into contact with gases, especially in the zone of infiltration. The solubility of gases is governed by Henry's law.

The volume of a gas that can be absorbed by a volume of water depends on the pressure and concentration of gases in the medium. Gas solubility in water decreases with increasing temperature and increasing salt concentration.

Most gases in contact with water have very similar solubilities, the order of magnitude only varying by a factor of two. Only certain gases, such as carbon dioxide, ammonia or hydrogen sulphide (H_2S), exhibit solubilities that are much higher.

The solubilities of the principal gases in contact with water are reported in Table I-XI.

TABLE I-XI — *Solubilities of some common gases in water.*

Gas	Solubility (mg/l)
Nitrogen (N_2)	23.3
Oxygen (O_2)	54.3
Carbon dioxide (CO_2)	2 318
Methane (CH_4)	32.5
Hydrogen (H_2)	1.6
Hydrogen sulphide (H_2S)	5,112

1.7.2 Chemical attack

The agents of corrosion and chemical weathering of rocks are quite complex; they involve the following main mechanisms:

— *Hydration*: penetration of water into the lattice of crystalline solids.

— *Hydrolysis*: plays an important role in the destructive action of water on silicates.

— *Oxidation*: a phenomenon of considerable importance in highly oxygenated infiltration zones. Its effect is felt to a lesser degree at greater depth. Oxidation particularly concerns oxides that are deficient in oxygen (magnetite), as well as sulphides such as pyrites and also organic matter.

— *Reduction*: corresponds to the reverse of oxidation, being equally important in groundwaters.

After the initial processes of dissolution and chemical attack, and often following a more or less prolonged residence underground, secondary phenomena can bring about modifications in the chemical composition of the waters. The most significant of these reactions are:

- Sulphate reduction.
- Base exchange.
- Oxidation.
- Renewed dissolution/precipitation of phases, leading to changes in solute concentration.

1.7.3 Physico-chemical properties of waters

The main objective of physico-chemical studies is to ascertain the origin of groundwaters as well as their potability or degree of pollution.

The following may be cited as examples of the physico-chemical parameters generally investigated in groundwaters:

— Physical parameters (temperature, pH, conductivity, etc.), which need to be measured *in situ*.

— Chemical parameters, major and trace element composition:

- The cations: Na^+, K^+, Ca^{2+}, Mg^{2+}
- The anions: HCO_3^- , SO_4^{2-} , Cl^-, NO_3^- ;
- Silica: SiO_2;
- Trace elements: Al^{3+}, Fe^{2+}, Mn^{2+}, Fe^{2-}, etc.

These parameters can provide some indication of geochemical exchanges with the reservoir.

a) Water colour

The colour of pumped out water is a parameter that can be readily estimated. It is measured by comparing the tint of the sample with calibrated known reference solutions. The results are expressed in colorimetric units or in degrees HAZEN.

$$1° \text{ HAZEN} = 1 \text{ ppm Pt-Co.}$$

In practice, coloured shades often replace the use of standard solutions.

b) Turbidity

The turbidity of a water becomes even greater with increasing amounts of colloid in suspension. In practical terms, turbidity is evaluated by means of a nephelometer which measures the intensity of light diffused laterally by the water sample. The unit of measurement is the IU (International Unit), which corresponds to 1 mg of formazine per litre. It is equivalent to the Jackson unit (used in English-speaking countries). In groundwaters of low turbidity, 1 IU corresponds to about 1mg/litre of colloidal matter in suspension.

c) pH

pH values encountered in water research grade from pH 0 (strongly acidic media) to pH 14 (strongly alkaline media).

The measurement of pH can yield some important information concerning the nature of waters.

— A pH of less than 7 corresponds to acid waters, which are consequently often aggressive and corrosive.

— Alkaline waters have a pH higher than 7 and tend to give rise to scale formation above a certain degree of hardness.

In reality, there is a precise value of pH beneath which waters become aggressive and above which the waters are encrusting. This value is called the equilibrium pH and is denoted pH_S.

From the actual pH of water and its theoretical pH_S, it is possible to define two indices commonly used in corrosion problems:

— LANGELIER'S index: $I_L = pH - pH_S$,

— RYZNAR'S index: $I_R = 2 pH_S - pH$.

The role played by these parameters in the prevention of corrosion phenomena is discussed in more detail in a later section.

d) Conductivity

Conductance is the reciprocal of resistance; it characterizes the property of a substance — in this case liquid — to allow the flow of an electric current. The unit of electrical conductance is the siemens (or mho), whereas the unit of resistance is the ohm.

The measurement of resistance or conductance is of great practical importance since it enables the rapid and precise monitoring of the slightest perturbation reaching groundwater bodies.

Furthermore, the conductivity (or resistivity) of water is a relatively faithful indicator of its mineralization. Conductivity increases with the concentration of dissolved salts and varies as a function of temperature. It is expressed in μsiemens/cm (or μmhos/cm), while resistivity is expressed in ohm-cm.

A simple relation can be used to convert conductivity into water salinity:

$$1 \ ppm = 1.56 \ \mu S/cm$$

The variation of conductivity gives important information on the evolution of water quality. In relation to conductivity, the mineralization of water may be represented as indicated on Table I-XII.

TABLE I-XII — *Relation between water conductivity and mineralization.*

Conductivity in µS/cm, at 20° C	Mineralization
< 100	very weakly mineralized water (granitic terrains)
100 - 200	weakly mineralized water
200 - 400	slightly mineralized water
400 - 600	moderately mineralized water (limestone terrains)
600 - 1 000	highly mineralized water
> 1 000	excessively mineralized water

e) Hardness

The hardness of water is due mainly to the presence of calcium and magnesium salts in the form of bicarbonates, sulphates and chlorides. It therefore represents the concentration of alkaline earth cations, which is measured in broad terms by titration (determination of water hardness).

Hardness is commonly expressed in degrees on the French scale as follows:

— 0 - 7°: very soft water

— 7 - 14°: soft water

— 14 - 20°: moderately hard water

— 20 - 30°: fairly hard water

— 30 - 50°: hard water.

— 50° and above: very hard water.

f) Alkalinity

This parameter gives the concentration of hydroxides, carbonates and bicarbonates of the alkali metals and alkaline earths. On the basis of the alkalimetric titre (AT) or the total alkalimetric titre (TAT), it is possible to derive the distribution of the three main groups of chemical species that give rise to alkalinity.

TABLE I-XIII — *Distribution of species responsible for alkalinity in water.*

Dissolved species	mg/l per degree of hardness	When AT = 0	When AT<TAT/2	When AT=TAT/2	When AT>TAT/2	When AT=TAT
(OH) CaO Ca(OH)$_2$ MgO Mg(OH)$_2$ NaOH	3.4 5.6 7.4 4.0 5.8 8.0	0	0	0	2TA-TAC	TAC
CO$_3$ CaCO$_3$ MgCO$_3$ Na$_2$CO$_3$	6.0 10.0 8.4 10.6	0	2 TA	TAC	2(TAC-TA)	0
HCO$_3$ Ca(HCO$_3$)$_2$ Mg(HCO$_3$)$_2$ NaHCO$_3$	12.2 16.2 14.6 16.8	TAC	TAC-2TA	0	0	0

g) Chemical substances present in water

The chemical composition of groundwater is highly variable. The water contained in an aquifer can undergo the effects of several different phenomena (solute concentration, base exchange, redox reactions) which are capable of causing partial modifications to the chemical properties.

In the same body, variations are generally observed from the upstream to the downstream side.

— Increase in the total concentration of dissolved salts.

— Increase in the $\frac{Cl}{SO_4}$ ratio due to the faster dissolution rates of chlorides compared with sulphates.

— Decrease in the $\frac{Ca}{Mg}$ ratio. In fact, the input of calcium from dissolution of CaCO$_3$ is rapidly cut off because the inflowing waters are already saturated from the start.

Sulphates

Mainly derived from gypsum and anhydrite, as well as from the oxidation of pyrites. Certain sulphates of magnesium or sodium can also be present.

Chlorides

Among other sources, these may come from salt formations in potassic evaporite basins. Since chlorides are not absorbed by the soil, they can be transported over long distances. In addition, they may originate from excessive pumping near the sea coast. A water containing less than 150 ppm of chlorides is good for all uses. It is still considered as drinkable up to levels of 250 ppm.

Nitrates and nitrites

Found mainly in surficial layers, resulting from the infiltration of surface waters. They are generally indicative of pollution.

Iron and manganese

Commonly present in groundwaters as reduced species (Fe^{2+} and Mn^{2+}). Their contents are usually of the order of 0.2-5 mg/l for iron and 0.05-2 mg/l for manganese. In contact with air, the ferrous ions (Fe^{2+}) are oxidized to ferric ions (Fe^{3+}).

Ammonium ion

The ammonium ion (NH_4^+) is very frequently present in groundwaters, resulting in most cases from the anaerobic decomposition of nitrogenous organic matter. Its concentration is generally in the range 0.1- 0.2 mg/l.

Dissolved gases

The solubility of dissolved gases varies in inverse proportion to the temperature and increases with pressure. The main gases encountered in solution in groundwaters are listed below:

— Oxygen, which is responsible for accelerating the corrosion of iron, steel, zinc and copper. Waters containing oxygen are even more corrosive at lower pH or when conductivity is high. On the contrary, a well oxygenated water rich in calcium bicarbonate will deposit a protective scale on conduits.

— Hydrogen sulphide (H_2S) renders water both aggressive and corrosive. Waters containing H_2S will attack steel and form iron sulphide encrustations. It is characterized by a smell of rotten eggs.

— Carbon dioxide (CO_2) is very abundant in groundwaters. The solution of this gas will be stable if the pressure is sufficiently high. Following a drop in pressure (e.g. due to pumping), degassing of the CO_2 is accompanied by the formation of calcium encrustations. Thus, it is preferable in this case to reduce the drawdown due to pumping and the speed of water entering the screens.

Other dissolved gases are encountered in groundwaters, but in smaller proportions. This applies, among others, to nitrogen, sulfur dioxide and methane.

1.7.4 Isotopic characteristics

Variations in the relative abundances of stable isotopes (^{18}O, ^2H, ^{13}C) provide information on the origin of groundwaters. These variations are due to the fractionation which occurs during phase changes or chemical reactions. In the course of the fractionation process, water vapour is always depleted in heavy isotopes with respect to the remaining liquid (or the condensate is always enriched in heavy isotopes with respect to the original vapour phase). Because of this fractionation, which depends on temperature, the relative contents of isotopes vary as a function of several parameters, i.e.:

— Evaporation.

— Elevation or altitude.

— Latitude.

— "Continentality" (degree to which a climate is affected by continental influences).

Radioactive isotopes (^3H, ^{14}C) enable the estimation of turnover rates and circulation flow rates for groundwaters, thus leading to the basic concept of residence time for waters in an aquifer.

In addition, enriched isotope spikes (radioactive or stable) can be introduced into a system on the small scale in order to carry out artificial tracer studies of the local flow pattern.

By measuring the isotopic composition of sulfur in the dissolved sulphate, it is possible to identify the eventual origin of this element. In the case of a natural source, it is derived from pyrites, evaporite formations or incursions of marine water. Artificial sources include atmospheric pollution from the combustion of fossil fuels and groundwater contamination arising from fertilizers.

Nitrogen isotopes provide information on the origin of nitrates in groundwater (industrial inputs, oxidation of human-derived organic matter or decomposition of organic matter in the soil).

1.7.5 Qualitative requirements

Groundwater is considered as polluted when it contains substances other than those associated with the natural structure of the terrain though which it has passed. In particular, pollution is said to exist if the concentrations of dissolved or suspended constituents exceed the maximum permissible levels according to national or international standards.

Apart from fulfilling the required physico-chemical criteria, drinking water should not have:

— A colouration corresponding to more than 15 mg/l platinum with reference to the Pt-Co scale.

— A turbidity greater than 2 TU (Turbidity Unit).

— Any odour or taste at a dilution factor of 2 (12° C) or 3 (25° C).

The main guiding principles concerning drinking water quality are set by the World Health Organization (WHO). For ionizing radiations, indicative recommended values are based on data established by the International Commission for Radiological Protection (ICRP), assuming a daily consumption of 2 litres per adult:

— Total alpha-emitter radioactivity (exclusive of radon): 0.1 Bq/litre,

— Total beta-emitter radioactivity (exclusive of tritium): 1 Bq/litre.

The dose contributed by drinking water should remain lower than 0.05 mSv/person/year.

Through the regulations 3954-87 of 22nd December 1987 and 2218-89 of 18th July 1989, European legislation on radiological protection defines the maximum admissible standards for radioactive contamination in liquids destined for human consumption.

— Strontium isotopes (^{90}Sr): 125 Bq/litre.

— Iodine isotopes (^{131}I): 500 Bq/litre.

— Plutonium isotopes and alpha-emitting actinides (e.g. ^{239}Pu and ^{241}Am): 20 Bq/litre.

— Any other radionuclide with a half-life longer than 10 days (^{134}Cs and ^{137}Cs): 1 000 Bq/litre, (exclusive of ^{14}C and ^{40}K).

1.8 Conclusion

Groundwaters make up only a very small proportion (about 0.31%) of the volume contained in different reservoirs at the Earth's surface. To make the best use of this resource, it is important to define as precisely as possible the different parameters governing the storage, distribution and circulation of groundwater:

— Type of aquifer (confined or unconfined).

— Nature of recharge and flow (groundwater balance).

— Permeability.

— Porosity.

— Storage coefficient.

— Transmissivity.

— Diffusivity.

The circulation of groundwaters is governed by Darcy's Law. Knowing the hydraulic conductivity of a reservoir, it is possible to calculate the discharge rate of an aquifer.

Estimations of water resource volume, as well as the rate and period of turnover in the reserve, also represent essential quantities in the management of an aquifer. Using these data, it is then possible to define an extraction strategy in order to plan, in a rational way, the abstraction of waters from aquifers.

The hydrogeological approach, which is used to identify favourable zones for the recovery of groundwaters, can be outlined in terms of four stages: preliminary studies, field studies, photo-interpretation and geophysical prospecting.

CHAPTER II

Well design and construction

*"If all those who believe they are right were
not wrong, the truth would not be far away"*
P. Dac, *L'os à moelle*

The well must be of the highest possible performance, allowing for the maximum water abstraction while taking account of the hydrogeological environment in which it is sited. Several fundamental parameters enter into the choice of drilling technique, the nature and dimensioning of the casings and the gravel pack, and also affect the development of the catchment structure.

This chapter presents the main techniques used in borehole construction and gives information concerning the choice of equipment for the catchment structure. It also addresses the capacity of the equipment to operate satisfactorily under the best technical and financial conditions bearing in mind the potential difficulties which might be encountered during implementation.

2.1 Drilling techniques

According to the geology (more or less consolidated sedimentary terrains, hard basement rocks, etc.), or the required depth for the well, several different drilling methods may be considered.

The drilling diameter plays an important part. The choice of a starting diameter for drilling should be made in relation to the depth of the structure and its final dimensions. In fact, the nature of the pumping equipment (notably the number and dimensions of the pumps) will vary according to the expected discharge and the total static head and remains dependent upon the diameter of the equipped borehole.

The main methods of drilling presented here are applicable to the implementation of both exploratory holes and operational water wells. An exploratory borehole is generally intended:

— To establish a geological cross-section of the terrain.

— To determine the different water-bearing levels.

— To define the best catchment method for the water-bearing formations encountered.

— To undertake sampling of the formations and/or waters.

— To undertake pumping tests if the diameter of the casing is large enough.

— To serve as an observation well and thereby enable calculation of the hydrodynamic parameters of the aquifer. The disadvantage of this type of borehole structure is the construction cost (around 20 to 30% of the total cost of the completed structure for a standard well).

However, an exploratory borehole having an adequate diameter with respect to the fixed objective can be subsequently equipped for water extraction use. Otherwise, the hole can be re-bored so that it can be fitted out with a larger diameter and transformed into a producing well. Alternatively, the exploratory hole is left as it is and used as a observation well for a nearby producing well.

2.1.1 Percussion drilling

Percussion drilling is the oldest established method used on drilling sites. It was employed by the Chinese more than four thousand years ago. The method consists of raising a heavy tool (churn drill bit) and letting it fall onto the terrain that is to be traversed. The height and frequency of the drop are varied according to the hardness of the formations.

Two kinds of percussion can be distinguished: winch percussion and cable percussion. The latter is the more common method. The bit is suspended from a cable which is tightened and slackened in turn. The movements of the cable tool are rapid and its work is accomplished more through a hammering effect due to kinetic energy than through the effect of weight as in winch percussion. A swivel enables the bit to rotate on itself with each stroke. The hole is cleaned progressively by the lowering of a bailer which allows the cuttings to be brought to the surface. Drilling with this procedure can be implemented without the use of water or mud.

— Advantages:
  • It is a simple and relatively inexpensive technique (investment is generally less than for the other drilling procedures).
  • Since there is no drilling fluid (mud), there is no risk of polluting the groundwater body.
  • The bit can be recharged, reforged and sharpened on site;
  • It is a highly suitable method for well bores of medium depth;
  • The results in fissured terrains are very good (no leaks).
— Disadvantages:
  • Rate of penetration is quite slow;
  • Ill-adapted for soft or unstable terrains which require feed pipes;
  • The bit has to be brought to the surface regularly in order to clean the shaft (loss of time);
  • It is difficult to control the inflow of gushing artesian water.

2.1.2 Drilling by cutting

Drilling by cutting is better known as the Benoto process. In this type of drilling by sand washing or cutting, the casings penetrate the formation through the effect of their own weight or the action of a hydraulic jack. A grab bucket progressively empties the inside of the casing as long as it is situated above the standing water-level. Below that level, the use of a bailer is recommended.

In the presence of coarse fragments or blocks, the use of a free-falling bit ensures the breaking up of obstacles. Hydraulic vibrators can also be used to facilitate the lowering or extraction of the casings.

— Advantages:
- Rapid penetration at shallow depths below the surface in soft formations, particularly in alluvium (without coarse-grained material).
- Creation of drilling structures with large diameters.
— Disadvantages:
- Inappropriate method in hard terrains.
- Friction of the casings on the feed.
- Difficulty in extracting the support casings after the screens and gravel pack have been installed.

2.1.3 Rotary drilling

Rotary drilling is the most commonly used procedure. It has been tried and tested notably in the field of petroleum exploration.

A tool (drill bit) is attached to the end of a string of drill pipes screwed into each other. This assembly is driven with a rotary movement at variable speeds and under vertical compression due partly to the weight of the drill string or by hydraulic pressure.

The rotary movement is transmitted to the drill string and to the tool by a motor situated at the well-head. The drill pipes are hollow so that mud can be injected into the bottom of the borehole.

The different tools used in rotation are drill bits of several types according to the hardness of the terrains encountered (bits equipped with blades or discs, revolving cutters, tricone bits, diamond-faced or metallic carbide-studded cutting tools).

One or more very heavy drill collars may be placed above the drill bit to increase the vertical pressure on the tool, in order to facilitate the penetration and ensure straightness of the hole.

In summary, the whole assembly or "drill string" is made up of the following elements, from top to bottom (cf. figure 2-1):

— a swivel;
— a square drill pipe (or square kelly);
— ordinary drillpipes;

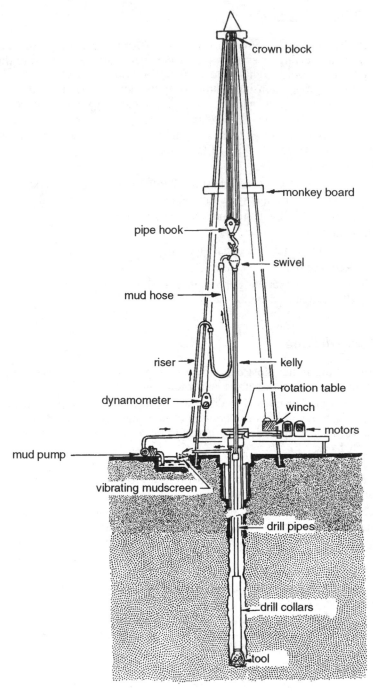

Figure 2-1
Diagrammatic layout of a rotary drilling rig (after A. MABILLOT).

— drill collars (which give extra weight and help maintain verticality of the hole;

— a tool (or bit).

All the string is suspended from a block installed in the derrick or else on a mast in the case of a self-supporting drill.

However, the rotary drill requires the use of a drilling fluid prepared on site which is injected continuously and under pressure into the hollow drill pipes of the string. This fluid emerges from ports near the bit and returns to the surface via the annular space between the pipes and the walls of the hole. The drilling fluid may be composed of:

— clear water;

— bentonite mud (type of clay mineral);

— mud with a synthetic biodegradable polymer base.

The main drawback in using bentonite mud is that it clogs the water inflow sections with an excessively thick "cake". This sometimes prompts the hydrogeologist to insist that drillers use clear water or biodegradable muds in water-bearing formations.

Once drilling is complete, the cake forms a more or less waterproof crust on the well walls. This crust must be eliminated in sections adjacent to the water-yielding zones. Sometimes, when the water pressure in the well is sufficient, the cake detaches itself. If this is not the case, hexametaphosphate is used to clear out the hole.

In certain special cases, notably the cleaning of holes with large diameters, the mud can be injected into the annular space to bring up the cuttings inside the drill string (this is called reverse circulation).

— Advantages of the rotary drill:
  - The depth reached can be considerable (several thousand metres in oil exploration). Moreover, depths of 300 or 400 m can be reached without casings in appropriate terrains.
  - The rate of penetration in soft terrains is high and can reach 100-150 m a day.
  - The drilling parameters (weight of tool, rotation speed, mud quality, mud injection speed) can be controlled in accordance with the formations to be traversed.
  - In soft terrains, rotary drilling leads to consolidation of the walls with a deposit of cake.

— Disadvantages:
  - Requires a drilling fluid and, as a consequence, the provision of water on site.
  - Possible clogging of the aquifer formations with bentonite mud (i.e. this type of mud cannot be used in water-well drilling).
  - Necessitates careful attention to the formations being traversed and also removal of cake.
  - Mixture of cuttings.

  • Risk of collapse in the event of a halt in drilling without cleaning the
    hole.

2.1.4 Down-the-hole (DTH) hammer drilling

This drilling method utilises impact accompanied by a thrust from the tool
which is itself rotating. The energy used to drive this equipment is compressed air
at a high pressure (10-25 bars). This process is of great interest for
hydrogeological work, mainly in hard terrains.

a) Conventional down-the-hole hammer drill

A pneumatic hammer fitted with drill bits is attached to the end of the drill
string. The percussive action is driven by releasing compressed air into the drill
string, whence the name *"down-the-hole (DTH) hammer drilling"*. The higher the
input air pressure, the smaller the risk of jamming. Most DTH hammer drills can
work at pressures between 4 and 18 bars.

The DTH drilling technique has been particularly developed for water
prospection in hard or fractured terrains.
  — Advantages:
    • Rapid rate of penetration.
    • Depths of around 150 m are commonly attainable.
    • Drilling fluids (air, foam, etc.) are well adapted to water-well drilling.
    • Good observation of cuttings (geological cross-section) and
      water-producing zones (hydrogeological monitoring).
  — Disadvantages:
    • Technique is ill-adapted to unconsolidated or soft terrains.
    • Risk of blockages due to cuttings, which lead to the need for frequent
      blow-out cleaning.
    • Depth of operation is dependent upon the compressor characteristics. It
      is advisable to have a very powerful compressor when working under
      considerable depths of water.

In certain cases, rotary and DTH percussion methods are associated together
in the same drilling operation.

In order to favour the holding up of the walls and the raising of cuttings, the
utilisation of drilling foam may prove necessary for DTH percussion drilling in
unconsolidated formations .

The drilling diameters for DTH percussion drills generally vary between 4"
(102 mm) and 15" (381 mm).

b) The Odex system for soft unstable formations

The perforation of cover terrains and alluvium is one of the most difficult
operations to carry out because of the quicksand-like and uncohesive nature of the
formations (i.e. running ground). In such cases, it is essential to case the walls of

the borehole as drilling proceeds. The Odex system (Copco-Sandvik Atlas) can be employed to achieve this purpose.

This method is based on the principle of an impact without rotation. It operates with a pilot drill bit having an excentric jig borer which enables the boring of holes with a slightly greater diameter than the outside diameter of the drill pipes. The casing is thus driven in progressively behind the jig borer by its own weight and the impact energy of the hammer. The drill pipes are joined to each other either by welding or by a screw thread. The excentric drill bit operates by clockwise rotation; an anti-clockwise rotation after completion of the drilling makes it possible to withdraw and raise the assembly to the surface. As in standard DTH percussion drilling, evacuation of the cuttings is ensured by a rising air flow, in this case between the drill pipes and the casing.

— Disadvantages:
  • When the excentric jams, it is often necessary to bring the tool back to the surface by extracting all the casing.
  • The use of foam is indispensable as soon as the depth is greater than 15 m.

The foam here has the same function as in the standard DTH percussion system. In addition, it lubricates and stabilizes the borehole thus facilitating the lowering of the casing.

Specially designed for the perforation of cover formations, this equipment can pass through extremely heterogeneous materials ranging from loose earth to homogeneous rock. The Odex system is even able to drill through blocks, either traversing them or pushing them aside.

In drilling for water, the Odex system is mostly used for casing from the surface down to the homogeneous bedrock, or for casing throughout the whole thickness of unstable formations. However, when drilling has to be continued in the underlying hard rock, the diameter of the telescoped pipes is limited by the diameter of the casing set at the bottom of the cased section.

The Odex system is equally useful for isolating upper groundwater bodies which are more vulnerable to pollution.

2.2 Drilling fluids

Drilling fluids play a crucial role in the execution of a drilling operation and particular attention should be paid to them. In the following, some basic information is given which is sufficient for "simple" drilling operations at shallow depths. For the implementation of deep drilling in complex geological formations, the reader is strongly recommended to consult the literature or take advice from specialist firms who will specify the composition of the mud and any modifications that need to be made as the drilling progresses. The choice and checking of drilling fluids are fundamental factors on which the success or failure of the structure may depend. In fact, the characteristics of the mud change with time and as a function of the materials encountered in the formations as they are being traversed. These modifications can cause serious damage: jamming of the drill string, total mud loss in the well, excessive cake clogging the aquifer, etc.

The drilling fluid has several functions, notably:

— Cooling and lubricating the drilling tool (drag bit, tricone bit, etc.).

— Sampling the geological formations encountered in a crushed form (cuttings), and raising them to the surface.

— Consolidating the borehole by depositing a clayey film (cake) on the bare well walls.

— Providing a counter-pressure against the entrance of flowing artesian water (by increasing the density).

— Giving useful information concerning the possible inflow of water or loss of head, owing to regular monitoring of the level in the mud pit.

In view of its multiple functions, the composition of the drilling fluid should be specified with the greatest care. Two main categories may be recognized:

— water-based fluids;

— air-based fluids.

The choice of drilling fluid will depend a large number of factors (type of terrain, possibilities for supply of water and chemical products, etc.). In general, water-based fluids containing clay or polymer additives are used in unconsolidated formations. On the other hand, the use of air-lift is reserved for drilling in compact or semi-consolidated rocks.

Whichever choice is made, success depends chiefly on the dosages, on the choice of additives and on the physico-chemical characteristics of the formations and the water in the drilled terrain.

2.2.1 Properties of drilling fluids

Control of mud condition involves maintaining certain characteristics in accordance with the fixed objectives as specified above. This is by no means a simple problem since the mud characteristics are continually being modified by the nature of the surrounding formations; fine sediments can cause coagulation of the mud, gypsiferous terrains will lead to flocculation, while a saturated aquifer formation will bring about dilution of the fluid, etc.

Mud is a colloidal mixture whose characteristics must be regularly checked and modified whenever necessary in order to maintain its required rheological properties: density, viscosity, filtrate strength, caking, pH and sand content.

a) Density

The density of pure water (at 4°C) is unity. The density of fluid mud can vary between 0.8 (if air is incorporated) and 2 (if barite is added; i.e. barium sulphate: $BaSO_4$, density 4.3). A fresh bentonitic mud has a density of 1.02 to 1.04, but this can vary during the course of drilling. The density is measured with a Roberval balance, or preferably a Baroid balance.

The density of the mud should be permanently monitored so that drill cuttings can be continually returned to the surface, while the condition of the walls is improved. Furthermore, it is able to balance any eventual influx of water (flowing artesian well). By using barite or water, it is possible to increase or decrease the density of the drilling mud (cf. figures 2-2 and 2-3).

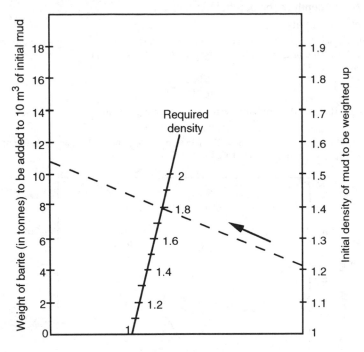

Figure 2-2
Weighting up of mud by adding barite (after A. MABILLOT, 1971).

In this example, 10.5 tons of barite must be added in order to raise the density of 10 tons of mud from 1.2 to 1.8.

The density of fluid mud is generally around 1.1, which represents a required dose of bentonite amounting to 3-8% or 30-80 kg per cubic metre of mud.

Variations in the mud volume are commonly observed, either being increased by dilution when a water-bearing layer is encountered having a higher pressure than the mud column, or being reduced due to losses into fissured rocks or zones of depressed groundwater-level.

When there is an increase in volume, it is advisable to stop the influx by increasing the density of the mud.

The hydrostatic pressure at the bottom of the borehole may be said to equal:

$$\frac{Hd}{10}$$

where:

    H : depth of the borehole in metres,

    d : density.

It follows that, for an influx produced at a depth of 150 m by a groundwater body whose residual pressure at ground level is 7 kg, the total pressure exerted by the aquifer at the bottom is equal to:

$$\frac{150}{10} + 7 = 22 \; kg$$

To counterbalance this pressure, the mud must be weighted up as follows:

$$22 \times (10/150) = 1.47$$

The weights of barite or volumes of water to be added are easily read from figures 2-2 and 2-3 (data from MABILLOT, 1971).

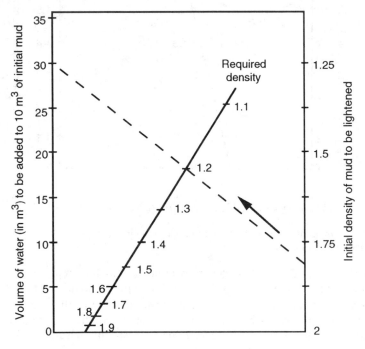

Figure 2-3
Lightening a drilling mud by adding water (after MABILLOT, 1971)

In this example, the density of 10 m^3 of mud can be decreased from 1.8 to 1.2 by adding 10 m^3 of water.

The causes of mud loss can be very variable in origin (e.g. a mud that is too fluid, a fissured or fractured terrain, a low-pressure aquifer, etc.). Each individual case requires an appropriate solution. The most problematic situation, albeit fairly frequently encountered, arises from total loss of mud which places the well structure in real danger. Numerous solutions exist but, here again, they must be adapted case by case.

b) Viscosity

An appropriate choice of viscosity ensures a clean bit as well as the efficient return to the surface and rapid settling out of drill cuttings. Furthermore, it leads to a reduction in pressure-head loss in the drill string.

The viscosity of a mud can be measured with a Marsh viscometer (on site) or a Stormer viscometer (in the laboratory).

c) Cake and mud filtrate

In permeable terrains, the drilling mud tends to drive water (mud filtrate) from the well towards the adjacent formation and leads to the build-up of a clay deposit (cake) on the well walls. In every case, the nature of the cake and the filtrate are controlled by the initial composition of the mud. As described above, these properties can change during the course of drilling, so it is essential to carry out regular checks on the cake and filtrate.

If the filtrate volume is excessive with respect to a given volume of drilling mud, then the cake is too thin and the walls are not held firm leading to a real risk of collapse. If the filtrate is too weak (cake too thick) there is a risk of the mud clogging the water-bearing formation.

Measurements are taken with a Baroid press, in which is placed a sample of the mud to be strained. A note is then made of the volume of filtrate collected in a given time as well as of the thickness and consistency of the resulting cake.

d) pH

The pH provides a measure of the acidity or alkalinity of the drilling mud. Acidic solutions have a pH in the range 0-5, neutral solutions in the range 6-8, while basic (alkaline) solutions have a pH higher than 8. pH measurements are of considerable interest since they reveal contamination by cement or by water from the aquifer if values are higher than 10-11. On the other hand, if pH is lower than 7, the risk of flocculation is high. The pH measurement should be carried out on the filtrate.

e) Sand content

It is clear that the presence of sand in the mud can be harmful because of its abrasive action (problems of wear of drill pipes, hoses, pumps, etc.). The presence of sand also increases the mud density and, where there are significant deposits towards the bottom, can cause blocking of the drill string. The permitted maximum sand content in a mud is generally evaluated at 5%. Sand content can be measured with an elutriometer.

The elimination of the sand is carried out using settling tanks or decantation pits, or by means of centrifugal desanders.

TABLE II-I — *Conditioning mud with bentonite (after A. MABILLOT, 1971).*

Average measurements and values	Apparatus used	Interpretation of the results, with consequences	Remedies and additives
Density (1.20)	Baroid balance	*Too strong:* - risk of mud loss - cake too thick -------------------------------- *Too weak:* - cake too thin - risk of wall degradation - collapses, artesianism	Dilution with water while controlling the other characteristics. Agitate vigorously. ---------------------------- Add bentonite (d=2.6) or barite (d=4.3). Stir vigorously.
Viscosity (40 - 45)	Marsh or Stormer viscosi-meter.	*Too strong:* - pumping difficulty - risk of jamming the tool during halts in the circulation. --------------------- *Too weak:* - risk of losing mud - risk of jamming by separating the grains from the mud.	Use of pyrophosphates (4kg for 100 l of water), tannins, lignites or ligno-sulphate. Keep a check on pH. --------------------- Add bentonite or starch
Filtrate volume (5-10 cm^3) cake thickness (5 mm max.)	Baroid filter press	Filtrate volume too large (cake too thin) - risk of collapse - risk of mud loss. --------------------- Insufficient filtrate (cake too thick) - risk of obstruction of in-coming water	Add starch or CMC (Blanose-Carb-oxymethyl-colloidal cellulose) 0.3-10 kg/m^3. Mix and stir. --------------------- Dilution with water. Control the other characteristics. Stir vigorously.
Sand (5% max)	Baroid screen	Risk of wear by erosion of mud pumps and flexibles	Use cyclonic desanders
pH (7 to 9.5)	pH paper	*pH > 11*: contamination by water or cement *pH< 7*: risk of flocculation	Use polyphosphates: acidic if the pH > 11; neutral if pH < 7.

f) Thixotropy

Thixotropy is the capacity for a mixture of particles in suspension to pass from a solid state (gel) to a liquid state while being agitated and to return to its initial state when the agitation ceases. It is therefore necessary to maintain fluid circulation during the drilling operation, even when the drill string is not turning, in order to avoid the mud solidifying and blocking the bit.

g) Conditioning the mud

Table II-I summarizes the main characteristics of the drilling mud and the means for modifying its condition.

It is estimated that the ideal characteristics for a fresh mud are the following:

— viscosity: 40-45;
— filtrate volume: 8 cm^3 for a mud sample of 600 cm^3;
— pH: 7-9;
— solids: 0.5%.

2.2.2 *Bentonitic muds*

Drilling muds are generally colloidal clay-based suspensions; the most commonly used type is bentonite. One gram of bentonite dispersed in water offers a surface area of 4-5 m^2.

Bentonite is mainly composed of a smectite clay called montmorillonite. This complex alumino-silicate mineral contains iron and magnesium ions which can be substituted with silicon and aluminium in various proportions to form bentonites with different characteristics. As a general rule, when the metallic ions of a bentonite are replaced by ions of another metal, its properties are modified.

For example, bentonite is hydrated in the presence of water, while undergoing a considerable increase in volume (12-15 times and sometimes 30 times). These variations arise from variably charged metallic oxide species.

It is noteworthy that bentonitic muds have the characteristic property of reacting and flocculating in the presence of nitrate-rich waters.

a) Bentonite

Bentonites are characterized by the Atterberg limits.

— Liquid limit: represents the water content below which a clayey paste behaves as a semi-liquid.

— Plastic limit: represents the water content below which a clayey paste ceases to show a plastic behaviour.

— Plasticity index: represents the difference between the liquid and plastic limits.

Several types of bentonite can be distinguished:

— Natural calcic bentonites.

— Natural sodic bentonites, which swell more than natural calcic bentonites; these are the clays used in drilling muds.

— Ion-exchanged bentonites, which are calcic bentonites transformed into sodic bentonites by the addition of sodium carbonate; the swelling of these bentonites can vary between 10 and 15 times.

— Activated calcic bentonites are ion-exchanged bentonites doped by the addition of polymers which improve their swelling capacity 10 to 25 times.

Table II-II shows the values of the Atterberg limits for clays and various types of bentonite.

TABLE II-II — *Atterberg limits of clays.*

Type of clay	Liquid limit	Plastic limit	Plasticity index	Swelling
Kaolin	20-50			1-2
Plastic clays	50-100	10-40	10-40	2-3
Bentonitic clays	80-150			3-6
Attapulgite Illite Sepiolite	80-150	30-40	50-110	4-8
Calcic bentonite	100-200		50-150	3-7
Sodic bentonite	450-550	50-60	400-500	12-18
Activated calcic bentonite	350-700		300-650	10-25

b) Special muds for swelling marls

Marly formations have the property of swelling in the presence of water. Hence, the risks of blocking the tool at the bottom of the well are considerable. A further problem is that the degree of swelling of marls increases with the alkalinity of the drilling mud.

To mitigate this problem, it sometimes proves sufficient to bring the pH of the mud to around 7.5 or 8 by adding pyrophosphate. Waterglass (sodium silicate), calcium silicate, or even starch can be added to the mud.

As a general rule, it is advisable to increase the density of the mud and the flow-rate of the pump, while traversing the marly formation rapidly so that hazardous zones can be cased at the earliest possible opportunity.

c) Muds with emulsified oil

These muds are obtained by adding diesel fuel (5-25%) and an organic emulsifier to standard mud. They lubricate the metal parts of the bit and increase the rate of progress of drilling. Care must be taken, however, to avoid pollution of the aquifer. For this reason, muds of this type are little used in hydrogeology.

2.2.3 Polymer muds

Polymers are chemical compounds with a high molecular weight resulting from the association of several simple molecules with low molecular weight. These substances can be used directly as drilling muds or as additives to bentonitic muds. They offer the following advantages:

— Drilling with reduced pressure at the bottom of the hole.
— Loss of fluids controlled without build-up of a thick cake.
— Torque and friction reduced.
— Samples are not contaminated by the fluids.

However, these polymers, both natural and artificial, have certain disadvantages:

In the case of natural polymers:
  • Bacterial proliferation takes place in a very short time.
  • Difficult to eliminate bacteria in the gravel pack.
  • The bactericides used can be toxic.

In the case of artificial polymers:
  • Risk of instability of the walls.
  • Risk of clogging.
  • Risk of pollution during destruction of the polymers by chemical action

a) Natural polymers

These compounds are generally organic products that are obtained from Guar gums. Their molecular structure enables them to produce, for the same amount of matter with the same viscosity, ten times more gel than a bentonitic mud.

The most commonly used brands are Revert and Foragum. The essential characteristic of Revert is the breakdown of its viscosity. In fact, after a certain time (around 2-5 days depending on the temperature of the water and the initial dose), the viscosity of the gel suddenly decreases and the mud becomes fluid and limpid like water. As a consequence, the drilling mud is cleared out quite naturally from the water-bearing formation and the cleaning and development operations are greatly facilitated. Furthermore, the breakdown of viscosity is accompanied by a change in colour of the mud which loses its dark blue hue and becomes colourless, thus making it easier to perceive the change of viscosity. However, if drilling conditions so require, the life of the viscosity can be prolonged by the addition of caustic soda (NaOH) to the mud. Inversely, the addition of chlorine stimulates the breakdown of viscosity. It should be noted that the pH must be fixed between 5 and 7 to allow optimal utilization of Revert.

b) Synthetic polymers

These substances can be used with bentonitic muds in combination with other polymers. The following products are used as ingredients in such muds:
— viscosifiers with a sodium acrylamide/acrylate copolymer base;
— biodegradable viscosifiers;
— biodegradable lubricants;
— foaming agents, biodegradable surfactant mixtures;
— thinners with a sodium polyacrylate base in a 30% aqueous solution;
— water-retention agents to control loss of circulation (minimum 200-fold swelling).

As mentioned previously, these polymer products can only be broken down by chemical action, leading to a real risk of polluting the groundwater body. The most commonly used chemicals for decomposing polymers are the following:
— sodium hypochlorite ($12°$ Cl): dose2-4 l/m^3, total destruction in one hour;
— hydrogen peroxide (110 vol.): dose 3-4 l/m^3, total destruction in about 10 hours;
— hydrochloric acid: dose 1-2 l/m^3, rapid partial destruction (90%);
— sodium carbonate: dose 10 kg/m^3, partial destruction (80%) in more than 24 hours.

Table II-III summarizes the main problems which might arise and the courses of action to be taken to resolve them.

c) Biodegradable polymer mud

Synthetic biodegradable polymers are only applicable if their life-span is longer than that of natural polymers, thus allowing for their elimination before the onset of the degradation process responsible for the proliferation of bacteria.

The proprietary product AQUA GS may be mentioned in this context, since it ensures a biodegradable mud resistant to bacteria and only becomes open to attack after 5-6 weeks into the drilling operation. It resists various types of contaminant (clay marl, dolomite, acid gas, ferruginous water, etc.) and is not destroyed by cement. Furthermore, it can be used at a low concentration (1-3 kg/m^3) with all types of water (soft, salt and hard). It limits the swelling of clays and leads to good sedimentation in the settlement tank thus assisting the circulation of drilling fluids. Nonetheless, it is advisable to operate with a neutral pH.

Another product — known as D 800 or AQUA J — offers similar characteristics with comparable doses.

TABLE II-III — *Utilization of GSP products to solve drilling problems.*

Rotary drilling 1 - *In cases of viscosity reduction, check:* • pH • salinity of the water • if groundwater body is being recharged 2 - *Good viscosity but well walls not holding:* • presence of clayey layers • presence of soft layers • borehole traverses a salt groundwater body • standing water-level near ground surface • buckling of drill pipes which strike the walls and break up the mud cake	 • pH should be higher than 5.5; it can be increased with caustic soda or sodium carbonate. • Above 8 g/l of salt, use a special product (GS 550 S). • Increase the density of the mud and reload it regularly every 2-3 hours. • Reduce dose of viscosifier and use a thinner, add potassium chloride. · Reduce the viscosity and cut high flow rates to facilitate the return of cuttings, otherwise progressively load the mud with salt. • Use mud containing GS 550 Sand salt to balance the hydrostatic pressure. • Add mud so that borehole is quite full, and salt to increase the density. • Reduce pressure on the drill bit.
Drilling with a DTH hammer drill • Halt without apparent reason, badly cleaned drill pipes forming a polymer film which blocks entry of the drilling fluid	 • Add solvent and leave for 10 mins before continuing drilling.
Core drilling Poor recovery of cores: • Uncohesive terrains • Core is destroyed by the mud, the core does not come up or falls back	Increase the concentration of GS 550. • Reduce the flow rate and decrease the passage of water from the crown. • Slow down rate of progress and increase the diameter of the core pipe. Use a shorter core drill.
Artesian drilling • Loss of circulation in the aquifer	Calculate the excess pressure and the density of the mud as well as the weight of the salt. Drill at a loss with light mud or inject a viscous mud containing a slow destruction agent (H_2O_2).

2.2.4 Compressed air

The use of compressed air as a drilling fluid offers a certain number of advantages:
— greater penetration speed in hard rock;
— reduction of weight on the tool;
— great capacity for clearing out the drilling debris;
— drilling facilitated in swelling clay formations;
— low water requirements.

a) Rotary drilling

Compressed air is efficacious in hard and stable formations (igneous, metamorphic or dense sedimentary rocks).

During drilling, the airflow rate is adjusted to maintain a velocity in the annulus sufficient for a satisfactory return of cuttings to the surface. In theory, the raising capacity of air is proportional to its density and the square of the velocity in the annulus.

b) Drilling with a DTH percussion tool

The higher the air pressure the smaller are the risks of jamming the tool at the bottom of the hole. Most percussion tools work between 4 and 18 bars. The choice of the compressor power depends upon the estimated consumption of air during drilling and blowing.

2.2.5 Stabilized foam

Foam is a mixture of liquid and gaseous components (i.e. water and dissolved chemical products, air). The foaming solution is sometimes supplemented with polymers or bentonite in order to increase its density and improve its viscosity qualities and also stabilize the borehole walls.

The dose varies in relation to the quality of the foaming agents, corresponding to between 0.2 and 2% of the weight of water used. The foam must maintain a certain consistency, comparable to that of "shaving foam".

A stabilized foam drilling fluid is used:
— With a rotary drill, when the use of mud is made difficult by the nature of the terrain, whether due to excessive fracturing or an insufficient supply of water to the bit.
— With an air drill, when there is excessive erosion of the borehole walls, in the presence of swelling clays or else when the evacuation of cuttings is made difficult by the influx of water into the well.

However, it should be stressed that this method may well prove inappropriate in the case of large water influxes into unconsolidated aquifer formation .

a) Foam for rotary drilling

In order to clean the borehole, a foaming product should be used at a dose corresponding to 0.3 to 0.5% of the water volume. This concentration can be increased slightly in the presence of an abundant inflow of water.

A viscosifier giving a stabilized foam is used for the consolidation of the well walls, at a concentration corresponding to 0.5% of the water volume. The characteristic airflow conditions are as follows:

— airflow rate for drilling: 350-400 litres/minute per inch of diameter to be drilled;

— injection pump discharge: a mixture of 0.1-0.2% in relation to the volume of air.

In soft terrains (high drilling speeds), the airflow rate and the volume of foaming solution are increased, but a proportion of 0.1-0.2% is maintained in the mixture. In hard terrains, on the other hand, the airflow rate and solution volumes are reduced.

b) Foam for air drilling

In the case of DTH percussion drilling, care is taken to observe the following recommendations.

— For small inflows of water or reduced air losses, the hammer should be placed at the bottom of the hole; firstly 0.3 to 1 litre of foaming agent then 5 to 15 litres of water are poured through the drill pipes before drilling is resumed.

— If an unstable zone is being traversed, 5 to 20 g of stabilizing polymers can be added to the water circuit.

— In the case of abundant water inflows, wall caving or heavy air losses, 0.2-0.6% of stabilizing viscofiers can be used with 0.3-1.5% of foaming product according to the amount of water influx.

It should be noted that the use of foam reduces the penetration speed by about 40% at 6-7 bars, 20% at 10.5 bars and 10% at 18 bars.

After use, the hammer should be blown and lubricated carefully. In the particular case of the Odex drill, injection of foam is not necessary.

2.3 Water well equipment

A water well is designed to extract the water contained in an aquifer formation. For this reason, whatever the drilling method adopted, the well structure always includes a lining tube maintaining the terrain in the non water-bearing upper part of the borehole and a screened section adjacent to the actual groundwater body (except in the special case of suspended screens).

It should be borne in mind that water well operation should take place under the best possible conditions, both from the point of view of quality:

— no pollution of the water around the catchment structure,

— no entrainment of solid particles,

and from the point of view of quantity:

— obtaining the highest capacity that is compatible with the aquifer and the proximity of other structures developing the same aquifer,

— seeking the highest possible specific capacity (i.e. yield per unit of drawdown).

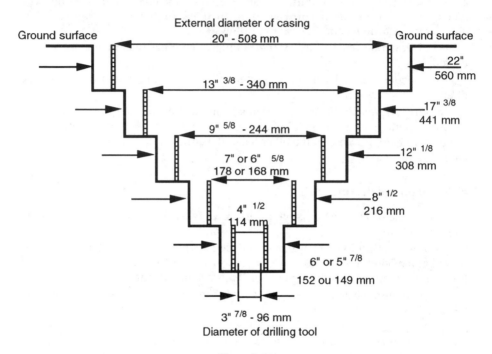

Figure 2-4
Relationship between diameters of the rotary tool and diameters
of the most commonly used casings.

Operating water wells are equipped with two fundamental elements:

— full (or blind) casings;

— screens or perforated casings.

The expected water-yield and the final depth of the borehole will determine the characteristics of the drilling tool and the diameter of the casings to be used (cf. figure 2-4). On the other hand, in relation to the desired yield, the choice of submerged pump will constrain the diameter of the casings.

The quality of the casings and screens is essential for the longevity of the well.

TABLE II-IV — *Characteristics of the principal casing diameters (API) after A. MABILLOT, 1971.*

External diameters		Thickness	Internal diameters	Average weight per metre with collars
(inches)	(millimetres)	(millimetres)	(millimetres)	(kilograms)
$4^{1/2}$	114.30	5.20	103.90	14.10
$4^{1/2}$	114.30	6.35	101.60	17.25
$4^{1/2}$	114.30	7.35	99.60	20.10
$6^{5/8}$	168.30	6.22	155.86	25.30
$6^{5/8}$	168.30	7.32	153.66	29.75
$6^{5/8}$	168.30	8.94	150.42	35.70
7	177.80	6.91	163.98	29.75
7	177.80	8.05	161.70	34.20
7	177.80	9.19	159.42	38.70
$9^{5/8}$	244.50	7.14	230.22	43.60
$9^{5/8}$	244.50	8.94	226.62	53.50
$9^{5/8}$	244.50	10.03	224.44	59.50
$13^{5/8}$	339.70	8.38	322.94	71.40
$13^{5/8}$	339.70	9.65	320.40	81.10
$13^{5/8}$	339.70	10.92	317.86	90.75

Some basic rules should be observed:

— Leave at least one inch (25.4 mm) of clearance between the pump and the internal diameter of the casing. This latter will therefore be around 5 cm larger than the outside diameter of the pump.

— Leave some clearance between the bare walls of the hole and the full casing, notably in anticipation of cementation of the annular space.

2.3.1 Casings

The diameter of the casing will be a function of the anticipated yield of the well. Table II-V enables the evaluation of casing diameter suitable for a maximum capacity under most circumstances.

The choice of casing type is then made in relation to the resistance to various forces or loadings (i.e. tensile stress, crushing, bursting and buckling).

TABLE II-V — *Casing diameter as a function of eventual well capacity.*

Minimal internal diameters of casings (inches)	Maximum planned yield (m^3/h)
4"	3
6"	50
8"	140
10"	250

Two main materials are used for smooth casings:
— PVC (plastic);
— steel, in particular:
  • black steel,
  • galvanized black steel;
  • steel covered with a plastic film;
  • aluminium chrome steel;
  • stainless steel.

The characteristics of the most common casings are as follows:
— length of sections: 3-6 m;
— thickness: 2-11 mm (steel), 4-16 mm (PVC);
— diameter: 100-2 500 mm (steel), 60-315 mm (PVC);
— connection: welded collars, threaded end-fitting (steel), screw-threaded (PVC).

2.3.2 Screens

The screen is the principal component in the equipment of a water well. Installed at the end of the full casing, and placed adjacent to part or all of the aquifer formation, the screens must fulfill the following functions:
— ensure the maximum yield of clear sand-free water;
— resist corrosion by aggressive waters;
— resist the crushing pressure exerted by the aquifer formation during its development;
— have the longest possible life-span;
— induce minimal pressure-head losses.

In the majority of cases, with simple artesian groundwater wells, the catchment structure is one of two types:
— borehole with "monolithic" fittings;
— telescopic arrangement with screens having a smaller diameter than the upper level lining tube.

There are several types of commercial steel screen (prefabricated without further modification on site):

— Screens with *round holes* are used in hard terrains, but have a low open area (10%),

— Screens with *oblong holes* have vertical rectangular slots whose width is at least equal to the thickness of the sheet steel and a standard length of 3 cm; they show open areas in the range 10-20%.

— *Louvre slot screens* have horizontal rectangular perforations in the form of a hood; they are mechanically resistant but have a low open area.

— *Bridge slot screens* are a frequently used type; they are constructed flat then rolled and welded, which provides good mechanical resistance owing to low metal removal; they show an open area varying between 3 and 27% according to the dimensions of the perforations.

— The *Johnson-type screen* has a continuous horizontal opening (or aperture) running the whole length of the screen, which is obtained by the spiral winding of a profiled casing wire welded onto vertical metal rods. The main advantages of such a screen are:

- the regularity and precision of the opening;
- the very low risk of clogging;
- he highest open area coefficient of all the screen types.

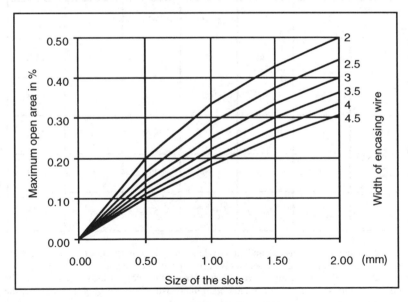

Figure 2-5
Variation of slot size, wire width and open area coefficient for Johnson screens (after documentation from Johnson Filtration Systems, 1992).

Table II-VI and figures 2-5 and 2-6 summarize the main characteristics of Johnson screens (diameter, open area, slot size, linear flow rate).

Prefabricated screens are also available in plastic (PVC). For DWS catchment structures, it is advisable to avoid manually produced screens wherever possible.

TABLE II-VI — *Diameters and weights of Johnson screens - Irrigator series (after documentation from Johnson Filtration Systems, 1992).*

Nominal diameter	External diameter (overall)		Internal diameter (slot size)		Resistence to crushing and weight of screen	
Inches	Inches	mm	Inches	mm	Resistence to crushing (bar)	weight (kg/m)
4	$4^{1/2}$	114	$3^{3/4}$	95	104	9.1
6	$6^{5/8}$	168	$5^{3/4}$	146	33	13.3
8	$8^{5/8}$	219	$7^{3/4}$	197	15	17.4
10	$10^{3/4}$	273	$9^{3/4}$	247	33	35.2
12	$12^{3/4}$	324	$11^{3/4}$	298	20	41.7
14	14	356	13	330	15	45.8
16	16	406	$14^{3/4}$	374	10	52.3
18	18	457	$16^{3/4}$	425	7	58.9
20	20	508	$18^{3/4}$	476	5	65.4
24	24	610	$22^{3/4}$	577	3	78.5

Note: the resistance to crushing and the weight of the screen is given for a zero open area; to obtain the weight of a screen with a given slot size (f) the weight (without slots) quoted in the above table must be multiplied by $l/l+f$, l being the width of the encasing wire. The same procedure should be followed to obtain the resistance to crushing, i.e. multiply the resistance to crushing at zero slot size by $f/f+l$.

The standard values of f are 0.5, 1, 1.5 and 2, while standard values of l are 2, 2.5, 2.75, 3, 3.25, 3.5, 3.75, 4 and 6.5.

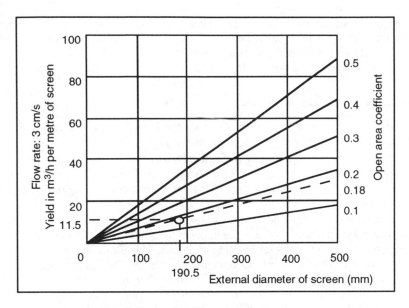

Figure 2-6
Flow rates, diameters and open area coefficients for Johnson screens
(after documentation from Johnson Filtration Systems, 1992).

Example: open area coefficient $C = \dfrac{f}{f + l}$

where f is the dimension of the slot between two coils of wire and l is the width of the encasing wire.

For $f = 1$ mm and $l = 1.14$ mm, $C = 46\%$, then with the same open area, we obtain:
- $C = 4$ to 6% for a louvre slot screen,
- $C = 16\%$ for a bridge slot screen.

The relation between the three quantities C, f and l (cf. figure 2-6) is conditioned by the flow rate at which the water enters the screen. The optimum speed has been defined experimentally as 3 cm per second. The equation linking flow rate, screen diameter and open area coefficient can be written as follows:

$$Q = \pi \times DC \times 0.03 \times 3\,600$$

from which we obtain:

$$Q = 340\,DC$$

where:
Q : low rate in m³/h or one metre of screen,
D : external diameter of the screen in metres,
C : open area coefficient.

For different reasons, due to the development of the well and its changing characteristics with time (obstruction of certain slots by fine particles, partial clogging, etc.), it becomes necessary to weight the theoretical flow rate with a reduction coefficient (e.g. around 0.5-0.75).

a) Open area and screen slot size

The specification of a screen slot is established in accordance with the interpretation of the granulometric curve of the aquifer formation.

The open area coefficient is a fundamental factor since it conditions the passage of water from the aquifer towards the well. Thus, it is usually necessary to rule out the use of screens having coefficients of less than 10% (round, oblong, louvred slots).

The screen slot size is defined in relation to the characteristics of the gravel pack. The gravel placed in the annular space between the walls of the hole and the screen constitutes an artificial filtering medium. The grain-size of the gravel - which generally shows a uniform distribution - is selected as a function of the grain-size of the surrounding formation. *The installation of a gravel pack enables the screen slot size to be increased and the entrance velocity of water coming into the screen to be decreased, which thus enhances the production yield of the well.*

To hold back a very fine homogeneous sand, a double screen may be used consisting of two screens, with one inserted inside the other. However, such a procedure is not to be recommended since it greatly reduces the pumped discharge of the catchment structure.

Insofar as the existence of fine sand is known before drilling begins, the borehole can be implemented with a large diameter. By this means, the size of the gravel pack can be increased so as to lessen the velocity field in the outer annular volume (perimeter of the pack), thus diminishing the flow of sand into the screens. This type of double screening has proved its worth in West Africa, where it has saved a large number of boreholes from sanding up.

b) Positioning of the screens

The screens should be placed adjacent to the points of greatest water inflow and, generally speaking, over the entire thickness of the aquifer. Their installation will therefore depend upon the geological cross section established from observation of the cuttings, and also from the speed of penetration and changes in mud quality. There is always a discrepancy between the cuttings observed at any given time and the depth of the tool; this discrepancy increases with the speed of penetration and the depth in the well, and is also a function of the discharge rate of the mud pump. The use of gamma-ray logs is often necessary to enable readjustment with the geological cross section. Logs of this type record the natural gamma radiation emitted by the clayey formations. When drilling takes place in sedimentary terrains, such logs allow the screens to be positioned with maximum efficiency.

The length of the screened section is a function of the thickness of the water-yielding zone, the degree of drawdown and the nature of stratification in the aquifer layer. The data necessary for positioning the screens are derived from preliminary survey work, namely:

— instantaneous logs, mud losses;
— core sampling;
— granulometric analysis of samples;
— geophysical logs.

It is recommended to select the screen lengths according to the following situations that may arise:

— An *artesian water-body in a homogeneous unstratified terrain.* In this case, 80 to 90% of the aquifer thickness is screened to ensure that drawdown does not fall beneath the roof of the water-bearing layer.

— An *artesian water-body in a heterogeneous stratified terrain.* Here, 80-90% of the more permeable layers should be screened.

— An *unconfined groundwater body in a homogeneous terrain.* In theory, in a layer less than 45 m thick, it is recommended to screen at least the lowermost third without exceeding 50% of the total thickness. With thicker layers, 80% of the aquifer section may be screened in order to obtain a greater specific capacity. In fact, the screen length to adopt is a compromise between the longest possible screen, which has the advantage of reducing the water inflow velocity, and the shortest screen placed at the base of the aquifer, which enables a greater drawdown. It is advisable not to draw down the groundwater-level below the screen.

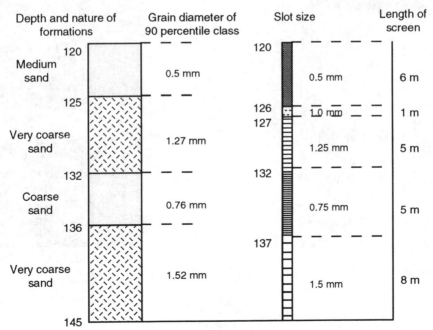

Figure 2-7
Choice of screen slots in a heterogeneous formation (after R. LAUGA, 1990).

1. Round holes

2. Oblong holes

3. Bridge slot

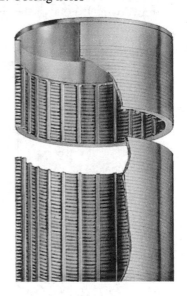

4. Johnson type

Figure 2-8
Differents types of screen (after: 1 to 3, data from TUBAFOR ; 4 A. MABILLOT, 1971)

— An *unconfined groundwater body in a heterogeneous terrain*. As a rule, the screen should be positioned in the most permeable layers so as to enable maximum drawdown under the best operating conditions. The screened section should be about one third of the total thickness of the aquifer. When the permeable layers are relatively thin, it is necessary to tap other less aquiferous layers with screen slot sizes that are adapted to each different layer (cf. figure 2-8).

To summarize, attempts should be made to screen the water-bearing formation to the maximum. It is useful to make provision for a pump chamber, a zone in which the pump can be positioned so as to redistribute the area of pumping depression and the velocity field of the water in the screens. As far as possible, it is advisable to reinforce the casings in the pump chamber since the successive switching on and off of the pump often provokes impacts between the pump and the casing leading to extra wear on the latter.

It is essential that the casings, and even more so the screens, should be positioned in the borehole with centralizers so that the gravel pack can be correctly placed.

2.3.3 Gravel pack

The pack of gravel pack should be composed of material that is:
— clean, without any clayey material;
— with rounded rather than angular fragments, to limit head (or entrance) losses;
— siliceous (and especially not calcareous), to avoid any risk of cementation or dissolution in contact with water during acidizing operations.

The thickness of the gravel pack should lie in the range of 3- 8" (75-200 mm), in accordance to the drilling diameter. It should generally be emplaced up to several metres above the roof of either the aquifer or the screened section. This surplus gravel is essential since it compensates for settling of the pack under its own weight.

Furthermore, it should be borne in mind that an overly coarse gravel pack in a fine sandy formation can cause sanding up of the catchment structure. On the other hand, a gravel pack with excessively fine grain size can lead to only partial development of the aquifer's potential and make it difficult to eliminate the drilling mud. This explains why bentonitic mud is not used.

Thus, the grain-size distribution of the gravel used in the pack is based on granulometric analyses of the aquifer formations that are to be tapped. These studies make it possible to define certain parameters such as the characteristic diameter, the fineness index or the sorting coefficient of the formations in question.

The characteristic grain diameter of a material is defined so that, in relation to the total weight of the sample, 10% of the grains are finer and 90% coarser than this dimension. It represents the x-axis coordinate of the 90-percentile grain-size class (cf. figure 2-9).

The median grain size is defined by the x-axis coordinate of the 50-percentile class.

The sorting coefficient (SC) is expressed as the quotient of the grain diameters of the 40- and 90-percentile classes on the cumulative curve.

The most commonly used experimental method is the following. A sample of material (about 100 g) is dried and then passed through standard mesh screens. After screening, the grains held back on the different screens are weighed. The weight in each case is then added to the sum of the preceding screened classes and the percentage of each class is calculated in relation to the total weight. The granulometric curve of the sample is then plotted with cumulative percentages along the y-axis and the dimensions of the screen meshes along the x-axis (cf. figure 2-9).

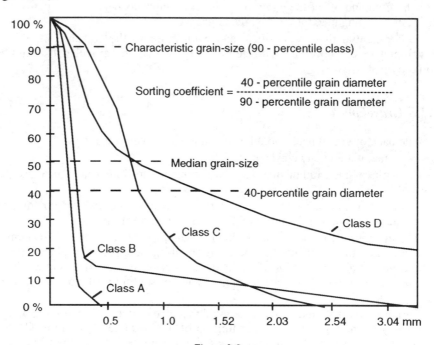

Figure 2-9
Granulometric curves characteristic of granular formations (after R. LAUGA).

By establishing the median grain size, characteristic grain diameter and sorting coefficient, it is possible to define the grading of granular materials:

— *Class A*: fine sand (grain size: 0.06-0.25 mm).

— *Class B*: medium or heterogeneous sand (grain size: 0.25-0.5mm).

— *Class C*: coarse sand (grain size: 0.5-2 mm).

— *Class D*: sand and fine gravel (grain size: 2-16 mm).

Outside these classes are the silts and clays (grains finer than 0.06 mm) and the stones and blocks (grains coarser than 16 mm).

It should also be noted that the shape of the grain-size distribution curve is also significant. Materials yielding a reverse S-shaped curve (classes A and C) generally have a greater porosity than those characterised by curves with a tail due to the presence of coarse grains (class D).

As a general rule, it is accepted that if the 90-percentile is less than 0.25 mm, the installation of an additional gravel pack should be envisaged. If it is greater than 0.25 mm, then it is considered that the formation can be developed naturally (self-development).

The sorting coefficient of satisfactory additional gravel should be situated in the range 2-2.5.

Finally, during the implementation of the catchment structure, the inclusion of a system should be envisaged which allows the further addition of gravel. In many cases, the gravel pack can settle or come to occupy the extra volume made available in the aquifer, particularly after acidification, and it then becomes necessary to add more gravel.

2.4 Cementation

This method consists of filling all or part of the annular space between the casing and the walls of the borehole with a cement-based mixture. Cementation is used mainly in the following cases:

— to block a cavity or large fissures which lead to heavy mud losses during drilling;

— to render the annular space impermeable and prevent the developed groundwater bodies being polluted by surface water;

— to fix the drive pipes (casing string) to the terrain thus protecting them from the corrosive attack of certain waters.

Figure 2-10 shows a diagram for determining the volume of water and weight of cement to use in order to obtain a given density and the corresponding volume of slurry. It should be noted that there are a fair number of rapidly-setting cements on the market. They can be used to limit the downtime of drilling operations.

The choice of cement quality and the eventual use of certain additives should be decided in accordance with:

— the nature of the formations and waters encountered;

— the temperature of the terrain (for very deep drilling);

— the setting speed of the slurry;

— the volume of cement to be used;

— the resistance of the slurry to contaminants which could be present in the drilling environment;

— the final resistance of the cement to crushing 7 to 28 days after setting.

Before undertaking cementation, the exact volume of the cement slurry to be used must be calculated. This volume is given by the following formula:

$$V = \frac{H}{2} \cdot \left(d_1^1 - d_2^2\right)$$

where:

   H : height of cementation,
   d_1 : drilling diameter,
   d_2 : casing diameter.

Cementation in a casing string must be carried out from end to end without interruption. A summary description is given here of the main types of cementation used in water-well drilling.

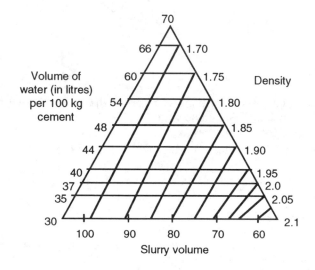

Figure 2-10
Triangular diagram showing the different parameters
in the preparation of pure cement slurry (after R. LAUGA , 1990).

2.4.1 Cementation through the outside of the casing

This method only applies to shallow boreholes. The quantity of cement slurry required to fill the annular space is introduced at the bottom of the well and the casing string to be cemented is emplaced immediately afterwards. The foot of the casing string is closed off by a destructible plug. The casing string is filled with water as it descends to prevent it from floating on the cement.

It is also possible to cement after the casing string has been put in place. In this case, the cement slurry is injected by a pump from the bottom and rises along the walls of the casing as the injection proceeds.

As a general rule, it is advisable not to carry out cementation via the outside of the casing when the implementation is difficult. However, if for technical reasons it is indispensable to employ such a method, the operation can be simplified by the use of an injection tube.

2.4.2 Cementation through the inside of the casing

In this case, the cement slurry is injected from the surface inside the casing and then rises along the outside of the walls. When cementation is complete, it is merely necessary to wash the interior of the casing string with clear water to flush out the cement.

There are several techniques for preventing the cement from rising up inside the casing.

a) Float shoe method

This method uses a special device, known as a float shoe, which is placed at the lowermost end of the casing string. The shoe consists of a plastic ball which acts as a plug preventing the passage of fluids from the bottom upwards. It is also in contact with the surface by means of a pipe screwed to the cementation shoe, thus allowing the cement slurry to pass into the annular space.

When the uncontaminated cement slurry appears at the surface, the filling of the annular space is complete.

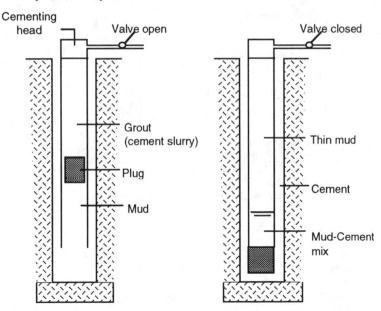

Figure 2-11
Emplacing a plug before injection of the cement slurry.

b) Free plug method

A destructible cemention plug (wood, rubber, plastic) is introduced into the casing string to be sealed before or after injection of the cement slurry.

— In the first case, the plug is placed between the mud present in the borehole and the cement injected directly behind the plug. After the cement has been put in place, a volume of mud under pressure is introduced. It drives the cement slurry upwards and enables it to rise into the annular space. The cement-mud contact should be situated within the casing (cf. figure 2-11).

— On the other hand, if the plug is emplaced after injection of the cement (cf. figure 2-12), it will be pushed directly by the mud or, eventually, by the water. When the cement appears at the surface, the operation is complete.

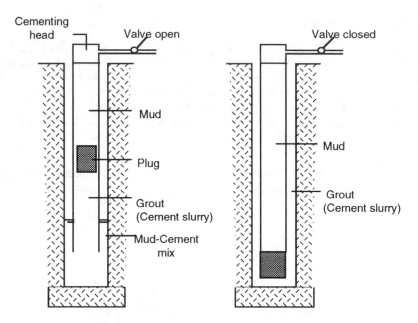

Figure 2-12
Emplacing a plug after injection of the cement slurry.

c) Cementation with a guided plug

This system is similar to the one described above, except that the cementing head is different (impermeable) and the casing includes a foot plug designed to hold back the cement plug as it moves downward.

The borehole being full of mud, the cement slurry is injected directly and, because of its density, it displaces the mud into the annular space. When the volume of cement (calculated beforehand) has been introduced, it is merely necessary to release the plug which then settles onto the cement. Following this, mud or water is injected which drives the plug and forces the cement into the annular space until the plug arrives at the bottom.

d) Partial cementation

It is possible to cement only a part of the cased section by carefully estimating the volume of cement to be employed. In this way, the lower or upper part of a catchment structure can be cemented. Several successive cementations can even be carried out separated by fine sand filters. This procedure is particularly useful for separating water-bearing horizons superimposed one above the other.

2.4.3 Well head

The well head should be built with great care since this determines the staunchness of the catchment works, which represents a very important factor in avoiding accidental pollution by surface waters or condensation. The well head should be constructed rapidly in order to avoid any accidents on site. During construction site work, it is the duty of the contractor to take the necessary measures to ensure temporary closure of the well as long as the equipping of the structure has not been completed. Conventionally, the well head is cemented to a height of at least 3 m above ground level. It is finished off with a concrete surface whose minimum size should be no less than the diameter of the borehole plus one metre.

The well head extends down through a full casing of at least 0.5 m length. This casing must be hermetically sealed so as to avoid any contamination from the groundwater body. It is evident that perforations of any kind cannot be tolerated in any of the full casings or joints.

Otherwise, the well head is generally part of a small civil engineering structure and should be designed with a view to optimizing water-yield. In particular, care should be taken to equip the structure with an anti-intrusion device and correct arrangements for easy removal of the pump and the catchment column. Finally, the well head should be sufficiently ventilated so as to avoid condensation which could be a source of pollution.

2.5 Development of water wells in alluvial formations

Well development begins after the catchment structure has been fully equipped (casing, screen and gravel pack). This operation consists of improving the natural permeability of the water-bearing formation around the screen. Development also aims to stabilize the aquifer in the catchment zone, to eliminate the cake or drilling fluid protecting the borehole walls and to increase the specific capacity of the well.

Evidently, even where no well development is actually carried out, the putting into operation of a catchment structure may lead to a certain degree of self-development. However, this latter is a slow process causing degradation of the pumping equipment, having no effect on sand bridging that can lead to sanding up of the well.

2.5.1 Cleaning

After drilling has been completed and the casing and gravel pack are in place, the structure must then be cleaned. This straightforward operation ensures clear and "clean" water is obtained after removal of the cuttings and fines that have accumulated in the borehole. Cleaning can be considered as the first phase in the development of the water well.

The cleaning of a water well is generally carried out by air-lift pumping, and is continued until clear water is obtained. It can lead to an increase of the discharge rate (by natural development).

In order to wash the whole thickness of the saturated terrains as fully as possible, it is useful to alternate periods of pumping with periods of rest. The successive fluctuations of the piezometric level enable an efficient cleaning of the aquifer formation.

2.5.2 Development

As described above, the development of a water well, among other things, involves improving the permeability of the aquifer formation surrounding the screen and also stabilizing the formation itself. It should be borne in mind that the immediate operation of a well without prior development will lead to a number of unfortunate consequences:

— It will not be capable of obtaining the optimal discharge from the aquifer.

— It will almost certainly bring about a considerable influx of sand (risking damage to screen and pump, clogging, settling of the gravel pack). Development is thus intended to finish off the cleaning of the borehole, screen and the gravel pack and to improve the hydrodynamic characteristics of the aquifer around the screen. It also has the objective of increasing the safe yield and producing clean water. The permeability of the terrain near the screen is also enhanced, particularly through the elimination of the maximum amount of fines in this zone and by the restructuring and stabilization of the gravel pack.

The methods of development are mainly applicable to loose or only slightly consolidated formations, as well as schist-type rocks and ancient granites. However, such methods are seldom valid for fissured or karstic limestones.

a) Methods of development

Development by overpumping
This is the simplest method of well development. It consists of pumping at a rate far above the estimated water extraction capacity:

— there are risks of irregular development as a result of vertical variations in the permeability of the terrain;

— this type of development can provoke a compaction of the fine sediments, which leads to a reduction in permeability;

— finally, there are risks that sand bridging formed by unidirectional flow (cf. figure 2-13) will cause reduced permeability.

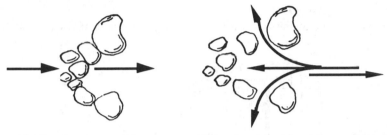

Unidirectional current flow
gives rise to sand gridging

Bidirectional (or bimodd) current flow breaks
up and destroys sand gridging

Figure 2-13
Formation and elimination of "sand bridging" at the screen face
(After A. MABILLOT 1971).

Development by alternating pumping.

This consists of alternating sudden stops and starts of the pump in order to create brief and powerful variations of pressure on the water-bearing layer, thus reversing the flow through the screen:
— it facilitates the disruption of sand bridging;
— there is a risk of wear and tear of the pumping equipment.

Development by surging

The vertical up and down movement of a surge block in the well creates a compression of the groundwater body on the way down, which drives the water and fine particulates into the formation, while depression on the way up sucks this material into the screen (from where the fines are recovered) (cf. figure 2-14);
— to be effective, the plunger must stay in the unscreened upper part of the well;
— inversion of the flow through the screen allows the elimination of sand bridging.

Development by high-pressure jetting

In this method, a tool with pressurized water jets is rotated as it is moved past the whole length of the screen (cf. figure 2-15):
— the fine particles penetrate into the screen where they are recovered by pumping or with a bailer;
— the efficiency of this procedure depends on the type of screen, being optimal for screens having continuous opening of the Johnson type.

Pneumatic development

This procedure uses the same principle as surging, by combining the forward and return flow of the groundwater around the screen brought about by the large volume of air introduced into the borehole. Pneumatic development is used both in consolidated and unconsolidated terrains. Two distinct methods can be adopted:

— the open hole method, which consists of alternating the pumping phases with air-lift and sudden blowing;

— the closed hole method, in which the casing is hermetically sealed with a joint.

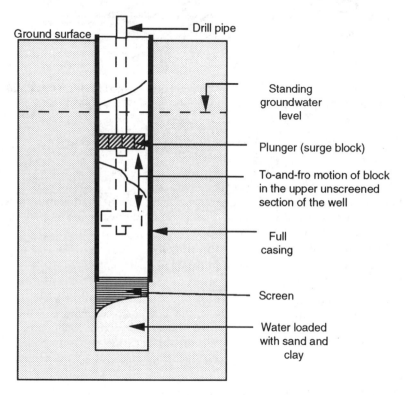

Figure 2-14
Development by surging.

Development by fracturing

This development technique makes it possible to widen existing fractures or create new ones in order to improve the specific capacity of the well.

Without going into the technical details of such a method, it can be pointed out that there are two types of artificial fracturing:

— hydraulic fracturing (or hydrofracting) in which an aqueous phase is injected into the borehole under pressure,

— fracturing by explosives, where the nature and charge of the explosive must be precisely determined in relation to the aquifer formation.

Chemical development

The chemical reagents used In this method comprise acids and polyphosphates.

— Acids are particularly effective in calcareous soluble rocks (limestones, dolomites). In general, dilute hydrochlororic acid is used. The volume of acid to be used increases with each successive acidizing operation, varying from two to five times the volume of the hole to be acidified.

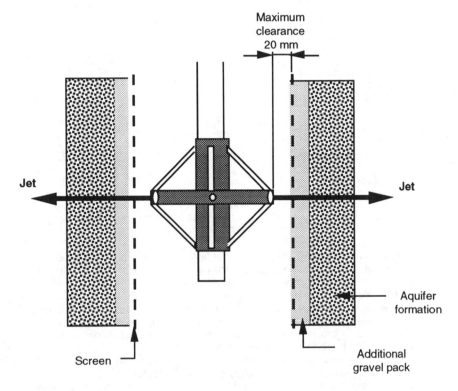

Figure 2-15
Diagram showing development by high-pressure water jetting.

Polyphosphates are used to disperse the drilling cake and also the clays derived from the surrounding formations.

Wive-wound screen

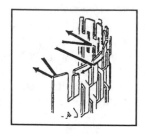

Bridge slot screen

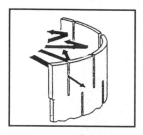

Ribbed screen

Louvre slot screen

Figure 2-16
Influence of the type of screen slot on the efficiency of development by jetting
(after F-G DRISCOLL, 1986).

Example of development by acidification. The following is a summary of a technical example involving acidification used to improve water catchment in a limestone formation. The example is taken from R. BREMOND [1962], and refers to a borehole tapping the Maastrichian aquifer at Taïba (Senegal).

Borehole description: the well is intended to extract water from two aquifers, a Paleocene limestone formation situated between 240 and 330 m and Maastrichtian sands at more than 400 m depth. Although the groundwater bodies are captive, there is communication between them. The borehole log is shown in figure 2-17.

The objective was to acidify the limestone formation between 240 and 330 m. The various pumping operations carried out to wash the formation showed the specific discharge of the well to be about 1 m³/h per metre of drawdown. A study of head losses made it possible to identify the presence of two flow regimes, one laminar and the other turbulent. It could thus be assumed that the laminar head loss was due to the terrain and the turbulent head loss was the result of increases in velocities around the borehole. This could have two causes: the limestone is slightly fissured, or the fractures had been plugged by the invasion of drilling mud.

In both cases, the terrain around the drill had to be treated in order to increase its fracturing or place the clay in suspension to separate it from the limestone. Acidification was performed in three stages (cf. tables III-VII and III-VIII) according to the following procedure:

1. Injection of acid.
2. Injection of water to drive the acid into the formation.
3. Injection of water to drive the acid into the formation.
4. Sudden decompression.
5. Pumping.

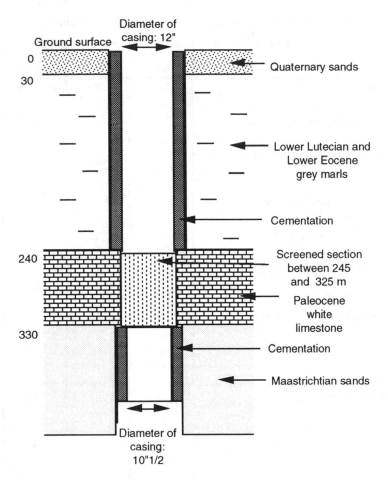

Figure 2-17
Taïba borehole log.

TABLE II-VII — *Nature and quantities of products used during acidizing operations.*

Product	1ˢᵗ operation	2ⁿᵈ and 3ʳᵈ operations
Commercial-grade acid (22 degrees Baumé):	5 000 litres	10 000 litres
Water:	7 000 litres	14 000 litres
Citric acid:	50 kg	100 kg
Corrosion inhibitor (IFF SSO):	15 l	30 l
Emulsifier (Cemulsol):	5 l	10 l

TABLE II-VIII — *Characteristics of acidizing operations.*

	1st operation	*2nd operation*	*3rd operation*
Volume of acid injected (l)	10 070	23 055	23 350
Injection time (mins)	9	18	13
Average discharge capacity of pump (m³/h)	67 130	76 850	107 750
Volume of water injected after the acid (l)	3 000	2 500	4 500
Water injection time (mins)	3	2	4
Injection time of compressed air into aquifer (mins)	6	7	8
Pressure obtained on aquifer (kg/cm²)	5	5	5
Decompression time (s)	40	40	40
Time between beginning of injection and repumping (mins)	56	78	72
Results obtained with air-lift pumping:			
Capacity (m³/h)	24	25	26
Corresponding drawdown (metres)	9	7	6
Specific capacity (m³/h/m)	2.6	3.8	4.3
% increase in capacity	160 %	280 %	330 %

The net results after all the acidification operations show that the specific capacity of the catchment structure increased from 1 m³/h/m to 4.3 m³h/m, which represents an increase of 330%.

Subsequently, prolonged pumping with agitation to facilitate cleaning of the borehole showed that the head losses due to turbulent flow in the groundwater body had been practically eliminated and the entire groundwater body could be considered as exhibiting laminar flow.

The flow rate obtained was 130 m³/h for a drawdown of 31 m, which corresponds to a specific capacity of 4.2 m³/h/m.

Development with air-lift pumping

This is the most commonly used and most efficient method of well development. It involves alternating air-lift pumping with phases of compressed-air injection, using a "double pipe" arrangement which allows delivery of air into the water column to produce a water/air emulsion (cf. figure 2-17).

— This method can be used either in open holes, with the possibility of displacing both the air line and the water pipe, or in closed holes with a fixed double pipe arrangement.

— Development with air-lift and scouring is continued as long as sand intrusion persists. As soon as the water is clear, the well can be prepared for water extraction operations by air-lift pumping continuously for at least one hour (cleaning).

Development of sandstones

Drilling operations in hard formations (sandstones in particular) nearly always cause plugging of the borehole walls. This can arise from the drilling mud (in rotary drilling), by smoothing of the walls (percussion), from the cuttings, clogging of the fissures or by blocking of the sandstone porosity with very fine sediment. In the majority of cases, the use of explosives is not recommended in sandstones, especially if the formation is brittle.

One of the more efficient development methods in this kind of terrain is a combination of air-lift pumping with the to-and-fro surging induced by the delivery of compressed air.

Another method consists of pressuring the borehole in order to lower the standing water-level, but without depressing it below the foot of the casing so as to avoid the penetration of air into the formation. The rapid opening of a surface gate valve will bring about a sudden rise of the water-level, thus carrying fine and loose sediments into the screen. These sediments can then be removed by air-lift.

b) Principle of development with air-lift pumping

Air supplied via a line is injected from the base of a tube submerged in the water well. The emulsion created in this way reduces the density of the water contained in the tube (cf. figure 2-18).

The atmospheric pressure on the water surrounding the tube causes a rise in the level of the emulsified water inside the tube. Provided that the required conditions are met, this rise can attain the same elevation as chosen for the tube outlet.

Required conditions:

For the water to reach the desired elevation, for example A in figure 2-19 (e.g. ground level), the length (BC) of the submerged part of the air line should be chosen in relation to the elevation of the well head (AB) or, more strictly, the total length of the air line (AC)

It is established that BC should be equal to 60% of AC:

$$BC = AC \times 0.60$$

Which can also be written as:

$$BC = AB \times 1.5$$

or otherwise:

$$AC = AB \times 2.5$$

Consequently, the total length of the air line must be at least equal to two and a half times the total height of the water column (including drawdown).

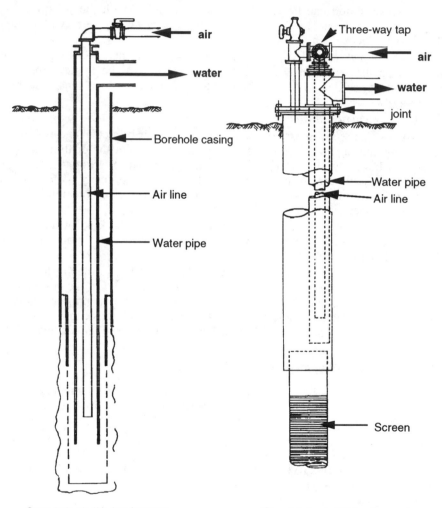

Open-hole air-lift development   Closed-hole air-lift development

Figure 2-18
Diagram of air-lift pumping equipment (after A. MABILLOT, 1971).

These conditions limit the use of air-lift development to catchment structures whose depth allows the introduction of such a length of tube below the drawdown level.

It should be noted that the standing water-level is not affected since it applies only at the start of operations.

The coefficients used in the three formulae given above (i.e. 0.60, 1.5 and 2.5) can be reduced slightly in the case of deep boreholes.

Measurements carried out on a large number of wells with elevations varying from 7 to 200 m have enabled A. MABILLOT to construct a nomogram for the setting up of an air-lift pumping operation (cf. figure 2-19).

The use of the nomogram presented in figure 2-19 is best explained by the following example taken from A. MABILLOT.

Let us consider a borehole of 300 m depth with a capacity of 50 m³/h for a drawdown of 90 m below ground level. The intention is to develop the well by means of air-lift pumping. The question is to determine the optimal dimensions for a device capable of delivering 50 m³/h at ground level. The head losses are considered to be negligible.

1) *Total length of the air line*: on the vertical depth scale on the right of the graph, the 90 m point indicates drawdown level for a capacity of 50 m³/h. From this point, a horizontal line is drawn which cuts the 45°-line labelled "drawdown level" at point B. A vertical line is then drawn passing through B, with length AB corresponding to a value of 90 m (the total elevation). The point C1 is defined as the intersection of the extrapolation of AB with the "minimum submergence" curve, and corresponds to the uppermost position of the foot of the air line.

The length AC1 measured on the depth scale is equal to 144 m. This represents the total minimum length of the air line. In this case, the submergence is BC1 (144-90 = 54 m), or about 37.5% of the total length of the air line.

By extrapolating the line ABC, an intersection C2 is obtained with the "maximum submergence" curve. The length AC2, measured on the depth scale, is equal to 192 m, corresponding to the maximum length of the air line. Beyond this value, no real improvement is acheived in air-lift performance. When extended to intersect the dotted line, the straight line ABC1C2 gives the length which should be given to the air line if a general rule is adopted fixing submergence at 60%, which yields 225 m in the present example.

The application of 60% submergence would require an extra 33 m of air line compared with maximum submergence and 81 m for minimum submergence. In addition to this saving of air line (and water pipe), it can be seen that the air-lift will function equally well in a borehole limited to a depth of about 150 m, whereas, with the 60% rule, at least 225 m of borehole depth are necessary. Thus, in the present example, there is a choice between 144 and 192 m for the total length of the air line. Any value between those two limits may be chosen. However, the water pipe should be made a few metres longer than the air line.

2) *Air pressure at the start*: this pressure corresponds to the weight of a column of water whose height is equal to the submergence of the air tube:

.   Minimum: BC1 =54 m or 5.4 bars
.   Maximum: BC2 = 102 m or 10.2 bars.

3) *Volume of air necessary* (reduced to atmospheric pressure). The horizontal line passing through C1 is prolonged towards the left, where it intersects the air/water ratio curve for "minimum submergence" at point D1. This value corresponds to an air-to-water volume ratio of 12.5, as read off the horizontal scale at the bottom of the nomogram.

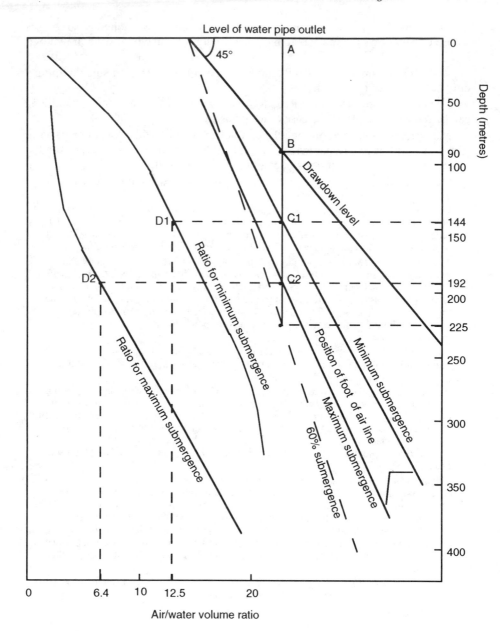

Figure 2-19
Air-lift pumping nomogram (after A. MABILLOT).

For the indicated capacity (50 m³/h of water), 50 x 12.5 = 625 m³/h of air (at atmospheric pressure) would be needed. In this case, a compressor is required that takes in at least 625 m³/h or 10 400 l/m of air, while releasing it at 5.4 bars (minimum). By extending the horizontal line from C2 further towards the left, another intersection point D2 is encountered on the air/water ratio curve for "maximum submergence". The air-to-water volume ratio at this point is brought down to 6.4. To obtain 50 m³/h of water, the compressor would have to take in 50 x 6.4 = 320 m³/h, but it would have to release it at 10.2 bars, as indicated for BC2 above.

In the example presented here (capacity: 50 m³/h), the water pipe should have a diameter of around 150 mm, while a $6^{5/8}$" casing would be suitable. The air line could be made up of a 2" (50.60) pipe weighing 6.4 kg per metre. For a length of 200 m, its weight would be approximately 1,300 kg. Even if the pipe itself could easily support such a load, having a metal cross-section of 900 mm², the joints would require careful attention, particularly the upper collars.

TABLE II-IX — *Diameters of the water pipe and air line in an air-lift pumping device.*

Pump discharge (m³/h)	Water pipe diameter (mm)	Air line diameter (mm)
6 - 12	60	20
12 - 20	90	30
20 - 30	100	40
30 - 50	125	50
50 - 90	150	65
90 - 170	200	65
170 - 220	250	65

c) Monitoring well development

Starting from the principle that good development makes it possible to eliminate the influx of sand or sediments during the operation of a water well, and also to improve its specific capacity, the following criteria must be checked.

— *Sediment content.* Five samples should be taken during the pumping tests for final acceptance, the first after 15 min pumping time and the others at 25%, 50% and 75% of the planned pumping time. The last sample is taken at the end of pumping.

It should be remembered that the maximum permissible limit for suspended particulate matter is 10 mg/l in waters intended for human consumption.

— *Specific capacity.* Even if the sand can be cleared out of a formation during development, the grain-size sorting may not be optimal. A check is carried out by comparing the drawdown of the groundwater body for a similar capacity (cf. figure 2-20) before and after development, or by measuring the hydraulic conductivity.

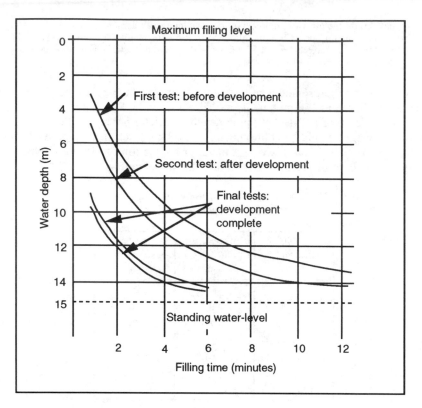

Figure 2-20
Monitoring well development by absorption tests (after A. MABILLOT, 1971).

2.6 Conclusion

The implementation of a groundwater well is a delicate enterprise which sets in motion a series of operations demanding the control of numerous specialized techniques. The success, productivity and life-span of the well are dependent upon these techniques. Any failure is inevitably translated into production problems.

It is not uncommon to notice - generally too late - that the drilling fluid is not suitably adapted to the physico-chemical or hydrogeological conditions of the terrain (cake difficult to eliminate or problems with collapse of the well). Moreover, the quality of the pumped out water may be short of excellent (inadequate equipment, insufficient or ill-adapted gravel pack and development), while the capacity may not come up to expectations (poor identification of the aquifers, inadequate catchment methods, clogging or obstruction of the water flow, etc.).

Consequently, the success of water well implementation is closely linked to a judicious choice of the different methods described in this chapter. In this framework, as discussed in the following chapters, the role of the project manager (or engineer responsible for running the site work) is of prime importance. In conjunction with the contracting company, the project manager should:

— Define the drilling techniques to be used (diameter of borehole and type of drilling).

— Choose a suitable drilling fluid and adapt it during the course of drilling according to the formations traversed.

— Resolve problems that are specific to the equipment, the installation of the gravel pack and the cementation.

— Determine the best type of development to be adopted in relation to the hydrogeological characteristics, while defining its principles and duration.

CHAPTER III

Well hydraulics

"Rather know nothing than half-know much"
Nietzsche

Pumping or aquifer tests should be carried out after the cleaning and/or the development of a water well. Such tests make it possible to determine:

— *the characteristics of an aquifer complex/catchment structure, by establishing a yield-depression curve, s = f(Q)*, which is a veritable identity card for the well system drawn up on the basis of *step pumping tests;*

— *the hydrodynamic parameters S (storage coefficient) and T (transmissivity)*, calculated from *long-term pumping tests;*

— the operating conditions of the well;

— the evolution of drawdown as a function of discharge rate and time for long-term water extraction (interference calculations).

Furthermore, pumping tests performed before the start of water abstraction or after a regeneration phase allow the satisfactory completion of cleaning and development procedures, including natural development, that concern the tested well.

3.1 Basic concepts

The parameters which characterise an aquifer are: permeability, transmissivity, porosity and storage coefficient. Moreover, Darcy's Law is known to be valid whatever the type of water circulation. However, it is not possible to describe in theoretical terms the flow of water towards catchment structures without firstly making a certain number of simplifying hypotheses and assuming a simple aquifer with ideal behaviour. Thus, since the setting up of formulae to describe groundwater flow is based on simplifying hypotheses, it is advisable to have a good understanding of them in order to make an appropriate choice among the numerous methods that are available for the interpretation of pumping test data.

3.1.1 Flow superposition

An ideally simple aquifer is assumed not to undergo recharge and has an infinite lateral extent. It is also homogeneous, isotropic and of constant thickness, being initially at rest and tapped throughout its entire thickness, while its water is released instantaneously during any fall in piezometric level.

In the following, it is demonstrated that such simplifications are not only justified but also realistic:

— A finite aquifer will always behave strictly as an infinite aquifer provided that the impulse set off by pumping is not reflected back from one of its boundaries towards the observation point, where it would produce a measurable perturbation.

This phenomenon sometimes takes so long to have any effect that many other perturbations of the system should first be taken into account:

— Although sedimentary media are usually heterogeneous in detail, on a scale of km^3 they are commonly more homogeneous than is generally believed.

— The heterogeneity itself (which, in fact, is expressed as $Kx \neq Ky$) only rarely plays a significant role, since in most cases the aquifer medium is strongly stratified, being defined by two hydraulic conductivities: one characterizing flow perpendicular to the stratification and the other flow parallel to the stratification. In general, permeabilities vary little within the horizontal plane, which may be considered as indistinguishable from stratification. The majority of natural groundwater flow takes place in this plane.

— Barring accident, the thickness of an aquifer varies very gradually, in such a way that an aquifer showing no abrupt changes in its thickness in close proximity to a borehole will behave as if it had constant thickness. The average thickness in the zone influenced by pumping can be assumed to be constant.

— Compared with a borehole screened throughout the whole thickness of an aquifer, a borehole tapping only part of the same formation will have a slightly lower yield. Groundwater flow is uniquely modified in the immediate vicinity of a borehole. It has been shown that this modification is only significant within a limited perimeter, which corresponds at most to 1.5 times the thickness of the aquifer. At greater distances, the flow lines have sufficient space to spread evenly throughout the entire thickness of the aquifer.

— As regards the release of water from the terrain following a lowering of the water table, it is well known that such an effect does not occur instantaneously. This phenomenon is only discernible if the aquifer formation has a very low permeability. In terrains of high permeability, most of the water is liberated very quickly.

— One of the starting hypotheses is that groundwater bodies are initially at rest. In reality, groundwater is never at rest, so flow must occur and the upper surface of the aquifer will be sloped.

All these difficulties can be overcome by applying the principle of flow superposition (cf. figure 3-1). The possibility that different water flows can be superimposed is a fundamental property of porous media. This means that the respective simultaneous effects of two different causes of change in groundwater-level at a given point can be added together algebraically. For example, if pumping is carried out during a period of natural lowering in water-level, this can be taken into account by subtracting the natural depression from the measured drawdown. Such an operation amounts to normalizing the piezometric level, not to the initial level, but to the natural level extrapolated to the time of the measurement.

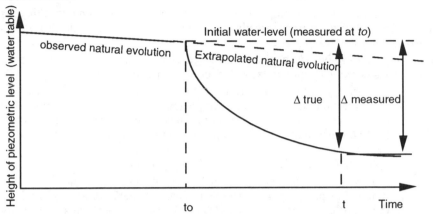

Figure 3-1
Practical application of the principle of superposed flow..

This correction is of practical interest particularly for long-term tests. Hence the need to keep a close watch on the natural evolution of the piezometric level before undertaking such a test.

3.1.2 Influence of starting up water extraction

The startup of water extraction from a well causes a cone of depression to appear on the aquifer, whose initial piezometric surface is considered to be flat. The cone is funnel-shaped, with its axis centred on the borehole.

The objective of pumping tests is to measure the dimensions of this cone at a given moment of its evolution, while maintaining constant discharge rate.

The geometry of the cone of depression determines:

— The drawdown, denoted as s, is measured by the drop in the water-level in the well or in a piezometer situated at distance x from the borehole axis. The level of water in the well is termed the dynamic or pumping water-level. The depth of the pumping water-level below the initial piezometric level, in an undisturbed regime, is known as the drawdown, s. This value corresponds to the pumping depression plus the head losses due solely to the presence of the well. The drawdown measured during rising water-level (recovery) is called the residual drawdown s_r.

— The radius of influence, noted R, is the distance from the well axis at which drawdown is nil or negligible.

At a regular discharge rate, three factors influence the dimensions of the cone of depression:

— *The hydrodynamic parameters* (transmissivity and storage coefficient).

In fact, the radius of influence is directly related to the transmissivity and indirectly to the storage coefficient.

— *Pumping time,* the dimensions of the cone increase with time until there is an eventual stabilization (quasi-steady state regime) in which the aquifer restores its water balance.

— *The flow regime.* At constant discharge, two concepts of the flow regime may be considered in relation to the influence of the pumping time:

- *The steady state* (or equilibrium) regime, in which, after a short pumping time (around 1 hour), the geometry of the cone of depression remains constant. This was the hypothesis put forward by H. DUPUIT (1863). It is in some ways a snapshot of the hydrodynamic behaviour of the aquifer.

- *The non-steady state* (or non-equilibrium) *regime* takes account of the observed fact that the size of the cone of depression increases as a function of pumping time. This is the basis of the expressions given by C.V. THEIS (1936) and C.E. JACOBS (1950). In strict terms, the steady state does not exist except under exceptional conditions. In reality, we accept the appearance of a quasi-steady state regime.

3.2 Metrology

Various external factors can have a considerable influence on groundwater bodies. As a consequence, it may be necessary to measure the temperature of the water and air, the pH and/or the conductivity of the water, the atmospheric pressure, the position of the tides in relation to the test times (boreholes near a sea coast), pluviometry, the spates of rivers, etc.

A device for the evacuation of pumped water should be put in place. This water should not be allowed to return to the aquifer and should be evacuated to such a distance that it does not interfere with the groundwater.

3.2.1 Choice and layout of observation wells (piezometers)

The number of piezometer depends upon the type of problem to be addressed and, of course, the available funding. The observation wells must have a very low response time. Their choice therefore will be decided by the nature of the water-bearing terrain.

Open observation wells can be utilised in coarse alluvium, where permeability is generally high. On the other hand, impermeable soils (clays, clayey gravels, silts) require observation wells having a constant volume to reduce the response time. This type of observation well should be used whenever there are doubts regarding the permeability of the soil, since a variation of one unit in the exponent of the hydraulic conductivity causes a tenfold increase in the response time.

In theory, the observation wells are set out along two rectangular axes centred on the control well and at increasing distances away from it, the distance of the last one being close to the estimated radius of influence (around 200 m in highly permeable terrains, and 30-50 m in terrains of low permeability). In many cases, the distances between the observation wells and the control well follow a geometric progression which facilitates the graphical interpretation. Whatever the arrangement, there should be at least four piezometers placed at different distances.

In a confined aquifer, the depth of the screens is of no great importance because the head variations are generally small in relation to the initial pressure-head, the equipotential (or piezometric) surfaces being more or less vertical cylinders concentric to the well axis. On the other hand, in an unconfined aquifer, the piezometric surfaces become increasingly curved as the well is approached. It is only at great distances from the well that these surfaces become comparable to vertical cylinders.

3.2.2 Measurement methods

a) Measurement of flow rates

Measurements of flow rate and water-level are fundamental to every hydrogeological study. They should be carried out systematically in a quasi-simultaneous way if they are to be useful.

A drawdown measurement without the corresponding discharge is of no practical use, just as a discharge without any mention of the corresponding drawdown provides no significant hydrogeological information.

There are numerous methods for measuring the pumping discharge of a water well. These can be indirect, such as reading a meter which converts flow velocity into discharge rate, or direct, such as timing the filling of a recipient with a given volume.

Although the *reading of a meter* is easy and rapid, it presupposes a well-calibrated meter. This calibration should be checked periodically against a more accurate direct measurement.

The most accurate method for measuring flow rate consists of timing the filling of a given volume.

The measurement of filling time can be very accurate if there is a stop-watch and a mobile outlet pipe which allows lateral shifting of the jet at any moment.

To measure the volume, it is useful to have a reservoir with a known horizontal section; a small amount of water is first poured into the bottom of the reservoir and its level is noted after stabilization; then the new level is noted after filling. The flow rate (or discharge rate) is obtained from the following formula:

$$\text{discharge rate in m}^3/\text{h} = \frac{\text{section in m}^2 \text{ x height difference in m x 3 600}}{\text{filling time in seconds}}$$

$$\text{or } Q = \frac{L \times l \times H}{t} \times 3\,600$$

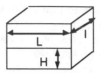

The pitometer (or Pitot tube) method is equally simple to set up. The assembly consists of a delivery main entering a rigid conduit with a diameter ensuring that flow takes place over the full cross section and along a minimum length of 2 m (and 15 times the conduit diameter D when this latter is >150 mm). At the end of the conduit, there is a measuring device made up of a section of tube identical to that of the intake with diameter D. The end of this tube is obstructed by a metal plate having at its centre a circular orifice of diameter d. On the side of the tube there is a pressure inlet attached to a transparent manometer tube fitted with a graduated measuring scale (cf. figure 3-2).

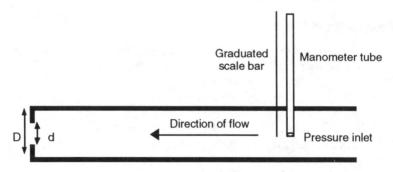

Figure 3-2
Cross-section of a typical Pitot tube.

The discharge rate of the pumped water is that which passes through the orifice of the plate. It is given by the Bernouilli expression:

$$Q = \alpha s \sqrt{2gh}$$

where:

Q : discharge rate,

α : coefficient of the experimental flow rate which depends on $\dfrac{d}{D}$

s : cross-section of the plate orifice $(\dfrac{\pi d^2}{4})$

h : height of the pressure-head.

This relation can be expressed in the simplified form:

$$Q = C d^2 \sqrt{h}$$

where $C = \dfrac{\alpha \pi \sqrt{2g}}{4}$ a coefficient which, like α, depends solely on the ratio $\dfrac{d}{D}$.

The numerical values of the C coefficient are given by the nomogram in figure 3-3.

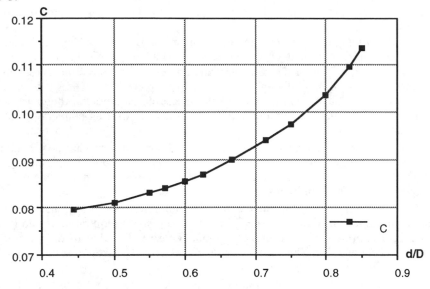

Figure 3-3
Determination of C coefficient (Pitot tube method).

b) Measurement of water-level

The aim of a pumping test is to furnish data whose interpretation will lead to determining the hydrodynamic parameters of the aquifer.

The data are essentially made up of measurements that are taken before, during and after pumping. They must be accurate, precise and usable independently of the operators who obtained them. Apart from the need to carry out the mesurements with the greatest rigour, it is also essential to use equipment that is in good working order.

It is advisable to measure horizontal lengths (distances from the control well to the observation wells), depths (groundwater level, borehole depth), time, discharge rates, pressures and various other quantities.

Direct method

— *Tapes with electric contact gauges* (electrical water-level gauges or dippers) are satisfactory for measuring the piezometric level. Insofar as the probe is of good quality, it enables measurements to the nearest centimetre for absolute values of 30-50 m. The apparatus is made up of an electric cable with two conductors having only slight elongation. The cable, calibrated or not, is wound round a drum. One end of the drum is attached to two isolated electrodes and the other end to an ordinary dry battery. A milliammeter or a high-resistance voltmeter is placed in the circuit, or alternatively, a relay is connected which supplies a bulb or an alarm. When the electrodes touch the surface of the water, the electric circuit is closed and the needle of the milliammeter moves, the bulb lights up or the alarm sounds. Numerous different models of this apparatus are available.

— The *analogue recorder*: the development of data processing equipment allows the rapid treatment of vast masses of data (automatic graph-plotting, conversion, preparation of data for interpretation, etc.), but all the measurements need to be digitized. Even though a set of analogue recordings makes it possible to follow continuously the evolution of the measured parameters, the charts so obtained require time-consuming and costly study, as well as conversion into digital form (digitization).

It is also necessary to consider the use of autonomous "automatic data acquisition systems" or loggers which place the digitized data in mass storage. The different modules are enclosed in waterproof packs and the current is supplied by 12 V cells or long-life batteries (e.g. lithium).

The continuing progress in electronics and microcomputing has led to increased miniaturization associated with minimum space requirements and low power consumption. The different components are vulnerable to damp, cold, excessive heat and dirt. This means that great care must be taken in setting up the on-site equipment.

We cannot be too insistent in advising the use of analogue recorders which provide very precise measurements with very short acquisition times (around one second) and which do not require staff to check water-levels day and night when meteorological conditions are unfavourable. Finally, by choosing a suitable data sampling step, the well effect and post-production effects can also be determined. It is preferable to have two recorders so as to be able to monitor a observation well at the same time as the main well.

Indirect methods manometer tube

This requires the installation of a waterproof piezometric pipe which goes down into the well and which is fixed along the pump column. The measurement procedure is as follows:

— Inflation of the standpipe with an air pump, or from a cylinder of compressed air/nitrogen fitted with a relief valve, until the pressure is stabilized.

— Reduction of the gas flow to the lowest value which allows a constant pressure.

— Reading the pressure in bars on a manometer connected to the standpipe

— Conversion of the pressure into metres of water (1 bar = 10.2 m of water).

The result corresponds to the height of the piezometric pipe submerged at the moment of measurement. It is therefore necessary to know exactly the total height of the piezometric pipe in order to calculate the required water-level.

At each step, with increasing discharge rate, the water height in the piezometric pipe falls (increasing drawdown), thus inducing a decrease in the pressure read off the manometer.

3.2.3 Implementation of a pumping test

Measurements of dynamic level and discharge rate (or yield) are made at regular intervals, being carried out during periods of falling (drawdown) or rising water-level (recovery after cessation of pumping), as shown in the following table.

TABLE III-I — *Schedule of measurements for pumping test.*

Time elapsed from the beginning of rise or fall in water level	Measurement of dynamic levels and drawdowns	Measurement of yield
0 - 5 min	every 30 s	every minute
5 - 10 min	every minute	every 2 min
10 - 20 min	every 2 min	every 5 min
20 - 40 min	every 5 min	every 5 min
40 min - 1hr30	every 10 min	every 10 min
1hr30 - 3 hr	every 15 min	every 15 min
3hr - 5hr	every 30 min	every 30 min
5hr - 8 hr	every hour	every hour
more than 8hr	every hour	every hour

The above measurement programme remains purely indicative. It is evident that, according to the response of the aquifer, a reduction of the interval between measurements may be deemed necessary.

3.3 Pumping tests in a steady-state regime

A step pumping test of short duration is used to evaluate the characteristics of aquifer and catchment complexes. These characteristics comprise: critical discharge, relative specific discharge, head losses in the well and its immediate vicinity, and the maximum operational yield. Such tests lead to the drawing up of a technical programme for equipping the well: casing, screens, gravel pack, pump power, etc.

The main aim of step pumping tests is to determine the characteristic *yield-depression curve of the well*, $s = f(Q)$, or the variation of drawdown as a function of the pumping discharge and *the critical discharge Qc.*

The yield-depression curve is particularly useful for adjusting the power of the pump before its installation.

3.3.1 Methodology

A step pumping test is generally carried out in four, or sometimes three pumping stages. The yield-depression curve is obtained by plotting the discharge versus drawdown relation on a graph.

From the methodological standpoint, there is a choice between several contrasting approaches:

— *Carrying out the test in a steady state regime.*

This is an ideal situation. Pumping proceeds with discharges of Q_1, Q_2, Q_3 etc. for a period of varying duration until stability is achieved (steady state). In this way, a discharge of Q_1 can be associated with a corresponding drawdown of S_1.

— *Carrying out the test in a quasi-steady state regime.*

This is the most widespread situation, but the closer the approximation to "quasi-steady state" conditions, the more frequent is the occurrence of pseudo-stabilization and the less meaningful is the test. With such tests, it is not possible to calculate the absolute values of the hydrodynamic parameters with an acceptable degree of validity. In particular, the specific discharge is only meaningful in a steady state regime [J. FORKASIEWICZ and J. MARGAT, 1966] and cannot be defined by this kind of test. The same authors consider that "the presentation of yield-depression curves constructed from pumpings without real stabilization, and the expression of specific discharge calculations on this basis, should not figure in serious hydrogeological reports".

In practice, each discharge yield is applied during equal time increments that are sufficiently long to justify the assumption that a quasi-steady state has been established. The same time increment is generally used for each step, while the aquifer is allowed to recover between successive steps for a period equal to the pumping time. Nevertheless, this technique can be used to obtain relative data rapidly in a well field were data are lacking.

The initial static water-level is measured before the pump is started up. Before the pumping test proper begins — the previous night if possible — the pump is switched on for about ten minutes with the pumping valve fully open so as to enable measurement of the maximum operational yield of the well, or Q_{max}.

The discharge (or yield) of each step can then be defined as follows:

Test with four steps		Test with three steps	
Step 1	$Q_1 = \frac{Q_{max}}{4}$	Step 1	$Q_1 = \frac{Q_{max}}{3}$
Step 2	$Q_2 = \frac{Q_{max}}{2}$	Step 2	$Q_2 = \frac{2 \, Q_{max}}{3}$
Step 3	$Q_3 = \frac{3 \, Q_{max}}{4}$	Step 3	$Q_3 = Q_{max}$
Step 4	$Q_4 = Q_{max}$		

For each step, the dynamic level , *DL*, and the yield, *Q*, are measured at regular intervals as shown in Table III-I.

It is also advisable to measure the height of any eventual marker bed in relation to the ground surface.

For all step pumping tests, it is preferable to halt water abstraction from the well to be tested at least one day prior to the test (a minimum of 10 hours before).

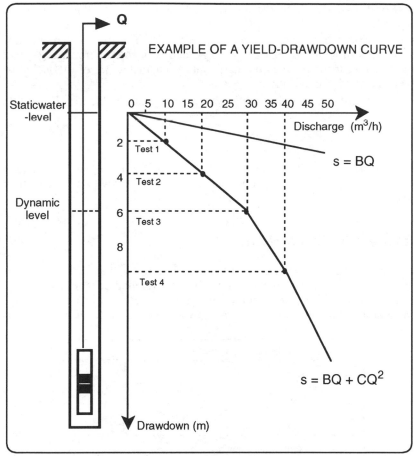

Figure 3-4
Example of yield-depression curve.

3.3.2 Interpretation

The results of a step pumping test are summarized for each step by two quantities: yield and drawdown. The drawdown is equal to the difference between the static water-level and the dynamic level. The stabilization of the water-level with pumping carried out at yields of Q_1, Q_2, Q_3... etc. corresponds to drawdowns of s_1, s_2, s_3... etc. The plotting of the representative points on a graph makes it possible to draw the yield-depression curve for a given catchment structure (cf. figure 3-4). The yield-depression curve is a fundamental feature of the well.

The drawdown measured in the well at time *t* is the sum of two head losses:

— *A linear head loss* is generated by laminar flow into the aquifer close to the well (Darcy's Law), and is denoted here as *BQ*. It is constrained by the hydrodynamic parameters of the aquifer and increases with the duration of pumping.

— A *quadratic head loss*, which is non-linear, arises from turbulent flow within the well structure, screen and casings, and is denoted here as CQ^2. It is solely a function of the pumped discharge and is characteristic of the technical equipment installed in the well.

The total drawdown, s, at time t is given by the expression due to C.E. JACOB:

$$s = BQ + CQ^2$$

This expression was established for a confined aquifer, but it can be extended to an unconfined aquifer on condition that the measured drawdown is less than $0.1b$ (where b is the thickness of the aquifer).

3.3.3 Concept of critical velocity

During pumping, the dynamic level in the well is lower than the piezometric level in the aquifer close to the structure. This difference corresponds to the height of the seepage surface denoted h'. It increases with the drawdown, attaining a maximum value when the drawdown in the aquifer is around $b/2$ (cf. figure 3-5).

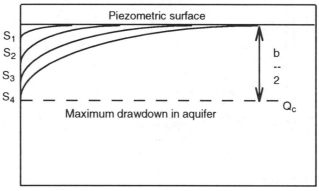

Figure 3-5
Diagram showing growth of the cone of depression.

Beyond the maximum depression of the seepage surface, any increase of drawdown in the well causes no further drawdown in the aquifer close to the well. The dynamic levels are stabilized and the yield ceases to increase any further as a result of the pumping. The area of pumping depression increases and the well is drained. Up to this limit, the depression curve is increasingly hollowed out to reach a maximum. Beyond a certain point, when the critical velocity has been reached, the laminar flow gives way to turbulent flow. The critical velocity corresponds to the critical discharge Qc. The turbulent regime causes an increase in the quadratic head loss, thus diminishing the discharge capacity of the well.

Moreover it provokes the drawing off of fine particles responsible for the clogging and/or silting up of wells.

In practice, the pumping discharge should always be lower than the critical discharge.

The yield-depression curve is a key element which cannot be dispensed with. It is essential to include it in all documents concerning the catchment structure. In fact, the yield-depression curve can be used later to detect any improvement or deterioration in well performance. Conventionally, the curve may show different types of characteristic shape (cf. figure 3-6):

1) Ideal well (straight line).

2) Real well after acidification (improvement).

3) Initial state of real well.

4) Real well after ageing (clogging).

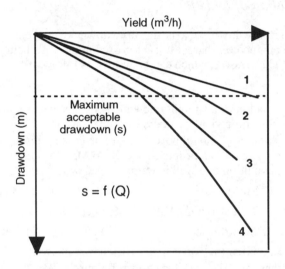

Figure 3-6
Characteristic types of yield-depression curve.

A straight line reflects the linear evolution of drawdown as a function of yield, without the appearance of head losses due to the well equipment (quadratic head loss). The quadratic head losses become all the more important as the curve becomes increasingly convex.

Under no condition can the curve present a concave-upwards shape. This would imply an invalid pumping test (incorrect measurements or the appearance of development during pumping).

3.3.4 Comparison of several tests on the same catchment structure

The carrying out of several step pumping tests on the same well during production (over several years) makes it possible to compare the yield-depression curves drawn up at each test with the initial yield-depression curve. The following information can be obtained from a study of such curves:

— *No change, the curves are identical.*

The structure has neither developed naturally nor become clogged.

— *The new curve plots below the initial yield-depression curve*

The structure has become clogged; the drawdown has increased for the same yield, so the specific discharge (discharge capacity) has diminished.

— *The new curve plots above the initial yield-depression curve.*

The catchment structure has developed during its service-life. For the same yield, the drawdown has diminished, so the specific discharge (discharge capacity) has increased.

3.3.5 Calculation of head losses

The specific drawdown, s/Q, is the drawdown level measured in the well divided by the pumped discharge. It is expressed in m/m³/h, being obtained from the equation of C.E. JACOB, which can be written as follows:

$$\frac{s}{Q} = B + CQ$$

This is the equation of a straight line which enables the formulation of certain simple expressions of the yield vs. drawdown relationship.

s/Q as a function of Q should be a straight line passing through the origin and having slope C. A method proposed by B. WALTON makes it possible to characterize the condition of a well in terms of its C value:

$C < 675$ m/(m³/s)²	Good well, with correct development
$675 < C < 1{,}350$ m/(m³/s)²	Mediocre well
$C > 1{,}350$ m/(m³/s)²	Clogged or deteriorated well
$C > 5{,}400$ m/(m³/s)²	Irrecoverable well

In general, there are four possible cases (cf. figure 3-7):

— straight line (1) passing though the origin, indicating that turbulent flow is strongly predominant in the aquifer and well ($s = CQ^2$);

— straight line (2) not passing through the origin ($s = BQ + CQ^2$);

— straight line (3) with zero gradient, parallel to the y-axis, indicating laminar flow with nil or negligeable loss of head in the screen and casing ($s = BQ$);

— concave-upward curve (4) ($s = BQ + CQ^n$).

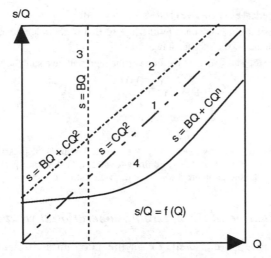

Figure 3-7
Graph of yield vs. specific drawdown showing cases discussed in text
(after J. FORKASIEWICZ, 1978).

The straight lines on a plot of specific drawdown vs. yield make it possible to determine the coefficients B and C in the equation: $\frac{s}{Q} = B + CQ$

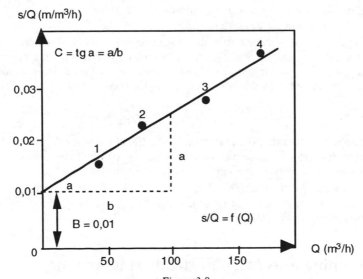

Figure 3-8
Linear relationship of yield vs. specific drawdown: calculation of head losses
(after J. FORKASIEWICZ, 1978).

In the numerical example given here $B = 0.01 = 1.10^{-2}$

- the B coefficient is obtained by the intersection of the representative straight line with the y-axis (specific drawdown).
- the C coefficient is equal to the slope of the representative straight line

$$C = tg\ \alpha = \frac{a}{b} = \frac{0.014}{100} = 1.4.10^{-4}$$

The equation of the representative straight line is:

$$s = 1.10^{-2}\ Q + 1,4.10^{-4}\ Q^2$$

The drawdown corresponding to each step is calculated with this expression. The values so obtained, when plotted on the yield/drawdown graph, coincide perfectly with the observed linear relation (cf. figure 3-8). Thus, the well test is correct.

3.3.6 Determination of the maximum operational yield capacity

The yield capacity of a well, Pr, is the maximum discharge that can be pumped over a given time without the pumping depression exceeding the maximum acceptable drawdown [J. FORKASIEWICZ, 1978]. The maximum acceptable drawdown is determined by:

— Physical and technical constraints of the aquifer complex/catchment structure, expressed by the critical discharge, Q_c and the corresponding critical drawdown, s_c, as measured from well tests (cf. figure 3-2).

In practice, if $Q_c = 150$ m^3/h et $s_c = 5$ m, the maximum yield, Q_{max} and the maximum drawdown should be lower by a factor of 5 - 10 %, i.e. $Q_{max} = 135$ m^3/h and $s_{max} = 4.50$ m.

— Socio-economic constraints, the main one being the production cost of the water at maximum operational capacity (Pr) which determines the depth of the dynamic level. The maximum drawdown adopted must therefore be equal to the maximum measured drawdown without going beyond the acceptable maximum limit.

$$Pr = Q_s \times s_{max} = Q_{max}$$

It should be noted that the maximum operational capacity can be higher than the critical discharge because a quadratic head loss can be tolerated on condition that it is not excessively high compared with the linear head loss.

In practice, Pr a function of the thickness of the unconfined aquifer ($s_{max} = b/3$) or, in the case of confined aquifer, of the water-level in the well before pumping, ($s_{max} = 0,75\ h$).

3.4 Pumping tests in a non-steady state regime

The main aim of pumping tests in a non-steady state regime is to determine the hydrodynamic parameters of the aquifer (transmissivity, T, and storage coefficient, S) and the optimal operational capacity of the well. In addition, account is taken of the well and aquifer characteristics, and also the eventual presence of other catchment structures in the vicinity (interference calculations).

Pumping tests in a non-steady state are of longer duration than those carried out in a steady state. They are performed with a single discharge step (constant discharge) for at least 48 hours, the optimal duration being 72 hours. The recovery of water-level should be measured for at least six hours and normally for a duration equal to that of the test. The pumping test discharge is chosen from the yield-depression curve determined by tests in the steady state.

The interpretation of water-level data (from pumping and recovery tests) is based on the use of hydrodynamic relations for the non-steady state set up by C.V. THEIS (1935) and later authors [L.K. WENZEL, 1942 and C.E. JACOB, 1950]. Theis's formula is presented in two forms: either an integral exponential equation or its logarithmic approximation. The latter formulation is easier to handle, and is the most commonly used.

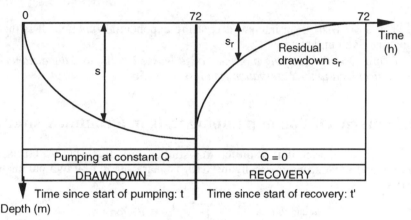

Figure 3-9
Long-duration pumping test.

A long-duration pumping test has three main aims:

— *Measurement of the hydrodynamic parameters*: transmissivity and storage coefficient.

— *Quantitative study of aquifer characteristics:* boundary conditions (confirmation of the distance from the well to the limits of the system, plugging of river banks) and structure (heterogeneity, leakage);

— *Evaluation of extractible groundwater resources* by direct "true scale" observations of the effect of production on the aquifer. Prediction of the evolution of drawdown as a function of pumping discharge.

The solution to C.E. JACOB's logarithmic approximation is obtained by drawing and interpreting the straight lines which represent drawdown vs. logarithm of pumping time or residual drawdown vs. logarithm of rise time [G. CASTANY,1967; J. FORKASIEWICZ, 1970 and 1972; C.E. JACOB, 1950; G.P. KRUSEMAN *et al.*, 1974 and TRUPIN, 1969].

Starting from a new concept of the hydrodynamic behaviour of aquifers, C.V. THEIS (1935) was the first to establish expressions describing groundwater flow towards catchment structures in a non-steady state regime. The basic general conditions for application of these expressions are those of the well test. Furthermore, the type of aquifer forming the basis of these calculations should satisfy three hydrodynamic criteria: it must contain a confined aquifer, be of infinite areal extent and have an impermeable substratum and roof. The general THEIS equation, which is applicable to all arrangements at the test station, is as follows:

$$s = \frac{Q}{4\pi T} \int_u^\infty \frac{e^{-v} dv}{v} \qquad \text{or} \qquad s = \frac{Q}{4\pi T} W(u) \qquad \text{with} \qquad U = \frac{x^2 S}{4 T t}$$

The term $W(u)$ is an integral decreasing exponential function. It is the well function given in tables.

In practice, the pumping time must be at least 42 hours and the distance from the control well to the observation well less than 150 m.

3.5 Interpretation of pumping tests in a confined aquifer

The case of a confined aquifer was studied initially by C.V. THEIS. This simple and ideal scenario (cf. figure 3-10) is characterized by two parameters T and S.

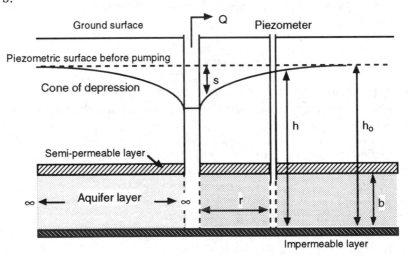

Figure 3-10
Perfect well in a confined aquifer.

3.5.1 Theis's bi-logarithmic method

This technique is used when the time t is small or when the distance x between the borehole and the observation well is too great. In such circumstances, the data are difficult to treat with the Jacob method.

The calculation of the hydrodynamic parameters T and S is performed using Theis's curve.

$$s = \frac{Q}{4\,\pi\,T}\,W\,(u) = \frac{0.08\,Q}{T}\,W\,(u)$$

where:

$$u = \frac{x^2\,S}{4\,T\,t}$$

$W(u)$ is a known and tabulated function.

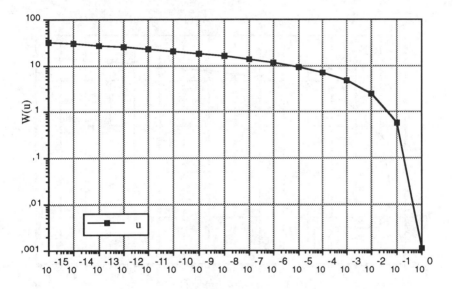

Figure 3-11
Standard Theis curve

For greater convenience in calculating, Y. EMSELLEM has suggested the use of the parameter $x = \dfrac{1}{u}$ proportional to the time t and to the function $F'(u') = W\,(u)$ whose representative curve on log-log coordinates is given in figure 3-11. In fact, this type of representation produces a symmetry in relation to the axis of the coordinates since $Log\ u = -\ Log\ (1/u)$. Thus, the standard curve does not change but the time increases in the direction of increasing x-axis coordinates, which is more explicit (cf. figure 3-12).

This gives:

$$s = \frac{Q}{4\,\pi\,T}\,F\,(u') = \frac{0.08\,Q}{T}\,F\,(u')$$

with

$$x = \frac{4\,T\,t}{r^2 S} = \frac{1}{u}$$

which can be written :

$$y = \frac{s}{0.08\frac{Q}{T}} = F\!\left(\frac{4\,T\,t}{r^2\,S}\right) = F(x)$$

or alternatively :

$$0.08\,Q\frac{y}{s} = T \quad \text{and} \quad S = \frac{4\,T\,t}{r^2\,x}$$

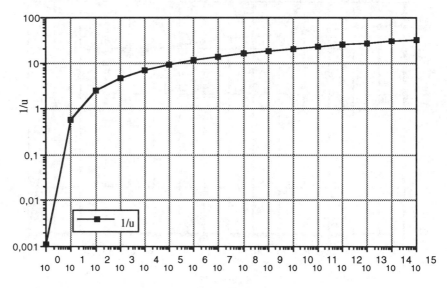

Figure 3-12
Standard Theis curve, normalized according to 1/u

In practical terms, it is advisable to carry out the following:

— Draw a Theis curve on the first graph using tables which give the $W(u)$ values. The $W(u)$ values are plotted on the vertical axis and $1/u$ on the horizontal axis. It is best to use log-log tracing paper.

— Represent the values observed from the observation well on a second graph with log-log coordinates. The vertical axis should be used for the drawdown values *s* and the horizontal axis for the time *t*. It is evident that this graph should be plotted using the same scale as used for the graph on figure 3-13.

The calculation of *T* and *S* consists of recording the observed values of *s* as a function of *t* on log-log coordinates, and then superimposing this curve onto the theoretical curve. The axes of the coordinates must be kept parallel. A common point is selected, and from that point the drawdown values *Δs* and corresponding times *t* are read off.

By superimposing the two graphs, it is possible to link up any point on one to a point on the other. Identification of the coordinates of this pair of points *(s, t)* and *[F(u'), u]* makes it possible to calculate *T* and *S* using the formulae:

$$T = \frac{0.08\,Q}{s}\,F\,(u')$$

$$S = \frac{4\,T\,t}{x^2\,u'}$$

TABLE III-II — *Well function for different values of u (after WENZEL, 1942).*

u	1	2	3	4	5	6	7	8	9
	0.219	0.049	0.013	0.0038	0.0011	0.00036	0.00012	38.10^{-6}	12.10^{-6}
$.10^{-1}$	1.82	1.22	0.91	0.70	0.56	0.45	0.37	0.31	0.26
$.10^{-2}$	4.04	3.35	2.96	2.68	2.47	2.30	2.15	2.03	1.92
$.10^{-3}$	6.33	5.64	5.23	4.95	4.73	4.54	4.39	4.26	4.14
$.10^{-4}$	8.63	7.94	7.53	7.25	7.02	6.84	6.69	6.55	6.44
$.10^{-5}$	10.94	10.24	9.84	9.55	9.33	9.14	8.99	8.86	8.74
$.10^{-6}$	13.24	12.55	12.14	11.85	11.63	11.45	11.29	11.16	11.04
$.10^{-7}$	15.54	14.85	14.44	14.15	13.93	13.75	13.60	13.46	13.34
$.10^{-8}$	17.84	17.15	16.74	16.46	16.23	16.05	15.90	15.76	15.65
$.10^{-9}$	20.15	19.45	19.05	18.76	18.54	18.35	18.20	18.07	17.95
$.10^{-10}$	22.45	21.76	21.35	21.06	20.84	20.66	20.50	20.37	20.25
$.10^{-11}$	24.75	24.06	23.65	23.36	23.14	22.96	22.81	22.67	22.55
$.10^{-12}$	27.05	26.36	25.96	25.67	25.44	25.26	25.11	24.97	24.86
$.10^{-13}$	29.36	28.66	28.26	27.97	27.75	27.56	27.41	27.28	27.16
$.10^{-14}$	31.66	30.97	30.56	30.27	30.05	29.87	29.71	29.58	29.46
$.10^{-15}$	33.96	33.27	32.86	32.58	32.35	32.17	32.02	31.88	31.76

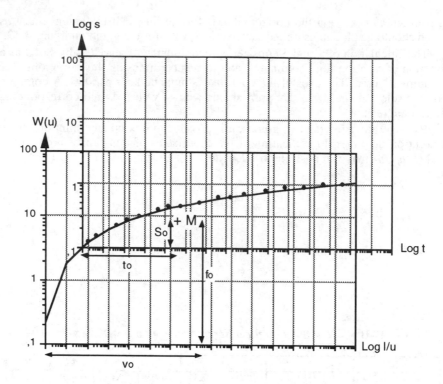

Figure 3-13
Example of application: Theis curve superimposed onto an experimental curve.

The linearity of the relations:

$$s = \frac{Q\ W(u)}{4\ \pi\ T}$$

$$t = \frac{1}{u} \times \left(\frac{x^2\ S}{4\ T}\right) = \frac{4\ T}{u\ x^2 S}$$

implies that the relation $t : s(t)$ can be derived from the relation $1/u : W(u)$ by an oblique translation whose components are:

$$\text{along } Os : log\ \frac{Q}{4\ \pi\ T}$$

$$\text{along } Ot : -\ log \left(\frac{x^2\ S}{4\ T}\right)$$

with $fo = \dfrac{so}{Q/4\,\pi\,T}$ and $vo = \dfrac{4\,T\,to}{s\,r^2}$ (cf. figure 3-13) we obtain:

$$T = \frac{Q}{4\,\pi}\,\frac{F_o}{S_o} \qquad\qquad S = \frac{4\,T\,t_o}{x^2\,v_o}$$

It is possible to simplify the problem. Indeed, since $T = 0.08\,Q\dfrac{W}{s}$, it is the $\dfrac{W}{s}$ ratio that really carries weight. On a log-log diagram, the shift of any length α corresponds to a multiplication by 10^α while the ratio $\dfrac{10^\alpha\,W}{10^\alpha\,s}$ does not change. It is therefore sufficient to find the correspondence between a simple fraction of s, for example 1 and the powers of 10, and the corresponding coordinates of W to avoid a double division. It is likewise for the storage coefficient.

Since $S = \dfrac{4\,T\,t}{r^2\,x}$ we must take $x = 1$, 10 or a simple number.

Figure 3-13 shows the superimposition of two graphs: the Theis curve and an experimental curve. This procedure enables the following data to be read:

$W(u) = 4,04$ (value of $\dfrac{1}{u}$ for $\dfrac{1}{u} = 100$)

$\dfrac{1}{u} = 100$

$\Delta s = 20$ cm

$t = 100$ min $= 6,000$ s

knowing the value of Q and the value of x, T and S can easily be calculated. If $Q = 120$ m^3/h $= 0.033$ m^3/s and $x = 45$ m we obtain:

$$T = \frac{0.08\,Q}{s}\,F(u') = \frac{0.08 \times 0.033}{0.2} \times 4.04 = 0.53 = 53.10^{-2}\,\mathrm{m}^2/\mathrm{s}$$

$$S = \frac{4\,T\,t}{x^2\,u'} = \frac{4 \times 0.53 \times 6{,}000}{(45)^2 \times 100} = 0.062$$

Beyond a certain characteristic seepage distance, the observation wells are stabilized in all directions.

3.5.2 Jacob's semi-logarithmic method

Starting from the Theis equation

$$s = \frac{Q}{4\pi T}\int_{u}^{\infty}\frac{e^{-v}dv}{v}$$

which can be expressed in the form of a limited expansion

$$s = \frac{Q}{4\pi T}\int_{u}^{\infty}\frac{1}{v} - 1 + \frac{v}{2} - \frac{v^2}{3!} + \frac{v^3}{4!} \pm \ldots\ldots dv$$

or by integrating:

$$s = \frac{Q}{4\pi T}\left[Ln(v) - v + \frac{v^2}{2,2!} - \frac{v^3}{3,3!} + \frac{v^4}{4,4!} \pm \ldots\ldots \right]_{u}^{+\infty}$$

it can be demonstrated that, if $u \to \infty$, the series between square brackets tends towards -0.577 (EULER's constant). On the other hand, if u is very small, the series between square brackets *is equivalent to Ln u*. It follows that, when u is small, Theis's integral exponential can be replaced by its logarithmic approximation given by C.E. JACOB, (1950)

$$s = \frac{Q}{4\pi T}(Log\,\frac{4Tt}{x^2 S} - 0.577216)$$

from which we obtain:

$$s = \frac{Q}{4\pi T}Log\,\frac{2.25\,Tt}{x^2 S}$$

After numerical solution and converting to decimal logarithms, this yields:

$$s = \frac{0.183Q}{T}log\,\frac{2.25\,Tt}{x^2 S}$$

As soon as a given point t becomes sufficiently large, the Theis equation allows a logarithmic approximation:

$$s = \frac{2.3\,Q}{4\,\pi\,T} \log \frac{2.25\,T\,t}{x^2\,S} = \frac{0.183\,Q}{T} \log \frac{2.25\,T\,t}{x^2\,S}$$

This equation, which gives the drawdown at a distance x from the well, is considered valid to:

— around 0.25% as soon as $\frac{1}{u} \geq 100$

— around 2% as soon as $\frac{1}{u} \geq 20$

— around 5% as soon as $\frac{1}{u} \geq 10$

— around 10% as soon as $\frac{1}{u} \geq 6{,}7$

It is estimated that an approximation of 5% is sufficient.

This is the equivalent of assuming:

$$t \geq \frac{10\,x^2\,S}{4\,T}$$

The calculation of T and S involves recording the drawdowns observed as a function of the logarithm of pumping time and then drawing the best fit straight line through the representative points. The slope of this line, $s = f\,(log\ t)$ is numerically equal to the increase of s per modulus 10 of t, while to is the time corresponding to the intersection of the straight line on the y-axis at $s = 0$.

In the logarithmic approximations formulated by C.E. JACOB, the first term is a constant, since both Q and T are invariable. In the second term, only the time is variable. The drawdowns grow as a function of the logarithm of pumping time. This condition is in conformity with the concept of a non-steady state regime.

The pumping data are recorded on semi-logarithmic graph paper. The drawdowns or water-level readings are expressed in metres and plotted from top to bottom of the graph in linear coordinates, while the pumping times are plotted in logarithmic coordinates. The initial piezometric level is indicated at the top of the graph. The scales are chosen in each case, in particular the time units (seconds, minutes or hours), so that all available space on the graph is used. The points so obtained define a mean straight line which is representative of the expression given by C.E. JACOB. At the beginning of pumping, the observed curve reflects the volume effect of the catchment structure which produces a turbulent non-linear flow. The intersection of the representative straight line with the initial piezometric level gives the initial fictitious time at the origin, denoted here as to.

During the pumping test, the following relations apply:

$$s = \frac{0.183Q}{T} \log \frac{2.25\,T}{x^2\,S} + \frac{0.183Q}{T} \log\ t$$

By substituting:

$$a = \frac{0.183 \ Q}{T} \ log \ \frac{2.25 \ T}{x^2 \ S}$$

$$b = \frac{0.183 \ Q}{T}$$

$$Y = s$$

$$x = log \ t$$

we obtain: $Y = ax + b$

This corresponds to the equation of a straight line with values of $Y = s$ on the y-axis and values of $log \ t$ on the x-axis. To facilitate plotting, use is made of semi-logarithmic graph paper (arithmetic ordinates, logarithmic abscise). The t values are plotted directly as x-axis coordinates. The establishment of this straight line makes it possible to determine T and S.

a) Determination of T

The slope of the straight line is represented by b.

$$\text{Since } b = \frac{0.183 \ Q}{T} = \frac{\Delta \ s}{\Delta \ log \ t}$$

$$\text{therefore } T = \frac{0.183 \ Q}{b} = \frac{0.183 \ Q}{\dfrac{\Delta \ s}{\Delta \ log \ t}}$$

b is determined graphically, being the increase of drawdown per modulus 10 of t.

Finally, since $T = KH$, a knowledge of T and H makes it possible to calculate the permeability K.

b) Determination of S

The intersection of the straight line with the abscissa (x-axis) is determined graphically, thus defining a point x_0 at time t_0.

For $s = 0$, that is to say $Y = 0$

$$\text{Let } s = \frac{0.183 \ Q}{T} \ log \ \frac{2.25 \ T \ t_0}{x^2 \ S} = 0$$

Since $\dfrac{0.183\,Q}{T}$ cannot be zero:

$$\log \frac{2.25\,T\,t_0}{x^2\,S} = 0$$

$$\frac{2.25\,T\,t_0}{x^2\,S} = 1$$

from which, knowing t_0, x and T, we obtain:

$$S = \frac{2.25\,T\,t_0}{x^2}$$

In conclusion, by observing the lowering of groundwater level during pumping at constant discharge, it is possible to determine the aquifer characteristics: transmissivity, permeability and storage coefficient.

An immediate consequence of determining these characteristics is the possibility of calculating the drawdown, s, of the groundwater at any distance, x, from the site of pumping, after pumping for time t with a corresponding discharge Q.

Finally, the above determinations were established from an observation of the drawdown in only one observation well and for a given discharge. They are equally valid for several observation wells which would give a better evaluation of T and S.

When the falling water-level is upset by variations of discharge or when several pumping discharges have been used, the representation s/Q log t can be used. Since the drawdown being a linear function of the flow rate:

$$\frac{s}{Q} = \frac{0.183}{T}\log\frac{2.25\,T}{x^2\,S} + \frac{0.183}{T}\log t$$

It can also be stated:

$$Y = \frac{s}{Q}$$

$$a = \frac{0.183}{T}\log\frac{2.25\,T}{x^2\,S}$$

$$b = \frac{0.183}{T}$$

$$x = log\ t$$

Where it is still the case that $Y = ax + b$

The determination of T and S is performed in the same way:

$$as\ b = \frac{0.183}{T}$$

$$we\ obtain\ T = \frac{0.183}{b}$$

where b is the increase of specific drawdown per modulus 10 of t,
and S is given by the same formula as previously.

The straight line obtained represents the evolution of the non-steady state flow which is observed at any point in the zone influenced by pumping whatever the discharge.

3.5.3 The critical radius

The critical radius, Rf, is the distance at which the drawdown, as calculated by the Jacob expression, falls to zero. It is a function of the transmissivity and the storage coefficient. This confirms the analysis of the factors affecting the dimensions of the cone of depression.

Therefore, it is in accord with:

$$s = \frac{0.183\ Q}{T} log \frac{2.25\ Tt}{x^2\ S} = 0$$

from which:

$$Rf = 1.5\ \sqrt{\frac{Tt}{S}}$$

It is noteworthy that Q does not appear in this formula, *so the radius of influence is independent of the pumped discharge*. The radius of influence is a function of the T/S ratio and time, increasing approximately as the square root of the time.

However, this fictitious distance, sometimes called the critical radius of influence, is purely theoretical since JACOB's equation is grossly inaccurate for slight drawdowns. It is better to evaluate the effects of a pumping step by calculating the drawdown at time t and at different distances using the Theis function (cf. section 3.5.1).

It can be seen that the convergence of $W(u)$ towards 0 is very rapid when u is increasing, as shown by the following values:

$$W(10^{-1}) = 1.82 \qquad W(1) = 0.219 \qquad W(5) = 1.1.10^{-3}$$

This means that, at a given distance r and beyond a certain time, the variations of drawdown are no longer perceptible.

It is apparent that, during the transition period of propagation of the cone of depression, its horizontal dimension is independent of the discharge. If the discharge is doubled, the drawdown is doubled at each point, despite the fact that the cone does not have a greater diameter. *The areal extent of the cone depends only on the properties of the aquifer and the pumping time.*

The cone's propagation speed is inversely proportional to the storage coefficient S.

3.5.4 Analysis of the Theis equation after cessation of pumping

As soon as the pumping is halted, the water-level in the well rises. This corresponds to the presence of a non-steady state regime which is evolving more and more slowly towards a quasi-steady state regime, which is also the regime of the natural flow from the aquifer. Account should still be taken of the natural variations in groundwater level. The level reached at the end of the recovery may be different from the initial level.

Recording the recovery in the control well and the observation well should be carried out with the same care as for the pumping water-levels. The analysis of these recovery curves provides data which are just as accurate, if not more so, compared with those obtained during pumping.

Observation of the recovery generally offers several advantages:

— no risk of perturbations due to irregularities in the running of the pump;

— a greater facility for accurate measurement of the water-level.

In accordance with the principle of flow superposition, the effects of a halt in pumping on the discharge $+ Q$ can be calculated, from the time of cessation of pumping, by assuming a virtual effect where pumping is continued at a discharge of $+ Q$ and , at the same time, an influx of water is set in motion towards the well at a discharge of $- Q$.

Indeed, a fictitious situation can be imagined where, instead of switching off the pump, it is left running at the same speed but with the water being poured back into the well rather than being carried away.

Let t be the time elapsed since the beginning of the pumping right up to the time of cessation and t' be the time elapsed after this cessation. The drawdown s due to the virtual continuation of pumping at a discharge of $+ Q$ is given by the following formula:

$$s' = \frac{0.183\ Q}{T} \log \frac{2.25\ T\ (t + t')}{x^2 S}$$

The recovery s'' due to a fictitious inflow of discharge $- Q$ is given by:

$$s'' = - \frac{0.183\, Q}{T} \log \frac{2.25\, T\, t'}{x^2 S}$$

In this way, the drawdown observed after the halt in pumping, or the measured residual drawdown s, is equal to the algebraic sum of $s' + s''$. From this, we can write:

$$s = s' + s'' = \frac{0.183\, Q}{T} \log \frac{2.25\, T\, (t+t')}{x^2 S} - \frac{0.183\, Q}{T} \log \frac{2.25\, T\, t'}{x^2 S}$$

Simplifying this, we can write:

$$s = \frac{0.183\, Q}{T} \log \frac{t + t'}{t'}$$

By substituting:

$$b = \frac{0.183\, Q}{T}$$

$$Y = s$$

$$x = \log \frac{t + t'}{t'}$$

once again we obtain the equation of a straight line: $Y = bx$

If s is plotted as the arithmetic ordinate and $log\ (t + t')/t'$ as the logarithmic abscissa, the slope b of the line enables us to calculate T.

A study of the recovery characteristics makes it possible to calculate the transmissivity. If the pumping has been carried out at a single discharge rate, the value of this discharge is then used for the calculation of T.

If there have been several successive discharge rates during the test the fictitious discharge can be used, for example:

1st step $Q = 20$ m^3/h for 24 h =   480 m^3
2nd step $Q = 40$ m^3/h for 24 h =   960 m^3
3rd step $Q = 60$ m^3/h for 24 h =   1,440 m^3

2,880 m^3

or a fictitious continuous discharge of $\frac{2,880}{72} = 40$ m^3/h

The really significant measurements are neither those taken during the initial stages of the recovery nor those taken at the end. These latter can be disturbed, for example, by the natural evolution of the aquifer. For the initial measurements, the second term of the Theis equation becomes applicable and the curve representing the water-level recovery is not a straight line. Therefore, the measurements taken between these extreme periods provide the data that should be used.

3.5.5 Anomalies encountered with the Theis graphical method

The following apparent anomalies are encountered when the Theis formulae are applied:

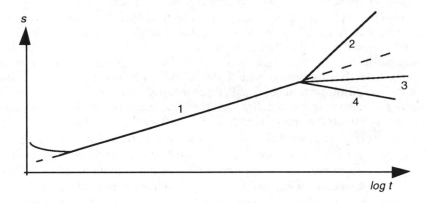

Figure 3-14
Different patterns of Theis curve

— On the one hand, as discussed above, the real curve obtained at the beginning of pumping lies above the theoretical log-linear relationship. The pumping time has to be sufficiently long before the second term of the Theis equation can be considered negligible.

— On the other hand, the upper end of the curve can exhibit variations of gradient (cf. figure 3-14).

Curve (1) represents the transient (non-steady state) flow under ideal conditions. However, under the effects of pumping, the cone of depression (in the aquifer) becomes increasingly widened out; the pumping depression may come to act upon a medium of different nature from the initial one.

— This second medium may correspond to an impermeable boundary interlayer, or a zone of lower or higher permeability. These variations can be recognized in the pattern of the curve.

— Lower permeability (2) will increase the drawdown s, and the curve will bend towards the ordinate (y-axis).

— Higher permeability (3) will have the opposite effect.

— Stabilization (4).

3.5.6 The skin effect and post-production

The interpretation of pumping tests in a non-steady state regime draws attention to both the skin effect and post-production. M. BONNET, P. UNGEMACH and P. SUZANNE have proposed a method for interpreting pumping tests which enables a definition of the hydrodynamic parameters of the aquifer based upon measurements taken in the well itself. In fact, observation wells are very frequently not available, so, in such cases, it is impossible to determine the hydraulic parameters of an aquifer by conventional methods.

It is generally accepted that conventional methods involving solutions to the Theis equation, under the hypothesis of radial circular flow in a homogeneous and isotropic aquifer of infinite extent, can be transposed to the case of real well. It is merely sufficient to assume that the distance r which appears in the formulae (distance from the observation point to the sampling point) is equal to the radius r_p of the well. However, in doing so, all perturbations arising in the well and its vicinity are implicitly ignored. But such phenomena can be significant and often make it very difficult to interpret water-level data from the well itself.

If we merely consider their influence on the water-level in the well, these phenomena, which are quite varied in detail, can be grouped into two categories:

— The skin effect, which is apparent during phases of pumping (depression).

— Post-production, which appears during the recovery in water-level.

A qualitative description of these "parasitic effects" is briefly presented here, along with an examination of their specificity and their interaction. These perturbations are expressed quantitatively by the generation of a conspicuous loss of head which can affect the drawdown levels measured in a given well. The method for evaluating this head loss is also summarized.

The skin effect

The hydrodynamic regime of non-steady state flow assumed here (circular radial symmetry in a homogeneous and isotropic aquifer) implies point sampling of the discharge without any modification of the aquifer characteristics. Indeed, the construction of the well and catchment facilities (equipment and completion) will disturb the flow in the vicinity of the borehole structure. A certain number of factors may explain this:

— Modification of the permeability of the natural medium near the well following drilling operations (clogging with mud or reworking of the formations). Improvement of the hydraulic conductivity by surging, washing with polyphosphates and acidification, or by the addition of a gravel pack.

— Presence of a screen modifying the pattern of flow and the head loss regime.

— Presence of pumping gear and existence in the rising main of turbulent flow deviating from the hydrodynamic regime typical of porous media.

— Otherwise, it should be noted that in the vicinity of the well, high flow velocities resulting from high discharges create a situation in which the assumption of laminar flow, which forms the basis of Darcy's Law, is no longer valid.

Changes in permeability due to the influence of drilling mud are commonly observed in oil-wells. These stuctures are almost always drilled with a rotary system and show productive capacities that are relatively low. In the oil industry, it is customary to refer to this phenomenon as the "skin effect".

In the case of water wells, the withdrawals are greater and the use of drilling mud is not widespread. Thus, it is likely that turbulence plays an important role around the well, across the screen face and within the rising main. For this reason, the less restrictive term of skin effect is also considered appropriate when applied to water wells.

Post-production

This phenomenon relates exclusively to the phase where the piezometric level rises following a halt in pumping. It involves the displacement of fluid towards the well in order to equilibrate the pressure. In other words, ignoring the compressibility of the water in the borehole (which is an acceptable hypothesis in the case of water wells tapping aquifers with a low pressure), this inflow ensures the filling of the well. Strictly speaking, post-production occurs throughout the recovery phase. In practice, it is only perceptible in measurements where there are considerable variations of volume in the well.

Thus, having taken account of the logarithmic formulation of water-level variations with time, the influence of post-production inflow is particularly well marked shortly after the cessation of pumping when there are rapid variations of water level (and consequently, volume) in the well.

Similarly, it follows that the magnitude of the post-production depends on the diameter of the well and the amount of drawdown attained at the end of pumping.

Through the intermediary of the amount of drawdown, it is shown that the well effect and post-production are phenomena that are closely linked to the beginning of recovery of the piezometric level.

Influence of skin effect on drawdown curves

In general, the different disturbances associated with the skin effect lead to the appearance, during the pumping period, of a conspicuous head loss ΔHp which is superposed upon the overall theoretical head loss which would be imposed by a porous medium supposedly tapped under ideal conditions.

However, it would be unrealistic to try to separate each of the components of the total head loss ΔHp. It evidently includes linear head losses (in the clogged drill bit, for example) and quadratic head losses (passage through the screen and rising main) of the CQ^n $(1 < n < 2)$ type. So ΔHp does not solely depend upon the discharge and, for a given discharge, it remains invariable in time provided that development phenomena do not occur during the test. In this respect, it is preferable to carry out tests on appropriately developed wells.

Therefore, it would appear that the actual observed drawdown (sr) during falling water-level will be equal to the sum of the drawdown of the aquifer

$sn = f(Q,t)$ and the head loss resulting from the well effect, that is to say $\Delta Hp = CQ^n = f(Q)$. *Therefore, if pumping proceeds at constant discharge*, the measured drawdown curve will be simply translated by a quantity equal to CQ^n with respect to the theoretical drawdown curve sn (cf. figure 3-15).

In the same way, the transfer of the real and theoretical drawdown curves onto a semi-logarithmic plot *(s, log t)* will yield two parallel straight lines. Thus, the calculation of T by Jacob's approximative method, which is based on the determination of the drawdown gradient, will in no way be altered by the presence of head losses. On the other hand, the calculation of the storage coefficient S using the formula:

$$ S = \frac{2.25\,T\,t_o}{r_p^2} $$

is only valid if to is given the value on the abscissa at the origin of the *theoretical straight line* $Sn = f(log\ t)$. Clearly, an erroneous result will be obtained if the same procedure is followed using the real straight line obtained directly from observed drawdowns.

Effect of post-production on recovery curves

In contrast to the skin effect, whose magnitude is a function of discharge alone, post-production inflow is a phenomenon which is variable in time. This gives rise to a shape distorsion and not just a displacement of the curves showing recovery of the piezometric level. The distorsion of the curves is particularly apparent at the beginning of recovery just when the post-production inflow is marked.

Furthermore, the importance of the head loss caused by the skin effect controls the amplitude and duration of the post-production inflow. This latter can prolong the skin effect when pumping is carried out during the recovery phase. The establishment of an undisturbed regime occurs only during the final stages of recovery. During the earlier stages, the real curve tends only very gradually towards the theoretical curve, starting first with the lower values. It may therefore be appreciated that, at this stage, the quantitative interpretation of the recovery curve can become highly problematic, especially when the catchment structure has previously been the site of a strong skin effect.

Quantitative interpretation

It is firstly necessary to assume that conditions are within the domain of validity of Jacob's equation. Furthermore, it may be considered that the pumping test is carried out at constant discharge and involves the regular observation of piezometric level during both the lowering and recovery of water-level.

Jacob's equation can be written as:

$$H_o - H = \frac{2.3\,Q}{4\,\pi\,T}\,log\,\frac{2.25\,T\,t}{r^2\,S}$$

where:

- Q : pumped discharge (m³/s)
- t_1 : duration of pumping (s)
- t : measured time from the beginning of pumping (s)
- t' : time elapsed since the cessation of pumping (s)
- Ho : height of the standing water-level (m)
- Hf : height of the dynamic level in the well
- H : height of the dynamic level in the well at the end of pumping at time t (m)
- T : transmissivity (m²/s)
- S : storage coefficient (dimensionless)
- rp : well radius (m)
- r : distance from the observation point to the well (m)

given that $\frac{r^2\,S}{4\,T\,t} < 10^{-2}$

If account is taken of head losses due to the skin effect (i.e. ΔH_p), then:

$$H_o - H = \frac{2.3\,Q}{4\,\pi\,T}\,log\,\frac{2.25\,T\,t}{r^2\,S} + \Delta H_p$$

At the end of the pumping $(t = t_1)$, the above equation can be rewritten:

$$H_o - H = \frac{2.3\,Q}{4\,\pi\,T}\,log\,\frac{2.25\,T\,t_1}{r_p^2\,S} + \Delta H_p$$

If post-production phenomena are ignored, the principle of flow superposition applied to the recovery leads to the formulation (for quite large values of $t - t_1$) :

$$H_o - H = \frac{2.3\,Q}{4\,\pi\,T}\,log\,\frac{t_1 + t'}{t'}$$

If H' = piezometric level in the absence of post-production inflow, and by eliminating Ho, we obtain:

$$H' - H_f = \frac{2.3\,Q}{4\,\pi\,T}\left[log\frac{2.25\,T}{r_p^2\,S}\cdot\frac{t_1\;t'}{t_1+t'}\right] + \Delta H_p$$

and if $t' << t_1$:

$$H' - H_f = \frac{2.3\,Q}{4\,\pi\,T}\left[log\frac{2.25\,T}{r_p^2\,S}\right] + \Delta H_p$$

Using this relation, it then becomes possible to calculate the head losses generated by the skin effect (ΔHp), provided that Hf, $t'H'$, rp, T and S are known.

In fact:

— Hf is known with accuracy (direct measurement).

— The determination of H' requires that t' is sufficiently small for the approximation to be justified. But, for low t' values, post-production inflow cannot be ignored. In order to remove this indeterminacy, it is necessary to make use of an artifice. This consists of extrapolating the linear part of the recovery curve (upper end) at time t' so as to determine H' graphically.

— The transmissivity T is obtained from the gradient of the drawdown test or from the linear part of the recovery curve. It should be noted in passing that the condition of constant discharge during drawdown (which implies the constancy of ΔHp) is indispensable for the determination of T.

— Insofar as the drawdown curve, which is influenced by the well effect, is not accompanied by recovery test data, any accurate determination of S is impossible. However, by the intermediary of its logarithm, an acceptable estimate of this coefficient can be obtained based upon the geological characteristics of the reservoir.

An example is presented here of the calculation of the well effect on the A 14 borehole in the Aubergenville well field (Yvelines, France), which is extracting groundwater from the Sparnacian Chalk aquifer.

Figure 3-15 shows the respective variations of the specific drawdown in relation to *log t*. The drawdown curve has a linear segment which makes it possible to determine the transmissivity $T = 2.2.10^{-2}$ m^2/s.

The calculation of the head losses linked to the well effect ΔHp scan be easily obtained from the pumping data:

$Q = 0.075$ m^3/s

$r = rp = 0.4$ m

Mean value of S in Chalk = 0.02

$Ho - Hp = 9.5$

$t_1 = 175,480$

From the equation:

$$H_o - H_f = \frac{2.3\,Q}{4\,\pi\,T} \log \frac{2.25\,T\,t_1}{r_p^2\,S} + \Delta H_p$$

we obtain

$$\Delta H_p = H_o - H_f - \frac{2.3\,Q}{4\,\pi\,T} \log \frac{2.25\,T\,t_1}{r_p^2\,S}$$

from which is derived

$$\Delta H_p = 9.5 - \frac{2.3 \times 0.075}{4\,\pi \times 2.2.10^{-2}} \log \frac{2.25 \times 2.2.10^{-2} \times 175{,}480}{0.16 \times 0{,}02} = 5.48$$

giving a numerical value of 73 for $\frac{\Delta Hp}{Q}$. This result confirms the existence of a strong well effect and makes it possible to quantify its contribution to total drawdown, which is 57% in this case.

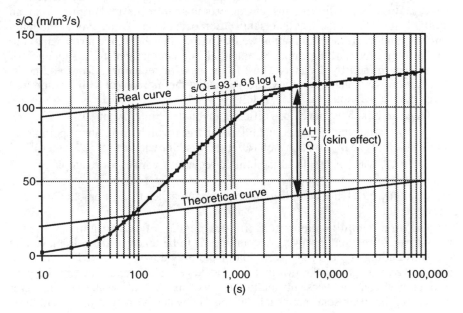

Figure 3-15
Drawdown of piezometric level in a water well

3.6 Interpretation of pumping tests in semi-confined (leaky) aquifers

In practice, real hydrogeological conditions are rarely exactly comparable to the ideal conditions defined above (cf. section 3.5). This places real constraints on the possibility of valid application of the Theis or Jacob methods described previously. So more complex models need to be devised along with corresponding methods of interpretation.

When pumping begins, the drawdown at any given point provokes an inflow of water within the water-bearing horizons situated above and below this point. This intake is expressed by a stabilization of the water-level, being either permanent or temporary according to the thickness of the aquitards, recharged aquifers or principal aquifer involved. The recharge can take two forms:

— Direct intake of natural water.

— An increase of pressure which amplifies the water input determined by the storage coefficient.

This phenomenon is therefore particularly sensitive in a confined aquifer with a weak storage coefficient; the characteristic feature of recharging from an aquitard stems from the fact that stabilization occurs simultaneously at the main borehole and at all the observation wells. On the other hand, recharge behaviour for an interlayer with an imposed potential is characterized by a sequence of events determined by distance. Thus, in the case of recharging from aquitards, observation wells that have not yet responded at the moment of stabilization will never show a reaction (see Jacob, below).

Three authors have developed different methods of interpretation:

— HANTUSH has tabulated the high recharges which bring about permanent stabilization of the observation wells after the first drawdown.

— BOULTON has tabulated the limited recharges which, for example, can be produced by a secondary aquifer.

— BERKALOFF has combined these two approaches to give practical rules of interpretation.

The first simplifications concern the degree of watertightness of the boundaries of the aquifer layer. Account needs to be taken of the role played by the surrounding layers in the evolution of the drawdowns observed in the main aquifer layer. It is well known that this "leakage" is of importance in the very frequently encountered case of "multi-layer systems", where a water-bearing layer is confined between semi-permeable layers. The water-bearing layer is said to be the principal aquifer because it is more transmissive.

The diagrams used to represent the influence of pumping in such systems are based on the principle that all the conditions of the Theis diagram (for the principal aquifer) are maintained, but are supplemented by conditions concerning the semi-permeable boundaries. The introduction of these conditions is reflected numerically by the definition of new parameters characterizing the leakage phenomenon.

It follows that the form of the solutions is evidently more complex. In particular, the graphical representations cannot be made with a single curve but require charts, or even a set of nomograms.

3.6.1 Semi-confined aquifers

Let us consider the case of a semi-confined aquifer in communication through a semi-permeable interlayer, which has negligeable storage, with an aquifer showing constant piezometric level. In this diagram, the third condition set by Theis (cf. section 3.5) is not supported, but all the others are satisfied.

The water-bearing layer tested here is in communication via a semi-permeable interlayer with an aquifer whose intake capacity is enough to maintain constant pressure despite a transfer of water towards the aquifer submitted to pumping (cf. figure 3-16). The amount of reserve specific to the semi-permeable interlayer is negligeable given the volumes of water passing through it. We can therefore disregard its storage coefficient.

During pumping, flow within the aquifer is increased by the delayed yield (coming from the groundwater body with a constant piezometric level) which is transmitted vertically through the semi-permeable interlayer. This discharge is proportional to the drawdown and is of a permanent nature.

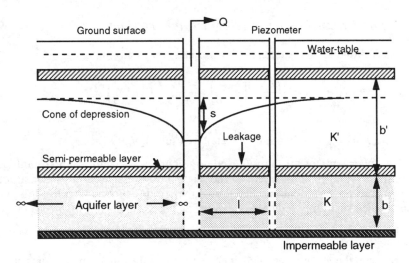

Figure 3-16
Ideal well in a semi-confined aquifer.

The parameters which define this transfer of water are:

$$\frac{K'}{b'} = \text{leakage coefficient}$$

$$\text{and } B = \sqrt{\frac{T\,b'}{K'}} = \text{leakage factor}$$

a) Leakage coefficient (or leakance)

This is a parameter of the semi-permeable layer which characterizes its capacity to transmit water vertically. It is defined as the discharge flowing across a unit area of the boundary between the tapped aquifer layer and its semi-permeable roof or footwall, caused by a difference of unit pressure between the tapped aquifer (also called the principal aquifer) and the groundwater body feeding the leakage.

The dimension of this coefficient is T^{-1}, while the values given by several authors range between 10^{-7} sec^{-1} and 10^{-9} sec^{-1}.

b) Leakage factor

This parameter characterizes the leakage effect in semi-confined aquifers. A high leakage factor corresponds to a short characteristic distance, and *vice versa*. In fact, although it was introduced as an aid to calculation, its values are rarely quoted in the literature. It has the dimensions of length: L.

3.6.2. The Hantush-Walton bilogarithmic method

The Walton method is intended for use in a non-steady state regime.

$$s = \frac{Q}{4\,\pi\,T}\,F\!\left(u', \frac{r}{B}\right) = \frac{0.08\,Q}{T}\,F\!\left(u', \frac{r}{B}\right)$$

where $F(u', r\frac{r}{B})$ is a function tabulated as a function of u' and $\frac{r}{B}$.

The upper ends of all the representative curves end with a horizontal step, while the lower part is identical to the Theis curve. Each curve is constructed from a set of observed time-related drawdowns, being plotted on a log-log graph which is then superimposed onto one of the standard curves (cf. figure 3-17).

The coordinates of any point in both axis systems $F(u', \frac{r}{B})$, u', s and t - together with the value of $\frac{r}{B}$ which enables fitting of the curve -make it possible to calculate the following parameters:

$$T = \frac{0.08\,Q}{s}\,F\left(u', \frac{r}{B}\right)$$

$$S = \frac{4\,T\,t}{r^2\,u'}$$

$$B = \frac{r}{\left(\dfrac{r}{B}\right)}$$

$$\frac{K'}{b'} = \frac{T}{B^2}$$

3.6.3 The Hantush-Berkaloff semi-logarithmic method

The Hantush method, which is also applicable in a non-steady state regime, consists of constructing a curve of s vs. $f(log\ t)$ for an observation well and then finding some points that are characteristic of this curve.

It can only be applied if the s value can be extrapolated in such a way as to find the value of s_{max} The Hantush-Berkaloff method makes use of the functions:

$K_o(x)$ and $e^x K_o(x)$ which are given in Table 1.

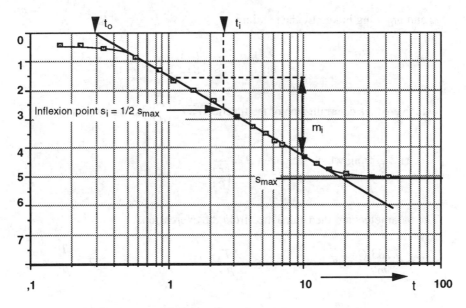

Figure 3-18
Plot of s vs. log t. Hantush inflection point method

The following procedure is adopted (cf. figure 3-18):
— 1. Plot the values of $s = f(log\ t)$,on semi-log paper and determine s_{max},

— 2. Determine the inflection point of the log-linear relation, where the drawdown s_i is equal to:

$$s_i = \frac{1}{2} s_{max} = \frac{Q}{4 \pi T} K_o\left(\frac{r}{B}\right)$$

— 3. Determine the slope m_i of the relation $s = f(log\ t)$ at its mid-point and the corresponding time t_i:

$$m_i = \frac{2.3\ Q}{4 \pi T} e^{-\frac{r}{B}} = \frac{0.183\ Q}{T} e^{-\frac{r}{B}}$$

while, at this same point:

$$u'_i = \frac{r^2 S}{4\ T t_i} = \frac{r}{2\ B}$$

s_i and m_i being linked by the relation:

$$\frac{2.3\ s_i}{m_i} = e^{\frac{r}{B}} K_o\left(\frac{(r)}{B}\right)$$

which enables determination of the function $e^{\frac{r}{B}} k_o\left(\frac{r}{B}\right)$ and we find in tables:

$e^{\frac{r}{B}}$ and $\frac{r}{B}$ from which we obtain $B = \dfrac{r}{\left(\frac{r}{B}\right)}$

The parameters are then calculated from the expressions:

$$T = \frac{0.183\ Q}{m_i} e^{-\frac{r}{B}} \qquad\qquad S = \frac{2\ T t_i}{B\ r} \qquad\qquad \frac{K'}{b'} = \frac{T}{B^2}$$

The values of the function $e^x K_0(x)$ have been tabulated by M.S. HANTUSH.

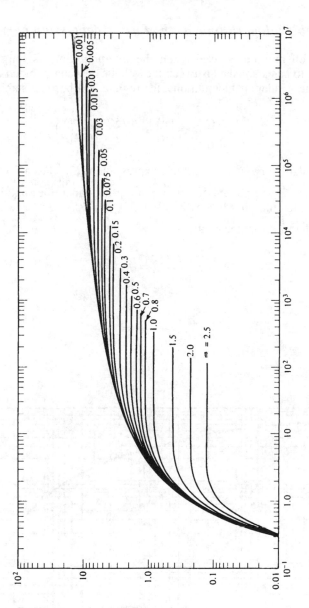

Figure 3-17
Standard curves for a semi-permeable aquifer in a non-steady state regime
(after WALTON).

3.6.4 Jacob's bilogarithmic method using depression curves

The Jacob method is used when the pumping time is long enough for equilibrium to be established between the volume of water pumped and the input from leakage. Under such conditions, the regime will become stabilized and we can write:

$$s = \frac{Q}{2 \pi T} K_o \left(\frac{r}{B}\right) = \frac{0.159 \, Q}{T} K_o \left(\frac{r}{B}\right)$$

The standard curve is a logarithmic expression of $K_o \left(\frac{r}{B}\right)$ as a function of to $\frac{r}{B}$ where values of s are plotted against r. On log-log paper, the superimposition of curves provides the coordinates $K_o \, (\mathsf{Y}(r;B))$, $\mathsf{Y}(r;B)$, s and r which serve to calculate T and $\frac{K'}{b'}$ from the relations:

$$T = \frac{0.159 \, Q}{s} K_o \left(\frac{r}{B}\right)$$

and:

$$\frac{K'}{b'} = \frac{T}{\left(\frac{r}{r}\big/ B\right)^2} = \frac{T}{B^2}$$

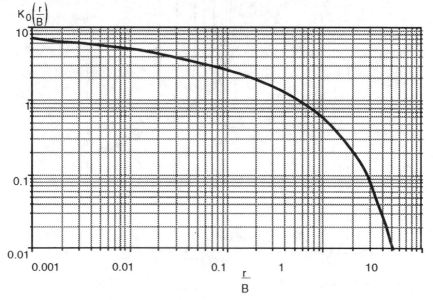

Figure 3-19
Standard curves for a semi-permeable aquifer in a steady state regime (after WALTON).

To apply this method, the data from several observation wells (at least three) are required.

3.7 Interpretation of pumping tests in unconfined aquifers

On the theoretical level, universally acceptable and rigourous solutions have not yet been found for the hydrodynamic problems of aquifers having a free upper surface.

Several methods have been developed recently that take account of the different conditions characteristic of unconfined aquifers, namely:
— Three-dimensional pattern of flow near the well.
— Variation in transmissivity with the amount of drawdown.
— Delay in draining of the water-bearing layer,
but none of these methods incorporate all three conditions together.

In the following, some general rules are reviewed and a summary is given of the Boulton method taking account of the phenomenon of leakage commonly observed in unconfined aquifers.

3.7.1 General rules

As long as the variations in water-level engendered by pumping are small in relation to the initial saturated layer thickness b, the vertical velocity component in the vicinity of the well and the reduction of T can both be disregarded. It is currently accepted that the equations valid for flow in confined aquifers can be applied to the flow in unconfined aquifers.

When this is not the case, it becomes necessary either to make use of corrected drawdown values or to apply specific methods for unconfined aquifers.

To sum up, there are three possible ways of interpreting pumping tests in an unconfined aquifer:
— *s < 0.1 b:* methods established for confined aquifers can be applied.
— *0.1b < s < 0.3 b:* methods established for confined aquifers can be applied on condition that calculations are carried out using corrected drawdowns:

$$s_c = s_m - \left(\frac{s_m^2}{2b} \right)$$

where:
s_c : corrected drawdown,
s_m : measured drawdown,
b : initial saturated layer thickness.

— $s > 0.3\ b$: specific methods for unconfined aquifers can be applied taking account of the vertical velocity component and the decrease in T.

3.7.2 Unconfined aquifer with strong drawdown

There is no entirely satisfactory theory in the case of ($s > 0.3\ b$).

One approach to modelling this situation consists of accepting that the release of water is not instantaneous. In fact, instantaneous release occurs at the beginning of pumping through decompression of the medium. Afterwards, release of the water takes place progressively through dewatering of the unsaturated part of the aquifer. This phenomenon is called dripping.

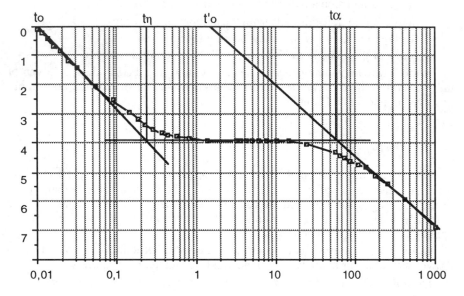

Figure 3-20
Draining in an unconfined aquifer.

In such cases, the drawdown curve has the general shape as given in figure 3-20, where three phases can be distinguished:

— a linear phase at the beginning of pumping, described by the equation:

$$s_1 = \frac{0.183\ Q}{T} \log \frac{t}{t_0}$$

— a linear phase at the end of pumping, with equation:

$$s_3 = \frac{0.183\, Q}{T} \log \frac{t}{t'_0}$$

— between these two phases, a horizontal step with:

$$s_2 = \frac{0.183\, Q}{T} \log \frac{t_n}{t_0}$$

If the storage coefficient, S, during the initial phase is low and of similar size to that observed in confined aquifers, and if S' is the storage coefficient of the final phase having a value close to the effective porosity of the medium, then, according to BERKALOFF, we can write:

$$T = \frac{0.183\, Q}{b}$$

$$S = \frac{2.25\, T\, t'_0}{x^2}$$

$$S + S' = \frac{2.25\, T\, t'_0}{x^2}$$

with:

$$t_0 = \frac{x^2\, S}{2.25\, T} \quad ; \quad t'_0 = \frac{x^2\, S'}{2.25\, T} \quad ; \quad t_\eta = \frac{0.561}{\eta \alpha} \quad ; \quad t_\alpha = \frac{0.561}{\alpha}$$

The following parameters can be obtained by calculation:

T : hydraulic transmissivity of the system,

S : storage coefficient of the immediately available reserves according to the Theis concept,

SS' : total storage coefficient ($S + S'$),

S' : delayed storage coefficient,

α : parameter expressing the rate of dewatering of the reserve S'.

3.7.3 Boulton's bilogarithmic method

Boulton's bilogarithmic method takes account of the draining phenomenon in an unconfined aquifer. It is applicable when the following conditions are satisfied:

— The aquifer is homogeneous, isotropic and of the same thickness in all zones influenced by the pumping.

— Aquifer of infinite lateral extent.

— Aquifer lies upon a horizontal impermeable substratum.

— The well is ideal.

— The well radius is negligeable.

— The pumped discharge is constant.

— Water release is not instantaneous; it is made up of two distinct phenomena:

   • instantaneous release of the water following decompression of the water-bearing medium and the water,

   • progressive release of the water due to draining of the dewatered layer. This second phenomenon is known as dripping (or draining).

The delay in the release of the water is observed through variations in the time of the storage coefficient. The water-sheet reacts to pumping in three different phases. At the beginning of the pumping, instantaneous release is preponderant because of the rapidity of its onset; the water-sheet responds like a confined aquifer with a weak apparent storage coefficient of the same order as that found in a confined aquifer. The second phase is very similar to the behaviour of a confined aquifer subject to recharge by leakage. Just as the pressure is falling rapidly, the water in the upper part of the aquifer becomes decoupled and shows only a very slow vertical movement, acting as a continuous recharge input until pseudo-stabilisation is achieved.

In the third phase, the pressure variation reaches the same order of magnitude as the eventual rate of lowering of groundwater-level. It may then be assumed that there is a constant storage coefficient which approaches the effective porosity of the aquifer formation.

Using such an interpretation of the behaviour of unconfined aquifers as his starting point, BOULTON formulated an expression giving the variations in drawdown as a function of time. From this, he derived a method for determining the aquifer parameters.

The parameters appearing in the Boulton method are as follows:

S': delayed storage coefficient

$B' = \sqrt{\dfrac{T}{\alpha S'}}$: drainage factor where $\dfrac{1}{\alpha}$ is an empirical constant called the delay index, B' is a factor characterizing the unconfined aquifer with slow gravity drainage. A high drainage factor corresponds to a rapid release of water, so if $B' = \infty$ the release of water is instantaneous. It has the dimensions of length: L.

The theoretical drawdown curve is S-shaped, with an equation that can be expressed in the form:

$$s = \frac{Q}{4 \pi T} \xi \left(u', u'_1, \frac{r}{B'} \right) = \frac{0.08\, Q}{T} \xi \left(u', u'_1, \frac{r}{B'} \right)$$

This equation is simplified in the first pumping phase to:

$$s = \frac{0.08\, Q}{T} E \left(u', \frac{r}{B'} \right)$$

where:

$$u' = \frac{4\,T\,t}{r^2\,S}$$

and, in the third pumping phase, this can be rewritten as:

$$s = \frac{0.08\,Q}{T}\,E_1\!\left(u'_1,\,\frac{r}{B'}\right)$$

where:

$$u'_1 = \frac{4\,T\,t}{r^2\,S'}$$

In its middle part (second phase), if $\eta = \frac{S+S'}{S} \geq 100$, the curve bends towards a horizontal line whose equation is:

$$s = \frac{Q}{2\,\pi\,T}\,K_0\!\left(\frac{r}{B}\right)$$

if $10 < \eta < 100$, the second part of the curve is not a horizontal line but has a slope noticeably less than the that at the beginning or at the end of pumping

Figure 3-21 shows two sets of standard curves: on the left, set A is comprised of standard curves $E\,(u',\,\frac{r}{B'})$ and, on the right, set B of standard curves $E_1(u_1',\,\frac{r}{B'})$.

The curves of set A serve to characterize the initial pumping phase, while set B is used to study the phase following the pumping step.

The method of interpretation consists of firstly superimposing the drawdown vs. time curve (plotted using logarithmic coordinates) onto one of the curves of set A. The coordinates of a point of coincidence are:

$$E\!\left(u',\,\frac{r}{B'}\right),\;u',\,\frac{r}{B'}\,,\;s \text{ and } t \text{ enabling the evaluation of:}$$

$$T = \frac{0.08\,Q}{s}\,E\!\left(u',\,\frac{r}{B'}\right) \text{ and } S = \frac{4\,T\,t}{r^2\,u'}$$

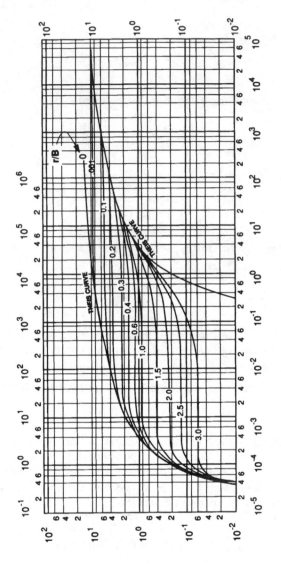

Figure 3-21
Set of Boulton standard curves (after Kruseman and De Ridder, 1970).

Next, the observed curve is offset to the right, parallel to the x-axis, in order to find the best fit with one of the curves of set B, while always maintaining the same $\frac{r}{B'}$ value. Knowing the double coordinates $E_1 (u'_1, \frac{r}{B'})$, u'_1, s and t we obtain:

$$T = \frac{0.08\,Q}{s} E_1\left(u'_1, \frac{r}{B'}\right)$$

$$S' = \frac{4\,T\,t}{r^2\,u'_1}$$

$$\text{and } \frac{1}{\alpha} = \frac{B^2\,S'}{T} : \text{ delay index}$$

The initial phase is sometimes not visible on the plotted graph, either because it is very rapid, or because it concerns an observation well which is too far away. In that case, the observed curve is superimposed only onto one of the curves of set B.

3.7.4 Berkaloff-Boulton semi-logarithmic method

This procedure, called Berkaloff's method, is used to determine the parameters included in the Boulton formula governing the evolution of drawdown with time. They can be easily obtained starting from values that are read off the semi-log diagrams in the case where $\eta = \frac{S + S'}{S} \geq 100$, that is to say, when a horizontal step is observed.

The drawdown curve can then be pictured as having three parts (cf. figure 3-22):

- a horizontal step with equation $s_2 = \frac{0.183\,Q}{T} \log \frac{t_\eta}{t_0}$
- two parallel segments which are asymptotic to the drawdown curve at the beginning and end of the pumping, on either side of the step:
  . the asymptote at the beginning of pumping has as its equation:
  $$s_1 = \frac{0.183\,Q}{T} \log \frac{t}{t_0}$$
  . the asymptote at the end of pumping has the equation:
  $$s_3 = \frac{0.183\,Q}{T} \log \frac{t_\eta}{t'_0}$$

By setting down:

$$t_0 = \frac{r^2\,S}{2.25\,T} \quad ; \quad t'_0 = \frac{r^2\,S'}{2.25\,T} \quad ; \quad t_\eta = \frac{0.561}{\eta\alpha}$$

we can also write: $t_\alpha = \frac{0.561}{\alpha}$

The interpretation consists of transferring the observed drawdown vs. time relation to a semi-logarithmic diagram and fitting the best straight line to the horizontal step, which corresponds to s_2 and drawing the two parallel straight lines corresponding to s_1 and s_3 knowing that the maximum differences for $t = t_\eta$ and $t = t_\alpha$ are numerically equal to *0.214 i*. Then, the following values can be read off the graph:

   t_0 at the intersection of the axis $s = 0$ with the asymptote corresponding to the beginning of pumping;

   $t'_0 = \eta\, t_0$ at the intersection of the axis $s = 0$ with the asymptote corresponding to the end of pumping;

   t_η and t_α at the intersection of each of these asymptotes;

   i the gradient common to both asymptotes.

As before, t_η is the same for all the observation wells.

The following parameters are derived by calculation:

— the hydraulic transmissivity of the system $T = 0.183\dfrac{Q}{i}$

— the storage coefficient of the immediately available reserve, according to the Theis concept $S = \dfrac{2.25\,T\,t_0}{r^2}$

— the total storage coefficient $(S + S') = \dfrac{2.25\,T\,t'_0}{r^2}$

— the delayed storage coefficient $S' = [\,(S + S') - S\,]$

— the parameter α expressing the drainage rate of the reserve S':

$$\alpha = \frac{0.561}{t_\alpha}$$

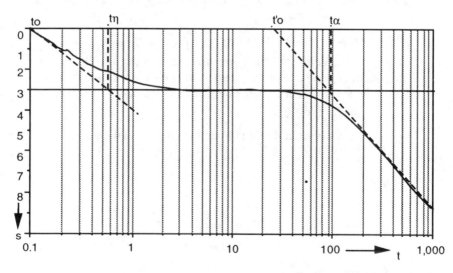

Figure 3-22
Curve s = f (log t) — Berkaloff method for unconfined aquifers.

Observation wells are indispensable when the existence of leakage is suspected. On inspection of the Boulton curve, it can be seen that if the pumping tests begin too late — i.e in the middle of the step — the observed curve cannot be distinguished from that of a classic Jacob semilog with head losses in the well.

3.8 Conclusion

This chapter attempts to present the main methods for the interpretation of pumping tests and sets out to derive the mathematical simplifications so the reader can easily make use of the equations. Nevertheless, despite the elegance of the mathematical devices proposed, their application will never lead to serious results if the problem is not properly stated in the first place. It is therefore fundamental to analyse the hydrogeological situation correctly before attempting to interpret the discharge tests. Under no condition can the refinement of the interpretation compensate for shortcomings in the data acquisition programme.

It should also be realized that the values found are orders of magnitude.

A deliberate choice is made here to restrict the presentation of interpretation methods to those that are well known and widely accepted. It should be understood that the analysis of discharge test data could be extended much further. The aquifer model can be considered as an overall interpretation of production tests that makes it possible to simulate the development of a well field according to many types of different scenario. The sophisticated analysis of head losses enables an assessment of well completion. It is then possible to check whether the work of the driller has been carried out according to the codes of practice. The interpretation of tests also leads to a better understanding of the geometry of the aquifer and provides evidence of the recharge limits, etc. Mathematical methods exist for studying all kinds of boreholes in aquifers displaying many different types of configuration.

If reliable tests are to be performed, a programme of correct pumping must be designed beforehand. This should include the careful setting-up of observation wells, and the implementation of well development and short pumping tests at different discharges. Pumping should then carried out for a sufficiently long duration for a steady state to be reached, not only in the well, but also in the observation wells, provided that the operation is in a recharged aquifer. It must be ensured that the operations on the ground are carried out conscientiously, while bearing in mind that these tests are difficult to implement and demand trained competent operators.

To cite H. CAMBEFORT once more, interpretations must be backed up by common sense and experience rather than by ingenious formulae. The mathematical aspects of the problem should not obscure our view of the qualitative analysis of the phenomena, which is probably the most important and often the most problematic part of the work .

Finally, the interpretation of a pumping test data cannot be reduced merely to the application of a few simple recipes. Data interpretation demands a critical mind, discernment and a certain flair for diagnosis. Natural conditions are often too complex to represent in an over-simplified conceptual framework. The temptation to apply formulae outside their domain of validity must also be resisted. Numerical results that are obtained under such circumstances run the risk of being illusory without this becoming apparent to the users. The role of hydrogeologists is thus of prime importance in the development of data acquisition programmes and interpretative procedures.

Notation used

α — parameter characterizing rate of dewatering of reserve (reciproced of delay index)

β — $\dfrac{r}{4B}\sqrt{\dfrac{S'}{S}}$ = factor included in the function $H(u', \beta)$

b — thickness of the aquifer
b' — thickness of the semi-permeable interlayer

B — $\sqrt{\dfrac{Tb'}{k'}}$ = leakage factor

B' — $\sqrt{\dfrac{T}{\alpha S'}}$ = drainage factor for unconfined aquifers (or leakage factor for semi-confined aquifers)

$\phi\left(u', \dfrac{r}{B}\right)$ well function for semi-infinite aquifers

$\zeta\left(u', u_1, \dfrac{r}{B}\right)$ well function for unconfined aquifers

$F(u')$ — $W(u)$ = well function for confined aquifers

$F\left(u', \dfrac{r}{B}\right)$ well function for semi-confined aquifers with leakage

$H(u_1', \beta)$ well function for semi-confined aquifers

$K_0\left(\dfrac{r}{B}\right)$ modified second-degree Bessel function of zero[th] order

K — horizontal hydraulic conductivity of aquifer

K' — vertical hydraulic conductivity of semi-permeable interlayer

$\dfrac{K'}{b'}$ — leakage coefficient

Q — pumping discharge

Q_c — critical discharge

r — distance of a given point to axis of pumped well

r_p — radius of pumped well

R — radius of influence of well

Rf — fictitious radius

s — drawdown

s_c — corrected drawdown

s_m — measured drawdown

s_{max} — maximum acceptable drawdown

s_r — residual drawdown (measured during recovery)

S — instantaneous storage coefficient of the aquifer layer

S' — storage coefficient of the semi-permeable interlayer or delayed storage coefficient of the unconfined aquifer layer

SS' — total storage coefficient $(S + S')$

T — transmissivity $(T = K b)$

t — time elapsed since start of pumping

t' — or t_r time elapsed since cessation of pumping (rise time) u $\quad\dfrac{r^2 S}{4Tt}$ = argument of well functions

u' $\quad\dfrac{1}{u} = \dfrac{4\,T\,t}{r^2 S}$ = argument of well functions

u" $\quad\dfrac{4Tt}{r^2(S + \dfrac{S'}{3})}$ = argument of well functions

u''' $\quad\dfrac{4Tt}{r^2(S + S')}$ = argument of well functions

u_1 $\quad\dfrac{4Tt}{r^2 S'}$ = argument of well functions

x $\quad$ distance of observation well from main well axis

CHAPTER IV

Supervision and final acceptance tests

"Power without control leads to madness"
Alain, Politique

This chapter presents what might be called a memorandum for water operators describing the main points that need to be checked by the person responsible for supervising the construction of a water well. Since it is impossible to modify the aquifer to any significant degree, it is essential to control the only parameter within our capability, i.e. the water well itself. Consequently, such actions should be undertaken according to a code of good practice. The construction of a well always runs up against unplanned factors and difficulties so that certain decisions must be made rapidly with a full knowledge of the facts. Hence the need to carry out the most rigourous checks.

The supervisor should be capable of establishing a climate of mutual trust with the contracting company which favours the smooth running of the work, while at the same time maintaining a rigorous checking procedure.

One of the inspector's principal concerns is to produce complete documentation on the work carried out. He or she should also possess indisputable records in the event of a claim or if problems arise during the service-life of the well.

It is quite clear that any defect in workmanship will sooner or later have repercussions on the structure, so it is necessary to keep archive records and report data in order to understand the origin and nature of the problems that might occur.

We can distinguish three main phases:

— checks during the setting-up of the construction site;
— day-to-day checks during the work;
— specific checks on the well which generally require the intervention of a specialist.

In general, an efficient check can help to avoid many problems and simplify the manner of reception of the well.

4.1 Preliminary checks before drilling

Preliminary checks are of great importance since they make it possible to prevent major problems at a later date. Indeed, it is always difficult — both from the technical and financial points of view — to stop the work schedule to discuss a problem, particularly if it could have been settled before the work began.

The checks to be made particularly concern the workplace environment (siting, access, etc.) and the conformity of the material provided by the drilling company.

4.1.1 The environment

As necessarily specified in the contract, it is important to respect the environmental constraints and the necessary means for operating in conformity with them. These issues should be addressed the earliest meetings with the drilling company.

As an indication, the following check-list gives the broad outlines of some environmental constraints that could apply:
— Access to the work site:
  • are the access roads or terrains robust enough for the movement of heavy machinery?
  • who is responsible for overseeing eventual reinforcement and even degradation?
    A certified report on the state of the site may be useful, depending on the nature of the potential degradation. This report could be entrusted to a bailiff so indisputable legal arguments can be made available.
— Installation of the drilling site:
On the site, it is important to mark clearly the zone reserved for the drilling company and define carefully the installation of the site:
  • is the ground liable to flooding?
  • should it be fenced off?
  • is there any risk to the local population from projected material?
  • are there any special hazards?

It is also suitable to remember and indicate that it is a work site (safety precautions to be taken).
— Connections with the road system and utilities during the work.

All the connection problems should be brought up, if only to compare them with clauses in the contract.
  • Mains electric supply or generator (who pays?).
  • Water supply.
  • Disposal of waste, which falls into three categories:
  . Water used for washing the equipment and developing the well (often being laden with mud or bentonite). Where can it be drained off without causing a nuisance?
  . Water from pumping tests often represents a considerable discharge rate, so it is essential to ensure an efficient disposal.
  . Water from acidizing operations can be evacuated into the environment only after neutralization.

— Nuisances:

Drilling work is noisy and dirty. It is therefore important to ensure that measures have been taken by the drilling company to limit these nuisances for those living nearby. In this regard, it will sometimes be useful to establish the noise level of the environment at the outset and to warn the local authorities and the neighbourhood of the nature and duration of the works.

It is also important to fix working hours at the site in relation to the acceptable noise level and to ensure that methods are provided for cleaning, upkeep and safety of the site. It is particularly important to respect the directives of the competent authorities in respect of the acceptable noise threshold.

4.1.2 Equipment

The drilling company must supply a list of the equipment and machinery which it is going to use. It is important to check that they conform to the specifications (power of the drilling rig, existence on site of all necessary equipment, etc.) and its effective presence before the work begins.

Discrepancies are commonly observed between the equipment stipulated in the contract and the equipment present on site. This can give rise to delays in the work as well as disputes.

Furthermore, it should be ensured that:

— the rig has sufficient lengths of full and screened casings available for fitting the borehole as soon as drilling has ended, whatever the choice of equipment arrangement. In broad outline, this arrangement will have been defined from the preliminary studies;

— the open area of the screens and the nature of the gravel pack (if necessary) are appropriate to the geology and assumed grain-size distribution of the aquifer (see Chapter III);

— the drilling fluids are suitable and present in sufficient quantities.

— the work site is properly supplied and can handle any drilling problem, etc.

4.1.3 The well site

The well site is determined by preliminary hydrogeological studies. It is decided in relation to the results of field work, photo-interpretation, geophysical studies, as well as environmental and production constraints (connection to existing resources).

Sometimes, for technical reasons, it may become necessary to move the structure slightly. It cannot be too strongly recommended to seek expert advice in this regard, particularly that of an accredited hydrogeologist. This is a crucial issue. Wherever possible, the hydrogeologist should be involved as early as possible in the project when the site is being chosen and the plant is being set out. This will ensure that the very best results can be obtained from the construction work. As a general rule, the advice of specialists should be sought.

In any case, it is preferable to avoid displacing the site layout. Regarding the nature of the terrain, two classic cases may be mentioned:

— *Location in an alluvial sedimentary zone.* The aquifers are continuous and the borehole can be freely sited inside a more or less extensive perimeter. In this case, for social or technical reasons, the drill site can be displaced in order to facilitate the work. The decision should be made in conjunction with the drilling company, the accredited hydrogeologist and the operator.

— *Location in a zone of crystalline basement or in limestone terrains.* The aquifers are discontinuous and closely linked to the fracturing in the rocks. In this case, the siting of the borehole, determined with great precision by the preliminary studies, should not be modified. If, however, a change of location proves absolutely necessary, it should not be effected until further studies have been carried out. Whatever happens, the drilling company should not interfer with the decision-making process.

Finally, displacing a drill site is to take risks concerning the sources of potential pollution, visible or invisible (notably sewerage pipes). With this in mind, a prior study of the inventory of pollution hazards is always very useful.

4.2 The "day-to-day checks"

The contract must indicate the content of the daily reports that are to be kept up to date by the drilling company and handed over to the supervisor. The careful filling in of these reports is indispensable, as they represent a surety for the acquired assets of the well. It is only on the evidence of these site record books that it is possible to settle eventual lawsuits regarding the sinking of a well.

One of the tasks of the supervisor is to ensure that these reports are pertinent and well-kept, and that the quality of their information is high (in particular the chronology of the operations, problems encountered during the work and the staff present on site). For these reasons, the work site record book must be signed and approved by the supervisor on a daily basis.

It is advisable to check a certain number of important points in the course of the work, notably:

— samples of formations encountered and their storage in bags or boxes;

— the depths drilled;

— determination of the volume of water inflows as well as their flow rate wherever possible;

— conformity of the equipment: blind pipes, screens (measurement of diameter, thickness, slot size);

— granulometry and composition of the gravel pack,

— details of every incident accompanied, wherever possible, by a written declaration;

— a check on the quantities of materials being used. The most important of these is the volume of gravel in the pack and, at a later stage, the amounts of products used in treatment.

4.2.1 Sample collection

To produce a relatively accurate geological section of the borehole, a sample must be taken at every change in the surrounding formation or, more generally, at every metre depth.

The taking of samples in gravely terrains makes it possible to establish grain-size curves which help in estimating the screen open area and the type of additional gravel to be emplaced given the hydrodynamic characteristics of the aquifer.

Samples of crushed rock (cuttings) enable the hydrogeologist to recognize the nature of the formations encountered and the eventual presence of fissures or underground water passageways (traces of oxidation).

The sample should not be washed If drilling has been carried out with mud. However, the composition of the mud should be noted at the time the sample was taken. All the material corresponding to a given sample is laid on a clean surface and then carefully mixed. The whole amount is spread out at a thickness of 2 or 3 cm, a square grid is traced and a small quantity is taken from each square. This method ensures the collection of a homogeneous and representative sample. It is then drained, dried and put in a bag with a label giving the sample characteristics:

— name and address of sender;
— name of the well and its geographical location (grid reference);
— size of sample taken;
— thickness of the formation sampled;
— date of sample collection.

In general, the cross-section established in the field by the hydrogeologist responsible for checking is sufficient and there is no need to keep the samples.

4.2.2 Depth drilled

The supervisor must keep an exact record of the number of drill pipes lowered into the borehole and know the position of the tool so as to be able at any moment to give the depth of the drill bit. This is essential for the noting of sample depths and locating any eventual inflow of water. Account should be taken not only of the length of the drill string but also of all the appliances that has been lowered into the hole (drill-collars, shock-absorbers, drill bit) so as to obtain a rigourously accurate calculation of the depth drilled.

Depending on the results obtained in the course of drilling, the depth at which drilling should be halted is determined by the supervisor in agreement with the hydrogeologist.

4.2.3 Water inflow

To determine the positioning of the screens it is essential to locate the depths at which water flows into the well.

— In the case of rotary drilling in sedimentary terrains, an examination of the drilling mud serves to identify the penetration of the tool into the aquifer. A lowering of the mud level in the well associated with a decrease in its density means that water inflow is occurring. This inflow is the more significant as modifications of the mud become visible and quantifiable. At this stage, measurements of the mud viscosity should be carried out..

A loss of mud signifies the presence of a highly permeable terrain. The same is true of drilling operated with air pressure when cuttings no longer come back to the surface because of air losses into fissures in the rock.

— In the case of DTH hammer or air-lift drilling in crystalline formations, this is no longer a problem since the water inflows are clearly visible. They are marked by a geyser at the surface which makes it easy to channel the water and measure its flow rate accurately.

4.2.4 Conformity of the equipment

During drilling, the supervisor must check the suitability of the equipment and appliances for the water-bearing formations encountered. It is strongly recommended to have a sufficiently wide range of material to meet any situation. Hence the advantage of specifications that are adapted to the circumstances.

In the case of soft sedimentary terrains, it is dangerous to leave a hole without equipment for an excessive period of time since the risks of collapse and filling in are too great. It is therefore essential to equip the borehole as quickly as possible after the drill string has been taken out.

In hard terrains (granitic crystalline basement, for example), the need to equip the drill is less urgent. In certain extreme cases (weak discharge) it may even be possible to leave the well as a "naked hole" so as not to hinder the inflow of water.

4.2.5 Particular incidents

The supervision also consists of noting events that are related to the drilling as and when they occur. These incidents must be recorded in the site record book.

For example, such incidents may involve the blocking of the drilling tool. This can sometimes occurs in the course of drilling, due to thickening of the mud (cake too thick) or the fall of stones (in hard terrains). In every case, the tool must be handled carefully so as to avoid blocking it permanently. It is important that circulation of the drilling fluid should never be halted while the tool is immobilized at the bottom of the hole, even for a short time.

There are also more serious incidents which, fortunately, are less common. An object may fall to the bottom of the borehole, thus interrupting the drilling. Among other items, this can be a light tool, a drill string or a casing pipe. An accident of this kind can have unfortunate consequences, in particular leading to downtime for the drilling team, and can even cause the abandonment of the borehole if retrieval is impossible.

In such a situation, the driller has to make a forecast assessment comparing the cost of retrieving the object (downtime for the drilling team, overrunning of estimated costs, failure, etc.) and the cost of sinking a new hole.

If retrieval is chosen, the first step is to identify precisely the object to be recovered: depth reading, exact position, shape of its upper surface so that an appropriate "fishing tool" can be produced. A cast of the upper surface of the object can be made to assist with this phase of the operation. It may also be possible to carry out an inspection with a video camera.

Once this phase has been completed, a suitable fishing tool (or one that has been specially made) should be used. The following is a list of some of these appliances (or devices):
— overshot (the most commonly used);
— grabbing tap and bell;
— cutter:
— fishing hooks or crow's foot;
— harpoons;
— spiders;
— sediment baskets,
— magnets, ...

4.3 Specific checks

A certain number of specific checks must be carried out during drilling. Some need special equipment, and they often require the help of a specialist.

The presence of a competent engineer on site is indispensable during the following essential operations:
— traversing the aquifer (characterization and trials);
— installation of equipment (screen, gravel pack);
— well development;
— pumping tests.

4.3.1 Determining the nature of the formations

This determination is essential in deciding all the technical details of the catchment work (position of screens, granulometry of the gravel pack, etc.)

As a general rule, it will be entrusted to a competent hydrogeologist who might envisage complementing the investigation with a lithological log. This operation is strongly advised so as to determine accurately the position of water-bearing layers.

— *Driller's log*: with this information, it is possible to obtain an immediate interpretation of the characteristics of the terrain being drilled (as a supplement to the geotechnical studies).

The parameters commonly recorded are the following (cf. figure 4-1):
—drilling rate or rate of penetration;
— pressure on the tool;
— reflected impact;
— injection fluid pressure;
— torque.

a) Drilling rate or rate of penetration

The rate of penetration of the drilling tool is linked to the weight exerted upon the drill bit (weight of the drill string plus pressure imposed by the drilling machinery). The driller's log measures the instantaneous rate of penetration of the tool into the formation. This rate can be calculated at intervals of between 1 mm and 20 cm in the drilled formation, thus enabling a determination of the lithological changes, fracturing, void spaces, weathered zones and clayey interbeds. The magnitude of the instantaneous rate of penetration can be modulated as a function depth, thus allowing the geologist to determine the nature of the formation and make correlations.

Calculating the drilling rate is not as simple as it might appear. Evidently, it will depend on the pressure exerted on the drill stem, but it is also affected by the tool's wear and tear, the condition of the circulation fluid, etc.

b) Pressure on the tool

This parameter can have a direct and marked influence on the drilling rate. By controlling the regularity of pressure on the tool, it is possible to facilitate the interpretation in certain cases by providing corrections on graphs that represent the rate of progress or penetration.

c) Reflected impact

This parameter leads to an assessment of the hardness of the rock. It can only be recorded when using the DTH hammer or the perforating hammer.

Hard rocks which require a strong impact will retransmit a percussive shock wave through the drill string to the surface, whereas soft rocks absorb a major part of the impact energy with weak transmission to the surface. When the drill encounters an evidently void space:

— the DTH hammer almost ceases to transmit a reflected impact, but the impact is felt extremely strongly once contact is reestablished;

— the percussion drill has a very strong reflected impact, but this will weaken if the fissure is filled with clay.

d) Injection fluid pressure

The recording of pressure of the circulation fluid makes it possible to detect the presence of permeable zones. It is an additional technique making it possible to confirm the presence of fractured zones with clayey fillings, for example.

e) Data processing

The data processing can be performed directly on site using a micro-computer. As an example, Table IV-I gives some information on the types of terrain encountered during drilling.

TABLE IV-I — *Evolution of drilling parameters as a function of the nature of the formation (after R. LAUGA, 1990)*

Type of terrain	Drilling rate or rate of penetration	Pressure on tool	Reflected impact	Torque	Injection pressure
Terrains requiring the use of percussion tools					
Compact crystalline rocks	Slow and regular	High	High frequency	Weak	Nil
Weathered crystalline rocks	High, linked to degree of weathering	Moderate or nil	Strong	Weak	Nil
Limestone (karstic zones)	Fast		Strong		
Clayed fissures	Fast	Low	Weak	Weak	Low
Chalk (in clayed zones)	Fast	Low	Weak	Weak	Low
Terrains requiring the use of destructive rotary tools					
Sands	High and regular	Regular and low		Regular	Low and continuous
Sand and gravel	Irregular	Increasing		Increases	Low
Marls	Slower than in sands	Higher than in sands		Stronger than in sands	Higher than in sands
Compact limestones	Slow	Very high		Regular	Depends on fracturing
Fissured limestones (open fissures)	Increases	Drops		Diminishes	Diminishes
Fissured limestone (closed fissures)	Increases	Drops		Diminishes	Increases

f) Torque

The torque exerted on the tool allows it to penetrate the rock. It is clearly influenced by the hardness of the different layers, by their capacity to "stick" (marls) or to swell with addition of water, being also affected by the pressure on the tool and by the injection fluid pressure. It serves as a complement to the other parameters.

g) Other parameters

Other drilling parameters, different from those summarized above, can also be mentioned:
— rotation speed;
— restraint (braking force);
— hammer drill acceleration (amplitude of the impact wave reflected by the ground at the base of the drill string);
— time interval between two measurements (giving average drilling rate as opposed to the instantaneous rate).

4.3.2 Measurement of permeability

Different methods can be employed to measure the permeability of a water-bearing formation:
— The Lefranc-Mandel test is mentioned here merely for information, since this method, which gives coefficients of point permeability in alluvial terrains, is hardly ever used in water-well drilling.
— The Lugeon test, which gives the permeability at a single point in the tested well. The permeability measured by this method is a fracture permeability. The Lugeon test is used mostly in boreholes with a diameter of less than 150 mm.

Principle of the Lugeon test

The Lugeon unit corresponds to the absorption of 1 litre of water per metre of depth in the borehole and per minute at a pressure of 1 MPa. A permeability of 1 Lugeon corresponds to around 10^{-7} m/s.

The Lugeon test is carried out on a section which must be isolated from the rest of the borehole. It involves injecting water in steps of 10 minutes duration at increasing pressures up to $P = 1$MPa (around 10 bars) and then returning to $P = 0$ using the same pressure values.

These two series of tests make it possible to establish two discharge rate vs. pressure curves which should theoretically overlap. In practice, it is apparent that the curves have different shapes, and their interpretation can be made in relation to the graphs shown on figure 4-1.

The Lugeon coefficient is calculated as follows:

$$N = \frac{2Q}{(P + 0.1\,H)\,L}$$

where:

  N : Lugeon coefficient
  Q : total quantity of water absorbed (in litres per 5 min)
  P : water pressure (bar)
  H : depth of the standing water-level
  L : length of test section (or slice)

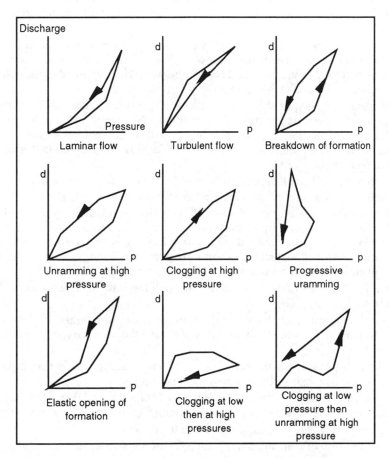

Figure 4-1
Shape of discharge rate vs. pressure curves(after R. LAUGA, 1990).

4.3.3 Downhole logging

Although data from cuttings are available after the drilling is finished, this information may prove insufficient in certain cases. It is then possible to supplement this information using logging tools which are lowered into the well in order to measure the different physical properties of the formations encountered in the borehole. The logging data obtained in this way are transmitted to ground level and recorded. Any recording of a parameter characteristic of the drilled formations as a function of depth is termed a log. By extension, the measuring device used to obtain a log is also called a "log" (or logging tool). The recorded parameters may of many varied types: physical (resistivity, radioactivity, acoustic velocity, etc.), geometrical (borehole diameter, thickness of mud cake, thickness of casings, screens, characteristics of cementation, etc.).

This practice, which is unfortunately rather uncommon, requires that the total benefits arising from the acquisition of physical and geometrical parameters should be included in the predicted cost of the well. The logging data so obtained will continue to be useful throughout the service-life of the well.

The data to be acquired are specific to the structure concerned. The amount and nature of these data are defined according to the type of catchment structure and the type of aquifer captured. Thus, a shallow catchment structure in an alluvial aquifer will require a different type of logging from a well that is several hundred feet deep.

As soon as well implementation is completed, the basic information obtained from logging can be used to provide a consistent data set.

Subsequently, this acquired data make it possible to adapt the running of well maintenance operations. It is evident that the more relevant information is available on the well, the easier it will be to detect any eventual malfunction. This information is crucial in enabling a sound diagnosis following a drop in yield and for the planning of well restoration.

Finally, it is desirable to combine the logging data with a video camera inspection:

— on completion of the structure, logging operations make it possible to reveal possible defects in workmanship and to take the necessary steps to remedy the situation;

— during production, the combination of visual and physical investigations facilitates the diagnosis of breakdowns that may occur.

At present, it appears that no more than 20% of new DWS structures are subject to complete logging checks before completion (Geotherma study, 1991).

With correctly chosen logs, it is possible to check:

— the quality of cementation (position, voids, thickness, etc.);

— the position and condition of casings;

— the condition and positioning of gravel packs.

The collection of such data (logs and video inspections) provides the fundamental historical basis defining the original state of the structure.

Figure 4-2 shows evidence for a resistant layer corresponding to dry limestones (dotted line) and a conducting layer, associated with negative response on the gamma-ray log, which corresponds to a water-rich calcareous bed (in black). In this example, the wet limestones are good conductors since the formation water contains a dry matter residue of around 2 g/l, thus accounting for the high conductivity of 2.5 ohm/m.

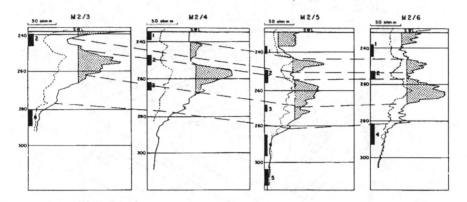

Figure 4-2
Correlation between electric and gamma-ray logs; Mikhili M2 well field, Djebel Akdar, Libya (after Arlab documents).

Numerous parameters can be recognized by well logging:
— logs characterizing catchment structure geometry;
— temperature log, useful with multilayer aquifers and deep wells;
— the flow rate log (or micro-current meter) which makes it possible to record the flow inside the undisturbed aquifer and during water extraction.

In most cases, these measurements are carried by out making use of electromagnetic, acoustic and nuclear properties. The recorded variables can be treated by numerous different types of data processing method which make it possible, through correlation of several variables, to obtain the relevant diagnosis.

a) Electric log

The electric log is used in order to determine the depth and nature of the aquifer formations encountered with a view to bringing them into production. The method consists of recording two depth profiles, one giving the resistivity and the other the electric potential of the drilled formations.

— The resistivity varies in response to changes in the nature of the rocks. Resistivity is inversely proportional to the quantity of water contained in a unit volume of the rock being considered and the conductivity of this water.

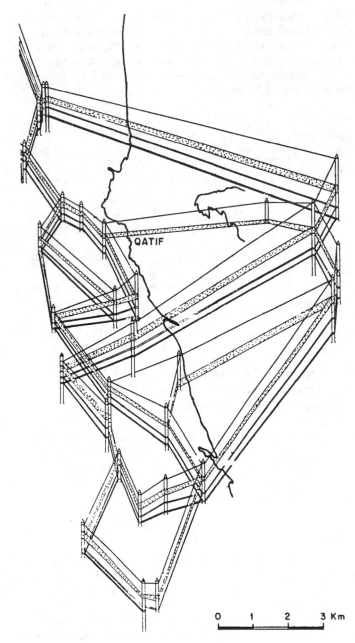

Figure 4-3
Correlation between electric log characteristics and observations in boreholes. Qatif zone, Saudi Arabia (after Arlab document). The electrically resistant layer corresponding to the permeable Eocene limestones is indicated by a dotted line.

— The self-potential (spontaneous polarization) varies as a function of two distinct phenomena: electrofiltration and electroosmosis:

- *Electrofiltration* is the result of the movement of waters towards the interior of porous layers, which produces a negative anomaly in the borehole adjacent to these layers and gives rise to an electromotive force. The direction of this anomaly, while showing positive potentials, can indicate a highly pressured layer discharging water into the well,
- *Electroosmosis* is the electromotive force proportional to the logarithm of the resistivity ratio between two different electrolytes.

b) Acoustic log

The acoustic log makes it possible to determine the structure of the rock, its lithology and the relative porosity of different formations. It can help to localize major influxes of water into semi-consolidated or consolidated rocks (sandstone, conglomerates, igneous rocks). An acoustic log may also enable the determination of the static level of the groundwater and lead to the detection of perched aquifers and possible fractures.

This type of log measures the propagation velocity of an acoustic wave generated by an electromagnetic source placed in the well. It should be realized that the rock or grains of a matrix transmit acoustic waves more efficiently than the pore fluids.

c) Nuclear methods

All geological formations emit natural radiation, including gamma rays. By introducing an artificial gamma-ray source (e.g. caesium 137 or cobalt 60) into the well, a dispersion of these rays can be created. This method, which is called gamma-gamma log, makes it possible to obtain a density versus depth profile for the investigated formations. In general, the higher the density the lower the porosity. The detectors are scintillation counters which are adjusted to detect the least energetic gamma rays. Radiation scattered by the Compton effect has lost most of its energy and therefore comes from a greater distance, but generally no more than 15 cm from the detector

Neutron logs provide an indication of the total porosity of a water-saturated medium or, alternatively, the degree of humidity of an unsaturated medium.

d) Borehole diameter measurement

By knowing the exact diameter of the borehole, it is possible to determine the degree of erosion in the course of drilling, the presence of swelling clays or resistant sandstone beds in a soft formation, or even the presence of fissures in limestone.

e) Thermometric measurements

The measurement of temperature in the well leads to the following information:
— detection of periods of seasonal recharge;
— localisation of a loss of injected fluids or inflow of water;
— investigation of fluid circulation in the formation on the other side of the casings;
— indication of the level of cement behind the casing after cementation.

The temperature at each point in the fluid depends on:
— the thermal conductivity of the adjacent rock;
— the distance from the surface;
— the geothermal gradient.

In general, the geothermal gradient is greater in formations with high hydraulic conductivity.

To detect losses of drilling fluid, a measurement is made after halting the circulation, then a second measurement is made after a new circulation of mud. This new mud is colder.

To detect an influx of water, part of the mud must be extracted from the well; it can then be seen that, below the influx of water, the temperature is lower than at the level of the influx.

f) Micro-current meter

This device serves to measure the velocities of vertical currents inside wells with high wall strength, but not in cased or screened wells drilled in granular formations. It is equipped with a screw that generates an electric current when it turns. This signal is transmitted and recorded at the surface. In relation to the number of rotations of the screw, and with reference to a preliminary calibration, a measurement of the current velocity is obtained. This is a direct function of the flow rate since the cross-section of the well is constant.

Circulation losses and the presence of more or less water-rich zones can also be identified.

The micro-current meter makes it possible to conduct static tests (measurement of the natural flow velocity of groundwaters) and dynamic tests during which the natural circulation velocities are accelerated by the injection of water or, in most cases, by constant discharge pumping at steady state.

The measurements are carried out at depth intervals of 0.10-0.50 m in the well during step drawdown and recovery tests. When the apparent current velocity is greater during drawdown than recovery, this means that the current is rising and *vice versa*.

The micro-current meter only gives an order of magnitude of the relative permeability of the different water-bearing layers. In operating wells, it can be used to reveal the productive zone of the aquifer.

g) Interpretation

Table IV-II defines the parameters which should be considered in correcting the raw geophysical data.

TABLE IV-II — *Geophysical parameters to be monitored.*

Type of log	Minimum diameter (mm)	Drilling fluid	Casing	Parameters to take into account
Resistivity and resistance	57	Necessary	Uncased	Resistivity of the mud, diameter and temperature for the qualities used.
Self-potential	57	Necessary	Uncased	Resistivity of the mud, diameter and temperature for the qualities used.
Gamma ray	57	Unnecessary but desirable	Cased or uncased	Resistivity of the mud and diameter for the qualities used.
gamma-gamma	57	Unnecessary but desirable	Cased or uncased	Resistivity of the mud and diameter for the qualities used, as well as the formation fluid and matrix density.
Acoustic	57	Necessary	Cased or uncased	Diameter of the hole, formation fluid and velocity in the matrix for the quantities used.
Neutron	57	Unnecessary but desirable	Uncased	As the gamma-ray log, with, in addition, the temperature, the salinity of the mud and the composition of the matrix.
Calliper	51	Unnecessary but desirable	Cased or uncased	None.
Temperature	51	Necessary	Cased or uncased	None.
Micro-current meter	51	Necessary	Uncased	Diameter of hole for the velocity log.

There are a great number of downhole logging techniques and only the more commonly used methods in water-well drilling are mentioned here. These highly sophisticated techniques are derived from the oil industry. In particular, great advances have been made in signal processing and also in the methods of data acquisition. It is now possible to inspect a well with different logging tools which continuously record a large number of parameters. The processing of these data on a calculator can enable the clear identification of the various formations encountered and their specific characteristics, while also facilitating correlations between the different wells in a catchment area.

Using the whole set of results so obtained, it becomes possible to interpret the observed anomalies, to determine the lithological composition of the layers and to assess the potential water-yield of certain formations.

4.3.4 Test for water-bearing layers

a) Method

When the lithological profile is established and the water-bearing zones have been recognized, it is possible to carry out a rapid water test. For a test in well-consolidated terrains (rock, conglomerates, hard sandstones) the mud is firstly thinned. The next stage consists of lowering the water-level in the well, after which the recovery is observed. This leads to a rough estimate of the discharge rate.

In softer formations, where the exploratory borehole is drilled as far as the bottom of the aquifer, the hole will be reamed as far as the roof of the aquifer layer. A test string is then sent down which consists of a blind pipe (for the unproductive zone) and a screen (for the water-bearing zone). The drawdown test can then be conducted, noting the volume of water withdrawn as a function of time and measuring the duration of the recovery. A rough specific capacity is thus determined which is lower than the real capacity but which is nevertheless useful in giving an overall idea of the discharge capacity of the well.

b) Relation between drawdown and discharge

While the variation of drawdown in relation to discharge is specific to each well, it is possible, with the help of a large number of examples, to establish an average drawdown vs. discharge capacity curve for two cases: a well in an unconfined aquifer and a well in a confined or artesian aquifer (cf. figure 4-4).

These curves make it possible to predict how a well will behave for different discharge or drawdown values, where the test has already yielded a drawdown value for a given discharge.

Let us consider a well in an unconfined aquifer 30 m deep with a static level of 10 m. During a test at 20 m³/h, the dynamic level is established at 12 m, which gives a drawdown of 2 m.

We wish to estimate the discharge for a drawdown of 5 m. If we consider that 100% of the drawdown corresponds to 20 m, then 2 m of drawdown is the equivalent of 10% of the total drawdown and 18% of the maximum discharge. A drawdown of 5 m would represent 25% of the total drawdown, which corresponds on the above graph to 42% of the maximum discharge. Thus, for a draw-down of 5 m, the well will be capable of supporting a discharge of:

$$\frac{42}{18} \times 20 = 46.5 \ \text{m}^3/\text{h}$$

The same well in a confined aquifer (lower curve) will give, for 10% of drawdown, 12% of the maximum discharge. For a drawdown of 5 m, or 25% of the total drawdown, we will obtain 30% of the maximum discharge. For a drawdown of 5 m, the well will therefore be capable of a discharge of:

$$\frac{30}{12} \times 20 = 50 \ m^3/h$$

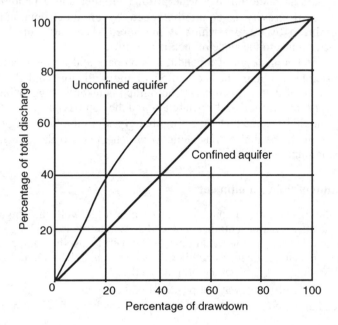

Figure 4-4
General drawdown vs. discharge relations for unconfined and confined aquifers

4.3.5 Equipment of the well

At the end of drilling and before completing the well, it is sometimes necessary to check certain parameters such as the verticality of the borehole or the presence of wall snags which could hinder the lowering of the casings.

— A special tool, called the drill reamer, is used to remove all undesirable asperities from the hole. This tool includes grooved rollers in very hard steel which are tilted at their outside edges. The diameters of these tools vary between 6" or 6 1/2" to 24" or 26".

— Checking the verticality of the hole is also very important before installing the flow string. A deviation of 0.25 to 0.50% is acceptable on condition that the straightness is satisfactory. Otherwise, problems may arise that could affect the running of the pump. In general, the final support pipes are allowed to have an angle, with respect to the vertical, of 1 sexagesimal degree per 30 m section. For information, some of the methods for checking verticality are listed below:

- Device with a plastic disk. The apparatus consists of a mast and a pulley whose axis is set at a determined height (3 m in general). A weight is suspended on the pulley from a cable which goes through a disk. Placed at the well head, this graduated disk will show the different inclination values of the cable at each operation (every 3 m).
- Apparatus made up of a waterproof container and a pendulum. The force of gravity is counterbalanced by an electromagnetic force produced by current which is measured. The value of the current corresponds to the angle of inclination (tilt).
- Apparatus made up of a pendulum-compass and a video camera. It is possible to obtain the inclination of the well bore and the compass bearing of that inclination. A photograph is taken at regular intervals to record the position of the pendulum and the compass needle.
- It should be noted that there are other types of clinometers (gyroscopic, chemical, thermal or mechanical) which measure the deviation of a pendulum.

a) Installation of lining equipment

Before the pipe lining equipment is lowered into the well, the lengths of the tubes must be checked and centralizers placed at regular intervals along them. The lining tube that is sent right to the bottom of the well is also be equipped with an end cap known as a decantation chamber. It enables the collection and storage of fine particles which might penetrate into the well.

There are three different ways of positioning the screens:

— placed at the bottom of the well, the upper part being free and protected by a casing;

— permanently suspended by means of a suspension cone;

— extended by a rising main (blind pipe) up to the surface.

Whatever the method used, close attention should be paid to the lowering of the pipe string, particularly the following points:.

— do not manipulate the tube without protecting its thread;

— inspect the treads, which must be clean and greases;

— avoid banging the pipes, and especially the screens, against metallic parts;

— begin by screwing slowly so the screw thread is correctly engaged;

— do not allow the pipe string to fall freely into the well;

— do not place the casings under pressure (breaking or bending of the string);

— during lowering of the string, keep a check on the length being introduced in order to know the exact position of the foot of the string.

If, for any reason, the casing string is stuck in the well, force should not be applied and the lining tubes must be brought back to the surface at once.

TABLE IV-III — *Choice of methods for the installation of screens according to the depth of the well and the presence of a gravel wall treatment (after R. Lauga, 1990).*

Deep well	Average well	Shallow well	with gravel pack	without gravel pack	Description of method
		X		X	Lowering of the screen to the well bottom protected by a casing and then raising the casing out of the aquifer formation.
	X	X		X	Lowering of the screen to the well bottom in a hole drilled without a lining tube in the aquifer formation.
		X		X	Emplacement of screen by bailing.
	X	X		X	Emplacement of screen by water jetting using a casing shoe with a clapper valve.
		X		X	Emplacement of screen without a valve shoe.
		X		X	Emplacement of screen using a plastic floating-ball valve.
	X	X	X		Emplacement of screen and gravel wall treatment between two lining tubes: screen extension tube and protection tube.
	X	X	X		Emplacement of screen and gravel pack between two lining tubes; the outler one is brought back to the surface.
X	X	X	X		Emplacement of screen on a suspension cone, introduction of gravel. Reverse circulation.
X	X		X		Emplacement of screen in a perforated casing using the wash down method.
X	X		X		Emplacement of screen in a perforated casing using inverse circulation.
X	X		X		Emplacement of screen in a perforated casing using cross-over method.
X	X		X		Emplacement of screen, open-hole gritting and reverse circulation.
X		X	X		Emplacement of screen, open-hole gritting by cross-over method.

Table IV-III summarises the different methods of installing the catchment screens, with or without gravel packs, in wells of different depth. It should be remembered that the screens must comply with certain essential characteristics.

— They must have the highest possible open area coefficient and be in contact with a highly permeable filter medium.

— The slots should have smooth lips, without rough edges. They should have a crescent-shaped cross-section in the direction of the current.

— The slot size should be a function of the grain-size distribution of the surrounding formation. The recommended entrance velocity of the water into the screens must not exceed 3 cm/s. Under such conditions, the risks of incrustation and erosion are greatly reduced.

— A quality of material must be selected that can resist both corrosion and the use of high-pressure jetting.

b) Emplacement of additional gravel

In order to be efficient and to allow well cleaning and development under good conditions, a gravel pack should have a minimum thickness of 3" (75 mm) over its operating radius. Experience shows that it serves no purpose to exceed 8" (200 mm).

TABLE IV-IV — *Volume of gravel in relation to the borehole and screen diameters.*

Diameter of borehole in inches	External diameter of the screen in inches									
	4	6	8	10	12	16	18	20	24	26
8	24.3	14.2								
10	42.5	32.4	18.2							
12	64.8	54.7	40.5	22.3						
14	91.2	81.1	66.9	48.7	26.4					
16	126.6	111.5	97.3	79.1	56.7					
18	156.1	145.9	131.7	113.5	91.2	34.5				
20	194.5	184.4	170.3	152.0	129.7	72.9	38.5			
24	283.7	273.6	259.5	241.2	218.9	162.1	127.6	89.2		
26	330.7	324.3	310.1	291.9	269.6	212.8	178.4	139.8	50.6	
28	389.2	379.0	364.9	346.6	324.3	267.5	233.1	194.5	165.3	54.7
30	447.9	437.7	423.6	405.3	383.0	326.3	291.9	253.3	264.1	113.5
36	648.5	638.4	624.3	606.0	583.7	526.9	492.4	454.0	364.8	314.2
42	885.7	875.6	861.4	843.2	820.9	794.5	729.7	691.1	602.0	551.3
48	1,159.3	1,149.2	1,134.9	1,116.8	1,094.5	1,037.7	1,003.2	964.8	875.6	824.9

Remarks concerning the table: 10% should be added to the given value to take into account the variations in borehole diameter and degree of compaction. The thickness of the gravel pack should not be less than 3". The gravel pack must extend at least 3 m above the uppermost screen.

The upper surface of the gravel pack must lie clearly above the roof of the aquifer layer or the top of the screen. The table given below indicates the volume of gravel (per metre depth in the well) to be emplaced as a function of the diameters of the borehole and the lining tubes (in inches). Conversion tables between the main international and US units are given in Annexes I and II.

Otherwise, unsatisfactory emplacement of the gravel pack can considerably reduce its efficiency. A problem that sometimes arises during gritting is the separation of coarse material, which falls to the bottom of the well, from the fines which tend to settle less quickly, thus remaining in the upper part of the filter medium. This will lead to the pumping of sand through the coarse material at the bottom while the upper part (finer particles) will only allow the passage of a very small flow.

This phenomenon occurs especially when the sorting coefficient of the gravel is higher than 2.5. It is also the case when the gravel is simply cast into the annular space from the surface. The separation brought about though the action of water or mud can produce — with the smallest obstruction — a gravel bridge which leaves a void space below. So this simplistic method of gritting should be avoided.

Two main ways of emplacing the gravel can be distinguished: by gravity and using a continuous circuit under pressure.

— *Gritting by gravity*: this method particularly concerns shallow wells. The gravel pack is put into place in the annular space by means of a conductor pipe fitted with a feed hopper at its upper end. The pipe takes gravel to the bottom of the well and is raised as the gravel is introduced. Water circulation (normal or reverse) can be employed to facilitate this type of gritting.

— *Gritting by continuous circuit flow under pressure*: for deep wells, the method currently employed consists of introducing the gravel in a light mud or water in reversed circulation mode. The gravel "suspended" in the mud or water is forced back by the pump, thus being carried into place from the bottom upwards around the screen.

Finally, in the case of large wells in aquifers that are likely to be acidized (chalk, limestone, dolomites), it is advisable to have available a mechanism allowing the addition of gravel after acidification and development.

c) Cementation

Once all the borehole equipment has been installed, it is often necessary to carry out cementation. Among other objectives, this operation aims:

— to seal the casing string to the walls of the well;

— to isolate the groundwater that is to be extracted from possible sources of pollution or prevent undesirable contact with other aquifers.

The quality of the cement should be chosen as a function of various parameters such as the nature of the formations and waters encountered, the temperature of the terrain, the volume to be emplaced, etc.

For water wells, which are generally shallow, Portland cement is used. Quick-setting cement can be added to limit the down-time for the drilling rig.

The technical aspects of cementation operations are discussed in Chapter III. Care should be taken to use the special cements that are suitable in view of the type of geological formations encountered (gypsum, aggressive waters, etc.).

4.3.6 Well development

It is unnecessary to repeat the different methods of well development set out in Chapter III. At this stage in the checking process, respect of the established protocol will be ensured: surging, air-lift, injection of chemical products, etc.

In particular, the correct setting-up of well development systems should be checked. Thus, for air-lift treatment, the following points need to be respected:

— The total length of the air line must be at least equal to 2.5 times the total height of the water column, drawdown included (see figure 3-18 and text referring to it).

— The water pipe should be several metres longer than the air line.

— It is advisable to check the air pressure at the start, as well as the volume of air available.

— Diameters of the water pipe and air line should be chosen in relation to the discharge.

— The specific capacity must be checked. A formation can be cleared of sand without achieving an optimal sorting of the remaining *in situ* particles. A check should be carried out by comparing the drawdown of the water-sheet for an identical discharge both before and after well development. Alternatively, this can be done by measuring the hydraulic conductivities.

At this stage in the checking, it is often useful to verify for a second time the height of the gravel pack. In fact, the gravel can often — if not always — settle under the effect of well development. It then becomes necessary to add further gravel up to the initially-determined level.

4.3.7 Pumping tests

The essential objective of pumping tests is to determine the operational capacity of the well and the hydrogeological characteristics of the aquifer it is tapping (i.e. values of transmissivity and storage coefficient). Knowledge of these parameters allows an assessment of the aquifer's evolution and the influence of productive pumping.

Discharge tests make it possible to construct experimentally the curves or straight lines which represent the functions governing the flow of groundwaters towards catchment structures.

The details of all the operations involved in a pumping test are presented in Chapter III. The following discussion is limited to defining the essential points of checks that are to be carried out during such a test.

— The test should always be performed on a completely developed and cleaned well, so that its properties may be characterized in a stable manner. At the end of cleansing operations, the pumped water must be perfectly clean and free of sand. Adherence to this rule must be scrupulously monitored by the supervisor.

— During cleaning operations, the well-borer is able to estimate an order of magnitude for the maximum yield of the structure; this value will make it possible to choose a pump that is best suited to the discharge tests.

— The piezometric level (standing water-level) should be systematically recorded before starting the pump .

— It is important to fix a certain number of discharge steps, using a regular schedule with increasing durations. The inspector should ensure both the regularity and the accuracy of the discharge rates. Excessively strong pumping should be avoided at the start of the test.

— The supervisor should also ensure that discharge vs. drawdown stabilization has been attained at the end of each step. The value of the dynamic level can then be noted.

— The recovery of water-level must be followed step by step as soon as the pumping stops and until it returns to its initial state (standing water-level).

— All measurements of drawdown and recovery must be recorded on a pumping data sheet which is given to the hydrogeologist responsible for interpreting the results. It should be noted that the use of pressure-sensor recorders has greatly facilitated this task.

— It is necessary to collect water samples at the end of pumping in order to carry out bacteriological and physico-chemical analyses. It is also recommended to make the necessary arrangements in advance and, especially, to notify persons from the official agencies who are in charge of taking samples for analysis.

This methodology must be imposed on the well borer (or driller) by the checking process so that, in every case, it is possible to:

— obtain an interpretation in a non-steady state regime and, in particular, establish a yield-depression curve for the well;

— define the critical discharge of the well from this curve and, following that, the operational yield of the well and depth of the pump;

— obtain an interpretation in a steady-state regime;

— obtain reliable comparisons concerning the evolution in the condition of the well and especially its possible future clogging.

4.3.8 Well data sheet

It is the role of the supervisor, while the catchment works is being constructed, to make a note of all elements which will serve to establish an "identity card", that is, all the geological, hydrogeological, mechanical and hydraulic characteristics of the water well.

These elements will be essential for the final acceptance of the structure.

If it is to be complete and effective, the well data sheet must above all contain the following information:

General observations:
- cartographic coordinates of the structure (latitude, longitude, elevation) eventually supplemented with administrative-type information (place names, map sheet number, etc.),
- dates of starting and completion of drilling,
- dates of starting and finishing of pumping,
- name of the drilling company,
- date of final completion of the structure,
- name of works inspector or inspection agency.

— *Studies carried out:*
- photo-interpretation (Y/N),
- geophysics (Y/N),
- other.

— *Geological cross-section::*
- highly detailed lithological log,
- nature of the tapped aquifer,
- nature of the fracturing, if any, and indication of the main strike direction,
- depths of water influxes.

— *Technical section::*
- depth of borehole, drilling diameters,
- different types of drilling used (rotary, DTH hammer, etc.) and nature of drilling fluids,
- penetration rate of the tool in the different formations,
- nature of the equipment: diameter of the casings quality, thickness, depth,
- description of the adapter bushings,
- presence or absence of cementation,
- type and description of mechanism for releasing the catchment (or flow) string,
- nature and position of centralizers in the catchment string,
- nature and granulometry of the gravel pack, depth of its top surface,
- type of screen, diameter, open area, slot size,
- depth of well as it is brought into production.

— *Hydraulic characteristics of the well:*
- discharge at completion of drilling,
- discharge and duration of well development,
- piezometric level of the aquifer(s), date of measurement,
- date and characteristics of pumping tests,
- discharge capacity vs. drawdown curve,

- low velocity log (micro-current meter),
- transmissivity, storage coefficient,
- specific capacity,
- authorised maximum yield.
— *Characteristics of the pumping equipment:*
  - date of installation of the pump,
  - make and type of pump,
  - characteristic curve of the pump,
  - depth of intake,
  - nature and diameter of the rising main,
  - operational yield.
— *Water quality:*
  - date of sampling,
  - results of bacteriological and physico-chemical analyses of the water.

It should be noted that software applications running on micro-computers (cf. Chapter VIII) can be used to follow up the implementation of a borehole and edit the drilling report.

4.4 Acceptance of the water well

Carrying out a water well project is a delicate and complex operation so, to obtain a structure offering the best possible guarantees, it is essential to ensure that it is submitted to a system of continuous checks.

This control must be handled by a specialist who, according to circumstances, may be a supervisor, a drilling engineer, an inspector or a hydrogeologist. The latter is an experienced individual who knows all the subtle details of the profession and will be capable of taking decisions in full conformity with the rules of the trade.

It should be realized that the implementation of a water well involves technical risks whose financial consequences are of considerable importance. The task of the works supervisor is consequently fundamental throughout the whole duration of the project. The supervisor checks the smooth execution of the technical programme and acts as an objective observer supplying the contracting authority with the relevant information for running the construction site.

As described previously, the supervisor's task mainly consists of:
— making sure of the technical means used by the drilling company;
— checking the conformity of the materials used;
— controlling the installation of the casings and gravel packs;
— checking the drilling parameters in the course of operation;
— directing the well development operations;
— supervising the well tests;

— ensuring that all the checking operations have been carried out in view of final acceptance of work.

The supervisor will also pay special attention to site cleanliness and the disinfection of equipment and fittings. Numerous bacteriological pollution problems can arise if the work is carried out under poor conditions of hygiene. The introduction of germs during the period of construction sometimes results in the seeding of bacterial cultures that are favoured by conditions such as water temperature in the aquifer or the intrinsic nature of the medium. Once these problems occur, they are difficult to combat, whereas it is straightforward to respect the conditions of hygiene while the work is still being carried out.

The supervisor is responsible for drawing up a report on completion of all work, in which each operation must be indicated in chronological order.

Finally, except in exceptional cases, it is difficult to know in advance the precise geological section of the well and the type of equipment that will be installed. The contracting authority thus will have an estimated price bracket for the site work, which is calculated on a theoretical basis and also as a function of the unit costs given on the company's priced bill of quantities. The real situation on completion of the work is always different and account must be taken of unexpected overheads: downtime of the site work, instrumentation time, development time, well tests, number of acidification operations, length of casings, etc. A contradictory effect sometimes arises where the cost estimates lead certain contractors to conceal eventual overspending on certain items by replacing them with others. In this way, wells exist whose characteristics on paper are quite different from those on the ground because it proved easier to inflate certain intangible items (drilling downtime, pumping tests, well development) in order to mask the overspend on lengths of casing, screens or some other kind of eventuality. These short-term "arrangements" result in the falsification of maintenance directives and evidently lead to erroneous rehabilitation diagnoses.

In the interest of the water operators and in the framework of sound management, it is crucial that work should be carried out unambiguously and that this should lead to the drawing up reliable reports. Non-information and petty financial arrangements reflect short-term views without any ethical foundations, and it is advisable to abandon such practices on a permanent basis.

4.4.1 Acceptance pumping tests

It is sometimes necessary to verify certain key points which guarantee the quality and longevity of the catchment structure. This is particularly true of acceptance pumping tests in which the characteristics of the well are once more investigated. To this purpose, a characteristic discharge rate vs. drawdown curve will need to be established making it possible to determine:

— the specific capacity;

— the critical discharge,

— the maximum yield.

These data will allow a comparative ("health") assessment to be drawn up when subsequent tests are carried out, and will indicate when it becomes necessary to restore the well.

Furthermore, the water quality must necessarily be carefully controlled. To achieve this, samples (minimum volume of two litres) are taken during the acceptance tests. The opportunity should be taken during this pumping test, if it is sufficiently long, to collect water samples for analysis by the European authorities.

4.4.2 Video camera inspection

Video camera inspection is advised for the acceptance testing of deep structures or in cases where particular difficulties have been encountered during construction. This is the only type of examination that permits the identification of eventual constructional defects (deformation, defective screwing, clogged or poorly-positioned screens, etc.).

It is desirable that the contractor is present during this operation. This examination can be accompanied by a measurement of flow rates using a micro-current meter which makes it possible to pick out the productive zones, for example in a karstic limestone aquifer. At a later date, this can enable comparisons of discharge capacity and deciding on well restoration procedures.

As far as possible, during the camera inspection, it is best to proceed with a pumping test using a discharge rate close to the operational capacity. This makes it possible to reveal any eventual upward flux of sand, but this is not always possible because of the difficulty of passing the camera cable alongside the pump.

4.4.3 Well logging tests

Well logging tests are to be recommended, even though they are costly, particularly when drilling problems have been encountered or when the final pumping is not giving the expected results.

A gamma-ray log makes it possible to reconstruct the geological section in case of doubts and check the adequacy of the technical and geological sections.

The gamma-gamma log is able to reveal the joins between casings and the location of screened sections. The position of the gravel pack can also be checked.

Finally, a probe can be used to check the depth reading at the bottom of the well.

4.5 Conclusion

The checking of the work in the widest sense represents an essential stage in the implementation of a water well. The engineering design department is responsible for the construction of the catchment structure. It is the direct

representative of the company and has responsibility for supervising a certain number of important points in the running of the works project, notably:
— collection of formation samples;
— recording depths drilled;
— determination of the position of water inflows and their discharge;
— conformity of the equipment in relation to the characteristics of the geological formations encountered;
— granulometry of the gravel pack, if required;
— follow up of all drilling incidents;
— checking the quantities of materials used;
— duration of well development and quality of the pumped-out water (samples);
— setting up of discharge tests and their interpretation.

In addition to its supervisory role, which is mainly intended to detect drilling problems and provide rapid remedies, the purpose of checking is to leave a written record of every stage in the construction work. From these notes, which are collected together in a work site record book and a well data sheet, it is possible at any moment in the service-life of the well to look up the original characteristics of the well. In this way, comparisons can be made which are indispensable in the efficient management of the structure.

Finally, no catchment of groundwater destined for human use should be carried out without filing a technical report upon completion of the construction work. This formal document should be sufficiently complete for the initial conditions of the well to be clearly identified. Once this information is available, maintenance of the catchment structure will amount simply to a question of political will on behalf of the contracting authority.

CHAPTER V

Water well protection

"Every representation of knowledge is an interpretation"
Terry Winograd and Fernando Flores

The considerable volumes represented by groundwater bodies, as well as their geographical distribution, flow characteristics and vulnerability mean that they are of fundamental importance in the management of water resources. However, considering the overarching nature of such resources, groundwaters cannot be separated from surface waters or dissociated from physical planning in the context of regulations which exist ultimately at the national, European or even worldwide level. In France, following the 1964 law, water has become an object of social policy negociation, and its management is tending to be organized in the manner of an acquired communal asset.

The control of groundwaters also involves making use of judicial means that include legislative and economic measures. It is necessary to speak in terms of a long-term vision, while taking account of the persistence of certain phenomena, particularly with regard to micropollutants which show a prolonged residence time. In the long term, it is advisable to incorporate these different aspects into a policy for national planning and development.

The extraction and management of groundwater resources entails the legal control of aquifers. The *groundwater right is* a right granted by law to take possession of groundwater and put it to beneficial use. In different countries, however, groundwater resources are governed by different laws. The evolution of groundwater law has taken such a tortuous path that confusion can arise about which law applies to a particular situation. In the USA, the demand placed on groundwater and the threat of groundwater contamination have led to major changes in the law in almost all parts of that country.

Groundwater rights are derived from two contrasting doctrines of law:

— the common-law or English doctrine of riparian rights, based on the premise that the water is the absolute property, in perpetuity, of the owner of the overlying land;

— the principle of prior appropriation, under the basic premise that the groundwater is the property of the state or the public.

Regulating agencies now evaluate the development of groundwater resources in the light of the public good.

Up to the present, groundwaters in France have been regulated by a certain number of official texts of limited scope that were enacted over the period from 1898 to 1973. The uses of water are now considered in the "public interest", added to which, the requirements arising from European directives have also led to a strengthening of the existing legislative framework. Taken together, these regulatory arrangements are tending towards the improved protection of groundwaters, particularly through the setting up of protected perimeters.

5.1 Catchment protection zones

Once the well has been drilled, it is appropriate to protect it against all external sources of pollution. To achieve this, the water well is surrounded by a certain number of protection zones.

The vulnerability of groundwater bodies to various types of pollution is conditioned by several factors:

— Filtering capacity of the reservoir.

— Thickness of the unsaturated zone of the reservoir.

— The flow rate of groundwaters, which has an influence on such phenomena as dilution, degradation and the absorption of certain pollutants.

— Natural protection of the aquifer, due to the presence of an impermeable layer above the aquifer which affords a barrier against surface pollution.

— Nature of the groundwater body, since unconfined aquifers are more vulnerable than the less accessible confined aquifers.

5.1.1 Groundwater regimes

A borehole does not simply tap groundwaters from the immediate vicinity of the well screen. According to the laws of hydrodynamics, a perimeter of influence of variable size is established within which the aquifer is subject to the effects of pumping. In this way, a radius of influence may be defined around each water catchment structure (see also § 4.5.3).

When the pump is started up, the water contained in the borehole is preferentially extracted and, as a result, there is an immediate lowering in water level. This drop leads to a pressure imbalance across the well casing, and the groundwater flows from the outside towards the catchment structure. Under these conditions, the piezometric surface is hollowed out and assumes the shape of a cone of revolution with curved faces. Two kinds of situation may arise:

— *Steady state regime*, in which both the form and extent of the cone of depression remain constant.

— *Non-steady state (or transient) regime*, in which these parameters vary with time.

At the beginning of pumping at constant flow rate, the water level in the well falls rapidly in a non-steady regime. The water level falls more and more slowly as pumping continues, and eventually tends to stabilize. A quasi-steady state regime is thus attained.

Although a strictly steady-state regime is the ideal limit towards which a non-steady state will tend to evolve, the flow of groundwaters into the borehole generally takes place under non-steady state conditions.

5.1.2 Effects of pumping on aquifers

As mentioned above, the piezometric surface grows hollow during pumping to form a cone of depression whose depth and extent increases with time (cf. figure 5.1).

a) Shape of the cone of depression

The form of the depression is not strictly a cone in the geometrical sense, but rather a convex-downward curved surface of revolution. The precise shape of this curve is determined by two parameters:

— Depth at the centre of the cone, which corresponds to the amount of drawdown in the well (also termed the dynamic or pumping level).

— Radius of the cone, which determines its aereal extent. This parameter corresponds to the distance at which the drawdown is theoretically zero or, in practice, too weak to be measured.

The depth-to-radius ratio therefore defines the shape of the cone. In turn, the shape of the depression profile depends on the pumped discharge rate and the aquifer characteristics (thickness and permeability), and thus the transmissivity of the medium.

Thus, for a given pumping rate, the drawdown at a particular location will be greater as the transmissivity decreases.

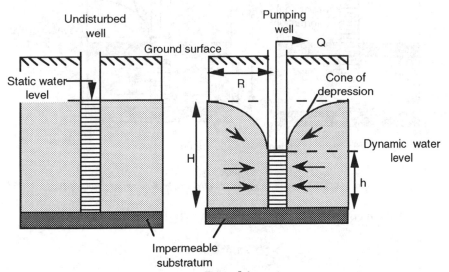

Figure 5-1
Principle of catchment from a unconfined aquifer (with a horizontal substratum).

In the case of an unconfined aquifer, the pumping discharge rate is given by the Dupuit equation:

$$Q = \frac{\pi K (H^2 - h^2)}{2.3 \log \frac{R}{r}}$$

where:

Q : discharge rate in m³/h,
K : hydraulic conductivity of the medium, dimensionless,
H : height of water table (or static water-level) above well bottom,
h : height in metres from dynamic level with respect with well bottom, so $(H-h)$ is the drawdown of the aquifer during pumping
R : radius of influence of pumping,
r : radius of well in metres.

When pumping is carried out in a confined aquifer, the cone of depression is no longer a limit of flow; it should be regarded as a theoretical cone of depression (see figure 5.2).

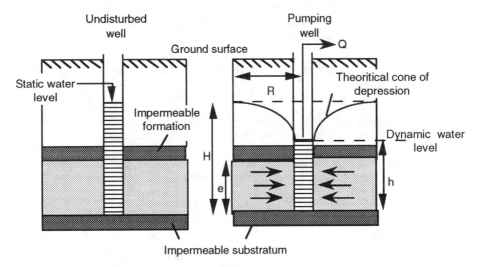

Figure 5-2
Principle of catchment from a confined aquifer (non-flowing artesian well).

The Dupuit equation can then be rewritten as follows:

$$Q = \frac{2\pi K e (H-h)}{2.3 \log \frac{R}{r}}$$

where:

Q : discharge rate in m³/h,

K : hydraulic conductivity of the medium, dimensionless,

H : height of static water-level above well bottom,

h : height en metres from dynamic level with respect with well bottom, so *(H −h)* is the drawdown of the aquifer during pumping,

R : radius of influence of pumping,

r : radius of borehole in metres,

e : thickness of the water-bearing layer in metres.

b) Growth of the cone of depression

During pumping at constant discharge rate, the cone of depression deepens and widens out. The growth of this cone progressively slows down with time, eventually attaining the radius of influence of pumping. The radius of influence is defined as the distance beyond which drawdown is imperceptible. Evolution towards apparent stability is more or less rapid according to the properties of the formations. Without discussing the technical details of the relations governing groundwater flow during pumping (cf. Chapter III), it can be simply pointed out here that, in principle, application of the non-steady state regime model and equations derived from it will lead to:

— Calculation of the numerical values for formation constants on the basis of discharge and drawdown measurements (provided essentially by pumping tests).

— Subsequent predictions of drawdown caused by pumping.

c) Groundwater flow under conditions of pumping

Under natural conditions, an aquifer is in a state of dynamic equilibrium. Pumping in a well will modify this equilibrium, thus bringing about a drawdown of the surface of the aquifer. This induces a flow into the catchment structure from its surroundings; two types of zone should be distinguished (cf. figure 5.3):

— The disturbed zone (or area of influence) is the area within which the water levels are influenced and therefore drawn down by pumping.

— The area of pumping depression is that part of the disturbed zone where all flow lines are directed towards the pumped well. This zone is included within the catchment recharge area, and extends upstream as far as the boundary of the system.

Any pollution taking place within the area of pumping depression will find its way into the catchment structure. The downstream boundary of this area is often taken as the outer limit of the near-field pollution control zone.

Determination of the area of pumping depression is based on the Theis equation, which is applicable to non-steady state regimes. Knowing the piezometric surface of the aquifer, it is possible to plot drawdown contours around the well as a function of the selected discharge rate. These curves intersect at certain points the contour lines on the free surface of the aquifer, thus enabling calculation of the modified values of head. From these points, it is possible to construct the lowering of the piezometric surface due to pumping. Tracing the flow lines leads to a definition of the area of pumping depression of a well.

The dimensions of these different zones vary as a function of discharge rate and are essential for establishing protected perimeters for water wells.

5.1.3 Protection zones

The determination of pollution control zones for drinking water capture can have important consequences on the community, equally as regards the health protection and economic aspects. Such studies are time-consuming and complex, and, in France, should be carried out by a hydrogeologist who is accredited in public hygiene matters.

The proposed perimeters should ensure an optimal security for the protection of water supplies, while not extending the area beyond a degree compatible with cost constraints.

As previously mentioned in Chapter II, a distinction is made between the remedial attenuation and well field management zones.

— The remedial action zone *(immediate protected perimeter)* is used to avoid the deterioration of water abstraction structures and prevent the spillage or infiltration of pollutants at or in close proximity to the catchment site.

— The attenuation zone (near-*field protected perimeter*) should provide effective protection of the catchment structure with respect to the underground migration of pollutants. It is determined in relation to:

- Characteristics of the aquifer and groundwater flow.
- The maximum discharge rate of the well.
- Absorptive and degradation capacity of the soil and subsurface formations with respect to pollutants.
- Dispersive capacity of groundwaters.
- Magnitude of drawdown during pumping.
- The duration and rate of transfer of waters between eventual point-sources of pollution and points of abstraction from the aquifer.

— The well field management zone *(far-field protected perimeter)* is an eventual prolongation of the attenuation zone to reinforce protection against permanent or diffuse sources of pollution.

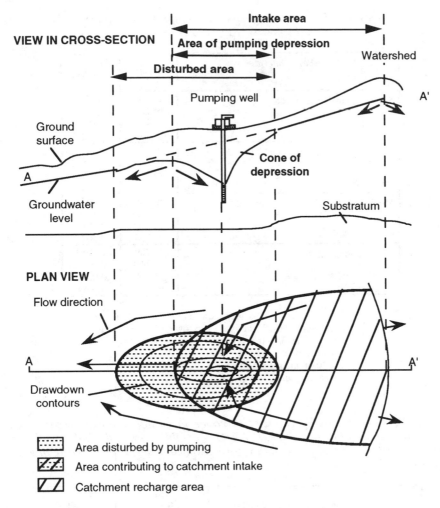

Figure 5-3
Diagram showing effect of pumping in a porous medium (after BRGM document).

a) Estimating the purifying capacity of terrains

The purifying capacity of the soil and the unsaturated zone is of importance in preventing the propagation of pollutants towards the aquifer. It plays a particularly crucial role in the context of environmental control (wastewater and solid waste disposal).

TABLE V-I — *Purifying capacity of aquiferous soil cover (after W. REHSE, 1977)*

Type of material	L (m)	I = 1/L
Silt-poor, sand-rich gravel	a) 100 b) 150 c) 170 d) 200	0.01 0.007 0.006 0.005
Fine to medium sandy gravel	a) 150 b) 200 c) 220 d) 250	0.07 0.005 0.0045 0.004
Medium to coarse sand-poor gravel	a) 200 b) 250 c) 270 d) 300	0.05 0.004 0.0037 0.0033
Gravel and pebbles	a) 300 b) 340 c) 360 d) 400	0.0033 0.0029 0.0028 0.0025

L is the horizontal distance required for purification.
The different transfer rates are as follows:
. a = less than 3 metres per day,
. b = 3-20 metres per day,
. c = 20-50 metres per day,
. d = more than 50 metres per day.

TABLE V-II — *Purifying capacity of cover formations (soil + unsaturated subsurface zone) (after W. REHSE, 1977). H is the thickness of the layer needed to achieve purification and I is an index (formation constant) characteristic of each type of terrain.*

Type of material	H (m)	I=1/H
Humus, 5-10% humus, 5-10% clay	1.2	0.8
Clay without mudcracks, clayey loam clay-rich sand	2	0.5
Clayey silt to silt	2.5	0.4
Silt, sands poor in silt or clay	3.0-4.5	0.33-0.22
Fine to medium sand	6	0.17
Medium to coarse sand	10	0.1
Coarse sand	15	0.07
Silty gravel rich in sand and clay	8	0.13
Silt-poor gravel rich in sand	12	0.08
Fine to medium gravel rich in sand	25	0.04
Medium to coarse gravel poor in sand	35	0.03
Pebbles	50	0.02

In 1977, W. REHSE proposed an empirical method for estimating the purifying capacity of a medium during the transfer of a pollutant from the ground surface vertically down to the aquifer and then circulating horizontally within the aquifer towards the water catchment structure.

The different soil categories may be classified in terms of their grain-size distribution. W. REHSE defined the soil thickness necessary under unsaturated conditions to ensure the purification of polluted waters. This author has determined the path length in different media — which is itself a function of the effective transfer velocity — required to achieve purification within the soil cover (cf. Table V-I) or the aquifer formation (cf. Table V-II).

H. BÖLSENKÖTTER *et al.,* (1984) added to this method by extending it to fractured media (cf. Table V-III).

TABLE V-III — *Purifying capacity of rocks*
(after H. BÖLSENKÖTTER et al., 1984).

Nature of medium	*H (cm)*	*I = 0.5/H*
Marls	10	0.05
Sandstones with clayey interbeds, clays, phyllites and micaschists	20	0.025
Basalts and volcanic rocks	30	0.017
Greywackes, arkoses, clayey or silty sandstones	50	0.01
Granite, granodiorite, diorite and syenite	70	0.007
Quartzites, cherty sandstones	100	0.005
Limestone	200	0.0025

b) Drawdown

The criterion of drawdown is linked to the concepts of area of influence and cone of depression as defined above. It is very important to delimit the area of pumping depression around a catchment structure since any pollution taking place within this zone will eventually lead into the water well.

c) Transfer time

This parameter is based on the time required for a pollutant to migrate from its point of entry into the aquifer to the catchment structure. It is also necessary to take account of the transfer within the unsaturated zone.

Within the recharge area for a water well, it is possible to establish contours of equal transfer time (isochrons). The more extensive is the zone of transfer, the better is the protection afforded to the water well.

The distance imposed between the water well and the boundary of the protected perimeter will vary according to the method adopted for calculating the transfer time (convection-related transfer time, modal time or arrival time). It should be noted that the most commonly used method is based on the convection-related time, which involves the effective flow velocity.

d) Distance

In certain cases, defining the protected perimeter depends on measuring the radius or distance between the well and the point under consideration. The disadvantage of this criteria is that it does not take account of flow and transfer processes affecting the pollutant.

e) Limits of water flow

The establishment of protected perimeters based on this criterion makes use of the physical, topographic and hydrogeological characteristics controlling the flow. These limits may correspond to a river, a canal, a watershed, a fault, a groundwater intake boundary or a impermeable aquifer boundary.

f) Current technical regulations in member-states of the European Union

This section summarizes the technical regulations that are typically applied in the different countries of the European Union. The compilation was drawn up by J.C. ROUX (1992). It appears that standards for sanitary well-protection zones (or protected perimeters) are regulated to a variable extent in different countries:

— In eastern Germany (ex-GDR), zone I (RAZ) is comprised between 5 and 10 m, while zone II (AZ) corresponds to a transfer time of 60 days and zone III (WFMZ) is subdivided into IIIA and IIIB with transfer times of 10 and 25 years, respectively.

In western Germany (ex-FRG), zone I should be fixed at least 10 m from the groundwater well, zone II should correspond to a groundwater transfer time of 50 days, while zone III should extend up to the limit of the catchment basin (it can be subdivided into zones IIIA and IIIB).

— In the Flanders region of Belgium, the abstraction zone should lie within 20 m of the well, zone-I should correspond to a transfer time of 24 hours and zone II to 60 days. The zone-II protected perimeter is set at a minimum of 150 m for artesian aquifers and 300 m for other types of aquifer. Zone III may extend to the limit of the catchment area, with a maximum radius of 2 000 m.

— In Denmark, no published directive exists stipulating the distance to the limit of the protected perimeter or the criteria to be used for its dimensioning.

— In Eire, the perimeter of zone 1A should be more than 10 m from the well, while zone 1B should have a radius of 10-30 m and zone IC a radius of 300-1 000 m.

— In Italy, the remedial action zone is never less than 5 m, while the attenuation zone is set at 200 m. In the Lombardy region, however, a transfer time of 60 days is applied for sanitary protection in the near field and 365 days in the far field.

— In the Netherlands, the first protection zone (or catchment area) corresponds to a minimum transfer time of 60 days, which is prolonged wherever possible to 100 days or 1 year. Due to the weak hydraulic gradients in the groundwater bodies, flow rates are slow (10-20 m/yr) and consequently zone I ranges from 30 to 150 m. The second so-called "protection zone" corresponds to a transfer time of 10 years and, finally, the third zone corresponds to a transfer time of 25 years. These two latter perimeters are situated at distances of 800-1 500 m from the groundwater well.

— In the United Kingdom, taking the example of the Yorkshire Water Authority, the protected perimeter corresponds to 150 days of fluid transfer, including 50 days within the unsaturated aquifer. For the Severn Trent Authority, this zone lies within a radius of 1 km around the water well. In the case of the Southern Water Authority, the protected perimeter corresponds to a transfer time of 50 days, which can be normalized to a distance of 0.5 km in porous aquifers and 1.2-5 km in the Chalk.

In a general way, the dimensioning criteria for perimeters are based on the purifying capacity in the intake area and a transfer time which is generally taken as 50 days in countries applying a standard. Whatever the circumstances, groundwater pollution control regulations and procedures for their application still remain rather disparate in the different member states of the European Union.

g) Summary

Table V-IV summarizes the procedures to be adopted and the methods that can be used to establish protected perimeters as a function of the selected criterion.

It should be borne in mind that each example of groundwater well protection is a special case that depends on the degree of vulnerability of the aquifer, the risks of chronic or accidental pollution, the transfer times and the boundary conditions. The prohibitions and regulations within the different protected perimeters are assessed by the accredited hydrogeologist and recorded in the Public Utility Declaration (DUP in French).

The measures to be taken within the three protection zones can be briefly outlined as follows:

— Inside the remedial action zone, the land should be purchased and enclosed by the developing authority.

— Within the attenuation zone, possible land use is more or less restricted according to the vulnerability of the aquifer and any prohibition should be the object of compensation.

— Although there is no question of prohibition inside the well field management zone, the regulations concern any activities, industrial plant or dumping that could represent a pollution hazard.

In practice, pollution control is merely confined to considering the well field management zone as a vulnerable zone within which general regulations (applicable throughout the national jurisdiction) should be prioritized with particular attention. However, raising general awareness not only of the necessity of protecting water resources but also the new water law will — in the long term — bring about modifications in the constraints for WFMZs. Orders are currently being drafted, particularly as regards the protection against diffuse pollution. It would appear that the trend is towards a better appreciation of pollution control based on a cross-fertilization of expertise between different disciplines (e.g. hydrogeology, agronomy, soil science, etc.). Finally, groundwater protection will be increasingly oriented towards a quantitative assessment of risks, so the resulting protective measures will necessarily become more extensive and constraining.

TABLE V-IV — *Procedures and methods for establishing protected perimeters according to selected criteria.*

Criterion	Procedure adopted	Method
Purifying capacity of cover formation	Soil survey Study of cover formations: test holes, permeability, mineralogy, chemical composition	Rehse method
Drawdown	Water-level measurement Pumping tests Hydrodynamic modelling	Graphical determination or calculation of the area of pumping depression, recharge modelling.
Transfer time	Water-level measurement Pumping tests Tracer studies Transfer modelling Study of cover formations	Calculation of transfer times or use of modelling curves for determination of isochrons
Flow limits	Water balance Tracer studies Chemical analysis Water-level measurement	Nomograms or type-curves
Arbitrary distance	Mapping Purifying capacity of cover formations (vulnerability maps) Water-level measurement Tracer studies	Multiple approach

N.B.: Environmental studies as well as mapping and chemical analyses are necessary in all cases.

5.2 Conclusion

On the global scale, it is known that groundwaters account for almost the entire inventory of available water resources. They correspond to about 600 to 700 billion cubic metres per year for the whole planet. In France, two thirds of the territory are covered by water-bearing terrains, from which are extracted 7 billion cubic metres per year, 60% being destined for human consumption. France is ranked seventh in the European Union for the proportion of groundwaters actually used.

Two thirds of the national territory of France is underlain by aquifers — or networks of aquifers — that are either confined or unconfined. A total of about 200 aquifers of regional importance (i.e. covering a surface area of more than 100 km^2) have been described and documented:

— 175 unconfined aquifers, with single or multi-layered geometry, commonly bounded by major watercourses flowing across the water-bearing formations. They comprise:

- 15 large-volume alluvial aquifers;
- 40 karstic aquifers;
- 20 multi-layered aquifers (at least in part);

— 25 deep confined aquifers, for the most part in communication with one of the aquifer types described above.

It is advisable to protect this resource. As regards the protection of water wells, the latest national report, carried out by the French Ministry of the Environment in 1986, estimates that 15% of wells are provided with effective protection (covered by a Declaration of Public Utility or statutory registration). Although this percentage is probably no longer representative of the real situation, the level of protection afforded remains low.

Finally, the carrying out of a Declaration of Public Utility is a long and exacting task for which it is advisable to seek assistance from specialists (engineering design departments, specialist organizations, geologists, hydrogeologists, expropriation agencies, etc.).

CHAPTER VI

Water well management

"He who knows nothing has no doubts"
P. Gringoire

The catchment of groundwaters is a key element in providing drinking water supplies for human populations. It plays a very important socio-economic role, and any shortcoming will have immediate consequences. In this respect, the protection, maintenance and monitoring of water catchment structures are all crucial factors in the management of the resource.

However, the ageing of water wells is inevitable and leads invariably to a decrease in yield. Therefore, all operators need to address certain questions concerning falling yield, i.e.: the aquifer may no longer be able to provide the initial discharge rate (modifications in standing water-level or conditions of aquifer recharge), or there may be a defect in the pumping system, or even an onset of clogging in the well.

At this stage of water well operation, it is advisable to check on the condition of the catchment structure in order to start any necessary rehabilitation work at the appropriate time.

6.1 Well operation

There are very few studies in the literature dealing with the operation of water wells. Some general remarks are to be found in monographs on water resource development (e.g. notes for water resource operators and water sanitation engineers), but very little of this material is specifically written with catchment works in mind. The major water companies have at their disposal various synthetical documents and policy guidelines destined for in-house use.

In order to develop a water well correctly, it is absolutely necessary to consider that the processes of water capture and pumping are inseparably linked. Under no condition would it be possible to manage one without the other. There are three essential conditions that ensure a suitable management of the particular type of system comprising groundwater catchment and pumping.

a) Adapting the pump to the hydraulic characteristics of the well

It is essential that the catchment structure should be equipped according to its specific characteristics, which are identified in the light of pumping tests and not in relation to the requirements to be met. The overdraft of a catchment structure will lead without fail to serious problems with sanding up, corrosion or clogging. It is appropriate either to carry out supplementary catchment works — whose operating conditions can only be fixed after performing tests — or increase the storage capacity by creating reservoirs.

The pump is a key element of the catchment work. It should be dimensioned in accordance with numerous criteria:

— Characteristics of the distribution system (direct connection of the well to the mains water supply after simple chlorination, or well supplying raw water to treatment plant).

— Equipment of the catchment structure, position of screens, location of the pumping chamber, diameter of equipment, etc.

— Local hydrogeological conditions, position of groundwater level, pumping (dynamic) water-level and predictable pumping regime.

— NPSH (net positive suction head) of the pump, particularly in the case of wells with very low pumping water-levels (thin aquifer, danger of vortex formation).

— Risk of interference with other boreholes within the same well field.

— Geographical location of the catchment work with respect to treatment stations; the fitting of flow limiters notably enables the minimization of groundwater level fluctuations during development.

b) Knowledge of historical well data

A knowledge of the historical (patrimonial) data is a crucial element in good well management. Production parameters for a catchment structure must be made available to the operators. With this in mind, the West Paris regional board of Lyonnaise des Eaux, which continues to manage a total of 150 wells, has made use of a database that provides access to full sets of historical data. All the variables concerning the boreholes within each well field can be made available through paper listings. In particular, these data comprise the following:

— A technical cross-section of the well structure.

— The main physico-chemical properties of the water.

— The standing water-level and the dynamic water-level at different discharge rates (yield-depression curve).

— The specific capacity of the well.

— The position of the pump and its characteristics.

— The maximum yield (limit not to be exceeded during water extraction).

An operator cannot correctly manage water wells without acquiring some information on the initial well field history. In the absence of this fundamental data, monitoring is impossible and, as a consequence, preventive maintenance cannot be carried out.

c) Technical equipment

In order to judge the satisfactory operation of a water well and detect any anomalies, it is necessary to provide the installation with a minimum amount of equipment which should also remain lightweight. In accordance with the general recommendations of water authorities, *the apparatus fitted to pumping equipment* should include:

— A water meter. Whatever its mode of operation (flow velocity meter, volume meter, ultrasound meter, etc.) and whether fitted with an emitter head or not, it should allow measurement of:

- The total pumped volume.
- The instantaneous discharge rate.

— An hours indicator (timer) for each pump, recording the number of operating hours. This serves both to check the useful life of the pump and to calculate the mean yield over a given period from the volume pumped in that time.

— An ammeter for each pump, measuring the current dissipated by the pump. Even when the pumping pressure head varies, this current always remains in the same range and is nearly constant for most of the time. As a result, an anomalous current reading will always be significant, whether it is positive or negative.

— A voltmeter is useful, especially in certain rural areas where the mains electricity supply may undergo drops in voltage from time to time. Insofar as a voltage drop may reflect a surge in current at constant wattage, a voltmeter can be useful for checking voltage stability.

— A manometer installed at or near the well head makes it possible to carry out various checks, notably concerning the satisfactory running of the pump. These checks involve taking readings at different points on the pump's discharge vs. pressure curve, and also ensuring that backflow conditions are normal.

— A device to protect the pump against dewatering. Although they can be of different designs (sensors, power relays, flow-switch, etc.), these devices, in addition to their protective function, also serve as good indicators of resource failure (abnormal frequency of shutting off due to tripping of the protection system).

— Sampling tap. For catchment structures that are not equipped with pumps (i.e. springs and gushing wells), water-level gauging is a very importance part of the checking procedure, along with the possibility of installing an automatic water-level recorder.

In addition, practical experience shows that it is necessary to equip the well with a branch pipe after the pump to allow the evacuation of pumped water without passing through the mains. This arrangement enables, in particular, the implementation of various pumping tests, sterilization of the well, evacuation of water, etc.

As regards the *catchment work* itself:

— A pressure gauge standpipe should be fitted which, provided it contains a level sensor, enables checking of the different levels of the groundwater (static and dynamic).

— A pressure detector (transducer) should be envisaged for transmitting the same hydrometric data as mentioned above, but in a manner suitable for computerized operation.

It is also advisable that the water well should be levelled with respect to the topographic base (NGF in France), in order to be able to place water-level readings in a consistent framework of data.

6.2 Well maintenance

Even when the work has been correctly implemented and protected against external pollution, ageing of the well is inevitable during it's service life. This phenomenon is generally expressed by a fall in yield which, in the absence any curative treatment, can lead to the pure and simple abandonment of the well. It is thus appropriate to attempt delaying the ageing as much as possible.

The diminishing yield of a catchment structure can have the following main causes, which may be independent of each other or linked:

— The degradation of mechanical parts in the well structure (well screens, pump).

— Clogging of the aquifer around the well screen and/or gravel pack.

— Failure linked to the water resource.

— Failure linked to water extraction problems.

Before presenting the causes of falling yield in water wells and the actions that are appropriate in attempting a remedy, it is suitable first to define the principles of good preventive maintenance.

6.2.1 Basic principles

It is only possible to ensure the management of a water well by knowing its main characteristics.

— *Design and construction characteristics.* The essential document in this regard is the technical and lithological cross-section of the borehole (see Chapter V for more details). Without going into unnecessary detail here, it should be recalled that the cross-section contains some important information on the geology of formations encountered during drilling, the various depths of interest in the well (drilled depth, treated or equipped intervals, depth to aquifer), the equipment of the catchment structure, nature and position of the pump, etc.

— *Hydraulic characteristics.* Obtained principally as a result of pumping tests (discharge rate, drawdown, transmissivity and, above all, maximum recommended yield).

— *Water quality.* In accordance with regulations, a water analysis must be carried out at the end of the long-duration pumping tests that follow the construction of a water well.

— *Temporal variation of parameters.* The collation of the initial data with current data (and eventually comparing them with intermediate data) leads to the creation of a database for the water well. If the initial data are lacking, supplementary studies of a very thorough nature must be carried out as soon as possible (downhole logging, pumping tests, etc.) which can then serve as a basis for later comparisons.

A record book with operational guidelines is attributed to each well. In all cases, this record is updated regularly with every incident or action concerning the well since its implementation (electricity cutoffs, influx of abnormal sand, electrical surveys, unclogging, changes of counter, well screens and pumps, etc.).

6.2.2 Regular maintenance

The maintenance of a water well concerns the catchment/pumping system as a whole rather than one or other of these elements taken singly. To achieve this objective, it is necessary to bear in mind a certain number of recommendations:

— All accessible parts of the structure should be kept in a perfect state of repair (protective masonry, enclosure, well head, water evacuation system, visible piping, etc.). In this context, any defect in maintenance represents a major source of pollution for the well.

— The access should be maintained in good condition. The remedial action zone should be kept clean and enclosed.

— Overhaul of the pumping equipment is just as essential as that of the catchment itself. One way of identifying malfunctions in the pumping/catchment system is by maintaining the well hydraulics in good order. In particular, the pump backflow column should be replaced on a regular basis. Complete checks are necessary at least every five years. Checking of the pump may be carried out at the same time (replacement of wearing parts).

— The maintenance of measuring instruments, such as manometers, ammeters, water meters, etc., forms part of the management of water-yield and operational reliability. In fact, the proper functioning of a water well cannot be thoroughly checked without precise metering.

Lack of maintenance is an ill-judged economic choice since it is always reflected in additional financial overheads. This because all actions not carried out at the appropriate time will in any case have to be done eventually, and often in a crisis situation.

a) Maintenance of the pumping system

It might be thought that maintenance of the pumping system has no direct effect on water catchment, but the real situation is entirely different. In point of fact, the jarring of the pump on start up and the number of submersible pumps falling into wells serve to illustrate the impact of pumping on water catchment structures. These commonly encountered problems bring about considerable damage to the well screens and casing. There are even some reported examples of pumps being found at the bottom of wells during downhole camera surveys, in cases where no-one suspected the presence of such objects.

A lack of pump maintenance can have another type of direct impact on water well development, by causing problems with the measuring instruments. If the water meter does not function or if the other instruments are out of order, then monitoring is difficult or even impossible.

Thus, it is necessary to keep to a strict procedure, with both economic and public health goals in mind, to carry out the appropriate maintenance of equipment.

Minor upkeep:

— Maintenance of all surface piping in good condition, as well as cocks, safety valves and fittings (clack valves, anti-water-hammer valves, etc.).

— Checking and carrying out regular changes of metering.

— Maintenance of sterilization facilities in an operational condition.

— Ensuring cleanliness and sealing of the well-head against insects and rodents.

— Cleaning the surroundings and enclosure.

— Regular checking of the electrical equipment used for monitoring and protection.

— Regular cleaning of settling basins and loading chambers in the case of spring discharges.

— Upkeep of drainage within the remedial action zone in order to avoid stagnation of surface water around the catchment structure.

— Checking wells bordering rivers after flooding to see that all is in order and scouring has not occurred around the catchment structure.

Major overhaul:

— Raising the pump every three years to check its condition, and especially replacement of the discharge column if necessary.

The period of three years between overhauls is only an average and it should be assessed on a case-by-case basis. For catchment structures with waters containing hydrogen sulphide or iron, for example, there may be an advantage in bringing up the pump every year. In all cases, a period of five years should not be exceeded, even for pumps that are only used infrequently, so as to prevent corrosion of the column (mostly in the region of the threading) or the falling of the pump down the well. Another factor to be taken into account is the ageing of the cable.

Raising the pump offers the opportunity of measuring the depth to the bottom of the water well. The systematic carrying out of this measurement at each raising of the pump can reveal eventual sediment or sand deposits, thus making it possible to programme the cleaning of the catchment structure (extraction of sediment material) and investigate the origin of the deposit.

— Restoration of masonry before its dilapidation, particularly in the case of spring catchment works, with special attention to access (doors and covers).

b) Maintenance of the catchment structure

The essential components of a catchment structure, apart from springs and large-diameter wells, are only accessible by specifically adapted means. Thus, any action undertaken within the structure generally necessitates a specialized type of operation. As a result, maintenance of the well structure is carried out at two levels, on those parts connected with the surface and those parts buried underground:

— The upper part of the structure requires the maintenance of a seal between the well casing and the well head. Minor problems at this level often give rise to bacteriological pollution. In such a manner, for example, corrosion of the casing near the well-head apron (barely visible due to moisture at the base of the well-head) will cause perforation of the casing and allow the influx of polluted water into the catchment work. Therefore, it is necessary to take care that all parts of the well connected with the surface remain water-tight (by painting, cementation, etc.).

— The buried part, that is, all equipment situated beneath the ground surface, can only be subject to periodic (scheduled) maintenance service.

This involves making the operator aware of the necessity of prevention rather than cure. The specification of the maintenance to be carried out is a function of the nature of the tapped aquifer, knowing that the operation will be less costly if the catchment structure is well designed, correctly developed and kept in good order.

With groundwater catchment from a limestone aquifer, acidification of the well every five or ten years may prove necessary in order to maintain the initial yield. In the case of catchment from a sandy aquifer, air-lift cleaning may be useful under the same conditions. Brushing and thorough disinfection is suitable in other cases where clogging is anticipated.

It is essentially the checking and monitoring of a well that make it possible to define in advance the nature and frequency of the maintenance work to be carried out, each catchment structure having its own specific characteristics.

Example of problem arising from lack of maintenance.
A submersible pump in a deep well was not checked for several years because it was operating correctly. One of the bolts on a flange gave way due to corrosion, and the other three did not stand up to the torsion caused by start-ups. The pump fell to the bottom of the well. The result is as follows:
— Major risk of deterioration of the well screen as pump falls.
— Danger of jamming on a casing collar during lowering of a new pump, thus making its installation impossible.
— Need to recover the fallen pump.
— Water-yield interrupted during the operation.
— Considerably higher cost than if the pump had been checked and changed during preventive maintenance.

6.2.3 Monitoring methods

It should be remembered that a single item of information on its own has itself practically no significance, even if it is anomalous. A diagnosis can only be drawn up from a set of data, hence the necessity of controlling maintenance on a continuous basis with a frequency that should be evaluated and optimized.

a) Quantitative monitoring

There are two types of data that are monitored quantitatively: those concerning the overall water-yield of the catchment structure and those which involve its discharge capacity.

When a water catchment structure is used to supply a community and the requirements subsequently rise, the duration of pumping must be increased since the extraction capacity is not expandable. Thus, it is necessary to follow graphically the changes in average daily pumping time. Two situations may arise:

— The *water extraction capacity* (or water-yield) remains constant. As long as the pumping time does not exceed 20 hours per day, the situation is not alarming. At more than 20 hours pumping, it is appropriate to try and boost resources. Experience shows that it is necessary to avoid pumping a water well for more than 20 hours out of 24. In fact, the aquifer is able to re-establish its equilibrium during a few hours recovery. However, if the pumping is uninterrupted over a long period, the entire area of the cone of depression remains dewatered. This phenomenon is sometimes so marked as to bring about oxidation or scaling.

— The *pumping time* increases without significant rise in water-yield. This corresponds to a reduction in the total capacity of the structure (water catchment or pumping).

Monitoring the evolution of water-yield is one of the key elements in catchment management. Moreover, it is essential to check and monitor yield since this measure reflects the state of the water catchment. In particular, yield is defined by its specific capacity; a steady value indicates that the catchment structure is maintaining a constant total capacity, whereas a falling value indicates a loss of total capacity.

As a general rule, it is considered that if specific capacity falls to a value more than 10% beneath the initial value, ageing of the catchment structure will set in leading to a fall in discharge capacity.

Even if there is no time-variation in specific capacity, it is nevertheless necessary to envisage a new step drawdown step pumping test every two years (which otherwise provides an opportunity to clean the pump and check its mechanical condition). It should be carried out at the same period in the year as the initial pumping test in order to eliminate as far as possible the effects of seasonal fluctuations in groundwater level.

The frequency of capacity measurements is adapted according to the use of the catchment structure and the extent of the well field. Among other parameters, care should be taken to note the following:

— The dynamic level (or static level if pumping has been halted due to a pumping problem or electricity cut off), with an indication of time, date and operator, always using the same measurement datum.

— Instantaneous discharge rate at the meter.

— Cumulative volume at the meter.

— Electricity mains current.

— Manometer pressure at head of pumping column.

— Standing water-level in any nearby disused wells, or recorded on water-level gauges (piezometer or observation wells).

It should be realized that a suitably monitored water well always provides some warning of future failure. In this context, discharge capacity tests carried out on a regular basis are the most efficient means of ensuring that the yield characteristics of the well are stable in time. The example given below (cf. figure 6-1) presents the yield-drawdown curve for a water well in the course of clogging.

Example of warning signs: During the second year of well development, it is noted that the curve indicates that, for a stable discharge capacity of 30 m³/hr, the drawdown increases from 2.40 m to 3.20 m; this is a worrying sign characteristic of a loss in total capacity of the well (increasing loss of head).

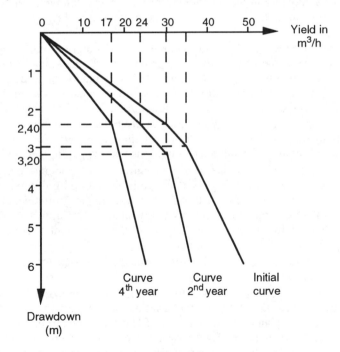

Figure 6-1
Variation of yield-drawdown curves for a well during clogging.

In the fourth year, it is no longer possible to obtain 30 m³/hr without dewatering the pump, which now lies 6 m beneath the standing water-level.

The well is then clogged. At the original pumping water-level of 2.40 m, the safe yield is only 17 m^3/hr.

In this way, owing to a simple graphical method, it is possible to monitor the evolution of a water catchment structure and to prevent a failure before difficulties arise.

b) Qualitative monitoring

It is necessary to collect samples of water for extensive physico-chemical and bacteriological analysis, with the date, hour, discharge capacity and duration of pumping prior to sampling. The frequency of these measurements should be at least three to four times per year, in accordance with the importance of the resource under consideration. In other respects, the frequency of measurements is subject to regulatory requirements (DDASS in France).

The regular monitoring of these characteristics offers a twofold advantage:

— On the public health side, it enables the detection of eventual changes in water quality and the identification of pollution indices.

— On the technical side, changes in certain parameters can lead to the detection of an anomaly in the functioning of the water catchment structure or its near-field environment, and provide a better understanding of its recharge mechanism.

The *bacteriological quality* of the sampled water is a good indicator not only of the natural protection of the groundwater but also the technical protection of the catchment structure. Any perforation of the casing or degradation of the cementation is reflected by an influx of impure surface waters. It should be borne in mind that several concordant analyses are required for a complete report.

The elements most commonly involved in groundwater catchment failure are iron and manganese. Thus, the evolution of their concentrations needs to be monitored as a function of operational conditions and the geological environment.

Above all, care is taken to check the concentrations of iron and manganese in the pumped water every time there is a change in the pumping regime (increases in discharge capacity) or during each abnormal variation in groundwater level. In fact, high concentrations of iron and manganese often appear under such conditions, thus leading to a risk of clogging of the catchment structure and the need to implement costly treatment procedures.

Two other parameters are of interest here, the electrical conductivity and temperature of the water.

— Conductivity generally shows little variation with time, and corresponds to the concentration of dissolved salts. Its variation can be taken as an indicator of intrinsic (structure-related) modifications of the well recharge. For example, communication between several groundwater bodies may occur at the same well due to corrosion of the casing or cementation. The monitoring of conductivity is of fundamental importance in the case of a water well that is developed in a coastal aquifer. A strong increase of conductivity is indicative of the intrusion (or encroachment) of brackish or even saline water into the well.

— The monitoring of temperature makes it possible to reveal changes in the origin of the water, but also contributes to an understanding of the behaviour of the well in relation to its environment, particularly with regard to problems of recharging.

Conductivity and temperature data are very easy to acquire, and do not represent a major investment in financial terms.

c) Checking of constituent materials

Groundwater wells are made up of various materials such as concrete, PVC, steel, etc., all of which undergo degradation. Because of this, regular checks are necessary. Thus, the frequency of such checks should be established at the time of construction. The design engineer is well equipped to know the mechanical, hydraulic and physico-chemical conditions to which the constituent materials of the catchment structure will be subjected.

These checks concern the following parts of the structure:
— Non-catchment components:
  • Condition of piping (distortion, tearing, perforations, thicknesses).
  • Condition of cementation (bond separation, disintegration, cracks).
  • Quality of bonding between different structural components.
— Catchment-related components:
  • Internal condition and position of well screen.
  • Homogeneity of the gravel-filter (pack).

It is important to check the state of the pump. To achieve this, a few discharge/pressure tests are required which also allow the detection of discharge capacity losses and pump wear.

Finally, the environment of the water catchment should also be monitored, particularly as regards human activities near the well and a strict application of the different protection zones.

6.2.4 Monitoring equipment

Failures of groundwater catchment are not easily detectable. As described above, the warning is given for the most part by indirect parameters (loss of discharge capacity, variation of physico-chemical and bacteriological characteristics). In such cases, it is often too late to avoid failure and the solutions become problematic. It is thus crucial to have a certain amount of equipment for monitoring:
— On the pumping system:
  • A water meter for measuring the instantaneous discharge rate and the cumulative pumped volume.
  • An hours indicator for each pump totaling up the number of operating hours.
  • An ammeter on each pump.

- A manometer.
- A water sampling tap.
- A sand trap to facilitate inspection visits.
- A bypass branch pipe to carry out discharging-well tests without backflow into the mains.

— On the catchment structure itself:

- A standpipe containing a water-level gauge.
- Electrodes for protecting the pump against dewatering.

— At the station:

- A chronometer.
- A water-level gauge.
- An accurate thermometer.
- A water analysis kit

— Off-site

- One or several water-level gauges (observation wells).

All the measurements should be reported in the record book at the station.

It is evident that the more data is available, the better is the possibility of detecting eventual anomalies. However, the frequency of measurements is a function of the size and importance of the catchment structure. In the case of a catchment field comprising several wells, the observations may be recorded on a daily basis. By contrast, weekly records are quite sufficient for a single well with a small discharge rate.

Certain types of check, which are more important by the nature of the work involved, should be carried out with a frequency of between three and five years. This applies, for example, in the case of raising the pump for maintenance and also changing the backflow column. The same applies to the video camera inspection and downhole logging surveys described in previous chapters.

The frequency of measurements should be established with care by a specialist (particularly taking into account the age and condition of the catchment work) and then scrupulously adhered to by the operators.

6.2.5 Frequency of measurements

In addition to regular observations, it is necessary periodically to carry out certain detailed investigations. It is broadly desirable to respect and, above all, make financial provision for a schedule of periodic checks on all developed wells in a particular field. In fact, the phenomena liable to effect groundwater wells do not manifest themselves in the same way in all cases, and they are not always easily detectable. The corrosion of part of the catchment structure, for example, is not always accompanied by warning signs before perforation occurs. It is only once the casing or well screens are perforated that certain indications make it possible to infer corrosion and perforation have actually taken place. So as to favour prevention rather than cure, it is advisable to take systematic measurements during periodic checks. A schedule of checks should comprise the following operations, although the list is not exhaustive:

Every year:

— A quantitative check with a pumping test.

— A qualitative check, involving a complete physico-chemical and bacteriological analysis of the water.

— A full report of water well development data, i.e.:

- Number of hours of pumping.
- Pumped volumes.
- Mean monthly discharge.
- Electric power consumption normalized to number of cubic metres pumped.
- Changes in standing and pumping water-levels.
- Comparison of pumping test results with initial data.

All these observations are analysed and compared with data from previous years in such a way as to identify possible trends. In particular, this analysis is characterized by the study of two fundamental parameters, the *total capacity*, whose stability or loss indicates a favourable situation or otherwise of the catchment work, and the *water quality*, which provides a similar type of indication.

Simple graphical methods are used, for example, to present the variation of specific capacity with time, and can be developed as tools for providing valuable information.

All these data may be entered into a well database, or failing that, archived so they are available and can facilitate maintenance of the catchment structure.

Every 3-5 years

According to the type of catchment structure and its exploitation regime, it is suitable to carry out certain checks every three to five years. The implementation of pumping tests, physico-chemical analyses, downhole logging and a video camera inspection would represent an optimum scenario.

Other types of action may prove necessary in certain contexts, i.e.: temperature profiles, acidification, addition of gravel, corrosion checks, sterilization, etc. The results of these checks are then compared with historical and archived data.

From the point of view of planning, it is a good idea to combine these operations with servicing of the pump. In financial terms, such checks represent an insignificant cost compared with the importance of the indications that they provide.

6.2.6. Summary

A water well that is functioning correctly is subject to minimal unplanned incidents (hazards) during its operation and has a defined maximum service life. To obtain the best operating conditions, it is simply necessary to respect the basic principles of good maintenance and management of the catchment structure:

— Adoption of a pumping schedule in relation to the characteristics and potential of the catchment work. As far as possible, it is advisable to avoid over-frequent start-ups and continuous pumping for uninterrupted periods of 24 hours.

— A pumping (or dynamic) water-level should be kept which at all times avoids dewatering of the well screen. The dewatering of well screens, which is a major cause of corrosion, may also facilitate the development of a clogging biomass.

— The maximum yield of a well should not be exceeded.

— Sufficient equipment should be available to ensure continuous monitoring.

— Regular upkeep of the well structure.

— Hasty conclusions should not be drawn concerning catchment works, and even more importantly, actions should never be undertaken in view of an anomaly without the advice of an experienced specialist.

6.3 Ageing of well structures

Despite all the precautions and the regular checking that are used to protect catchment structures, it is impossible to maintain them in good condition indefinitely. Thus, ageing is an inevitable phenomenon that is accompanied by several effects:

Corrosion:

— Electrochemical corrosion.

— Bacterial corrosion.

Clogging:

— Mechanical clogging;

— Chemical clogging.

— Biological clogging.

An enquiry carried out in France by GEOTHERMA in 1991 shows that the most common causes of well degradation are sanding up, scale formation and corrosion. Moreover, an analysis of the results of this enquiry reveals that 25% of all problems arising in water wells are due to the decay of the structure, while a further 25% are linked to abnormal development and 25% are caused by poor design. This means that *factors of human origin are involved in 75% of problems* and only 25% are associated with the characteristics of the water or aquifer. Finally, it should be pointed out that 30% of wells are less than 10 years old, as compared with 70% which are between 10 and 50 years old.

As a whole, water wells in France are old and in decay. A study carried out by the BRGM in the Nord-Pas-de-Calais region shows that the average age of wells is 46/47 years. In addition, it may be noted that out of a total of 1,136 recorded domestic water supply wells, only 582 are currently being operated. In this latter group of wells, 443 (76%) are more than 30 years old. This ageing of wells is also seen in the area managed by the Artois-Picardie Water Authority, where many of the catchment structures are more than 60 years old.

The age of wells is a worrying problem, especially since it does not seem to have been taken into account by a national policy for renewal. This state of affairs is even more disturbing because water industry professionals are agreed that structures subject to preventive maintenance run a serious risk of failure which increases with ageing.

6.3.1. Corrosion phenomena

The corrosion of water wells is far more insidious than clogging since its effects are often less detectable. On the other hand, its consequences are just as spectacular and important for the lasting quality (endurance) of the structure in question.

Corrosion can be an electrochemical phenomenon, although in some cases it takes place by purely chemical dissolution or even by biological processes. It should not be confused with the physical phenomena of erosion/abrasion or cavitation (separation of the fluid from the casing, formation of gas pockets, entrainment of fine bubbles which then implode to produce ultrasonic shock waves that tear up the metal surface).

— Erosion/abrasion is due to the kinetic energy of particles of sand or other detrital matter present in water which leads to the continuous disintegration of solid metal through regular and homogeneous abrasive action. The process is at the same time both mechanical and electrochemical.

— Cavitation is due to local changes arising locally in the vapour pressure which cause the release of water vapour bubbles followed by their sudden collapse. The implosions so produced take place at extremely high pressures, which accounts for the heterogeneous and cavernous disintegration of the metal.

The velocity of water circulation plays a very important role in this type of degradation. Corrosion is an extremely widespread phenomenon. According to its origin, a distinction can be made between electrochemical and bacterial types of corrosion. It can also appear in many different forms (e.g. developed over wide areas, pitting, hollowing out of the metal, cracks, etc.). If not detected in time, corrosion will lead to very serious constraints regarding rehabilitation, and even in some cases the complete reconstruction of the catchment work.

Corrosion is a physico-chemical phenomenon tending to break down a material that is out of equilibrium with its ambient medium. In water wells, corrosion can affect all the metallic and non-metallic parts (particularly those components based on cement). Only organic, plastic or bituminous materials are able to resist corrosion.

Corrosion is a complex phenomenon that is generally attributed to several concurrent causes:

— Presence of corrosive water within or just outside a well.

— Electrolytic cells set up between different parts of the same metallic structure in contact with waters of different composition, or between different metals connecting together but in contact with the same water.

— Activity of siderophilic and sulphate-reducing bacteria.

These corrosion phenomena can lead to the following effects:

— Simple dissolution of the metal over a wide area (uniform corrosion).

— Attack of metal localized on very small surfaces, in some cases at depth, giving rise to perforations (pitted corrosion).

— Attack of metal in troughs (hollows), appearing especially on screens composed of rolled (laminated) sheet or drawn tubes, with or without punched holes (intergranular corrosion).

— More or less branching linear cracks (cracking corrosion).

The casing and well screens are subject to different types and degrees of corrosion, being a function of their different physical characteristics and position in the well. Four zones of differential corrosion can be distinguished in water wells, i.e.: atmospheric zone (casing above the standing water-level), oscillation zone (between standing and pumping water-levels), submerged zone (beneath the pumping water-level) and external zone (casing in contact with the exterior of the well).

— In the *atmospheric zone*, condensation of water vapour represents the main factor of corrosion, which is also promoted by the presence of carbon dioxide and oxygen. Recent experiments have shown that the addition of small amounts of copper (0.2%) to the carbon steel used in the casing will increase resistance to corrosion by a factor of two.

— The *water-level oscillation zone* is the main site of corrosion, which arises essentially from differences in oxygen concentration evidently produced by alternating periods of submergence and emergence in the well. The fitting of stainless steel casing provides the best possible protection against corrosion. Steel type 304 (an alloy with 18% chromium and 8% nickel) has proved its qualities in this respect, whereas type 316 (with 18% chromium, 12% nickel and 2.5% molybdenum) is to be recommended.

— In the *submerged zone*, corrosion is generally less intense and alloy steel casings are not always necessary. Of course, in the case of aggressive waters it is advisable to fit stainless steel casings. Well screens are more exposed to corrosion, in view of their surface area and the velocity of water circulation, so it is recommended to use a stainless steel of type 304 or 316.

— Corrosion of the casing in the *external zone* can be avoided by carrying out cathodic protection, or alternatively, the problem may be circumvented by means of a cement casing.

Several types of phenomenon are involved in corrosion, including the electrochemical and bacterial processes described below.

a) Electrochemical corrosion

Taking account of the underlying thermodynamic principles which control corrosion (e.g. redox potential), the rate of corrosion is influenced by both the chemical and physical properties of the borehole water. Electrochemical corrosion is produced or favoured by the presence of gases and dissolved salts. The main physical factors involved are temperature and flow rate of the water. Dissolved gases represent the most important chemical factor. Among the three dissolved gases most commonly encountered, oxygen has the highest activity in terms of its effect on casings.

Generally speaking, electrochemical corrosion is brought about or promoted by the presence of one or more constituents in the water, which include the following:

— Acidic waters (having a pH of less than 7).
— Dissolved oxygen.
— Hydrogen sulphide.
— Carbon dioxide.
— Chlorides.
— Calcium sulphates (gypsum, etc.).

TABLE VI-I— *Electrochemical series: standard oxidation-reduction potentials.*

Reaction	Eo (V)
$Au = Au^{3+} + 3e^-$	+1.498
$O_2 + 4H^+ + 4e^- = 2\,H_2O$	+1.229
$Fe^{3+} + 1e^- = Fe^{2+}$	+0.771
$4OH^- = O_2 + 2H_2O + 4e^-$	+0.401
$Cu = Cu^{2+} + 2e^-$	+0.337
$Sn^{4+} + 2e^- = Sn^{2+}$	+0.150
$2H^+ + 2e^- = H_2$	0.000
$Pb = Pb^{2+} + 2e^-$	− 0.126
$Sn = Sn^{2+} + 2e^-$	− 0.136
$Ni = Ni^{2+} + 2e^-$	− 0.250
$Fe = Fe^{2+} + 2e^-$	− 0.440
$Zn = Zn^{2+} + 2e^-$	− 0.763
$Al = Al^{3+} + 3e^-$	− 1.662

N.B. The equilibrium potential Eo is measured with respect to a standard H_2 electrode, at 25°C (metal immersed in a solution of one of its salts).

When waters are corrosive with respect to the metallic parts of the well, these will function as an anode and acids present in the water will tend to dissolve the metal over its entire exposed surface. The main acid species responsible for this are H_2S and H_2CO_3, while HCl may sometimes be formed from other reactions. Although attack by H_2S is usually limited by the low concentration of this compound in natural waters, the activity of sulphate-reducing bacteria may produce larger quantities that can lead to corrosion by pitting.

As pointed out in the studies of P. MOUCHET, metal corrosion generally arises from the fact that metals are unstable in water and can thus be expected to undergo dissolution. As a result, metal cations situated at the interface tend to migrate towards the fluid phase, leaving electrons behind in the solid metal phase to balance the positive charges on the dissolved metal species. As the process continues, the water accumulates a positive charge and the metal surface a negative charge, which renders the migration of cations increasingly difficult (since they are repelled by the water and attracted by the metal). At the end of the process, an equilibrium is established and dissolution ceases; the potential set up between the water and the metal is termed the equilibrium potential of dissolution. A scale of potentials can be established with respect to a standard hydrogen electrode which enables a ranking of metals according to their capacity to pass into solution. This scale, otherwise known as the Nernst electrochemical series, is summarized in Table VI-I.

The Nernst equation is expressed as follows:

$$E = E_o + \frac{2.303\ RT}{nF} \log \frac{\left(Fe^{2+}\right)^2}{[O_2(aq)]\left(H^+\right)^4}$$

where:

Eo : standard electrode potential (V),
R : universal gas constant,
T : absolute temperature (°K),
n : number of electrons transferred,
F : 96,494 coulombs per mole (Faraday's constant).

In a general way, the equilibrium potential of dissolution E can be expressed as a function of the standard electrode potential Eo, as defined in the Nernst equation and given in the right hand column of Table VI-II, using the following simplified relation:

$$E = Eo + \frac{0,058}{n} \log c$$

where:

n : valency of considered metal ion,
c : concentration of metal in solution.

In the case of iron, we obtain:

$$E = -0,44 + 0,029 \log (Fe^{2+})$$

However, the situation does not remain as simple as suggested by this equation, otherwise corrosion would be restricted to an infinitesimally thin layer of metal. In fact, the process of metal dissolution described above may act in opposition to another process which takes place on more unreactive (or passivated) patches of the same metal immersed in the same electrolyte. The passivated electrode determines the setting up of a potential difference in the medium and produces an electric current which maintains the dissolution reaction — or corrosion — of the metal electrode.

Corrosion processes proceed in contrasting ways according to whether oxygen is present or absent.

Corrosion in an de-aerated medium (hydrogen corrosion)

The antagonist process is an electrode reaction in which hydrogen gas is formed from H^+ ions present in the water (cf. figure 6-2). This electron-accepting process is able to sustain the iron dissolution reaction described above, since it is itself an electron donor reaction. The overall reaction can be written as follows:

$$Fe^o + 2 H_2O \rightarrow Fe(OH)_2 + H_2$$

In a strongly acidic medium, the hydroxide-forming reaction cannot take place, so we obtain:

$$Fe^o + 2 H^+ \rightarrow Fe^{2+} + H_2$$

This process of hydrogen corrosion, which is incorrectly called chemical corrosion, is thus essentially electrochemical in character. It causes the formation of a positive and a negative pole (arbitrarily termed cathode and anode, respectively) between which an electric current will flow. Metal dissolution or oxidation occurs at the anode, while the cathode is protected against corrosion.

Dissolution under these conditions can only affect metals whose effective half-cell potential in the electrolyte is lower than the potential of the hydrogen cathode that forms the other half-cell. Thus, hydrogen corrosion can become established:

— On a more noble (electropositive) metal than the anode.
— On foreign impurities (oxides, dirt).
— On irregularities in the crystalline texture of the metal, where cold-rolled or work-hardened domains become anodic with respect to the metal itself.

Corrosion will continue to develop indefinitely as long as it is not limited by a lack of H^+ ions at high pH or by saturation of Fe^{2+} in solution. This latter process leads to the precipitation of ferrous hydroxide, forming a protective deposit that theoretically stops the corrosion.

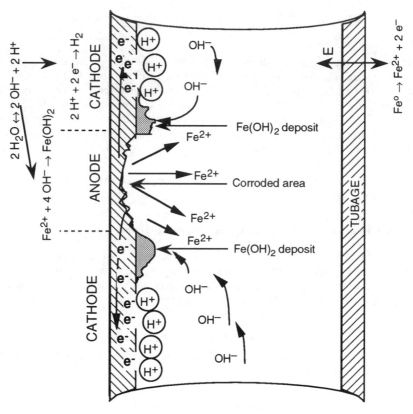

Figure 6-2
Electrolytic corrosion process in a de-aerated medium.

Since flow of the electrolyte will carry off the precipitate, hydrogen corrosion, in principle, can only be halted if the water is stagnant. E_{Fe} increases because the iron passes into the solution, while E_{H_2} diminishes due to the liberation of hydrogen gas. The reaction should cease when $E_{Fe} = E_{H_2}$, which, using the equations given previously, gives the expression:

$$-0.44 + 0.029 \log (Fe^{2+}) = -0.058 \, pH$$

After rearranging, this gives:

$$\log (Fe^{2+}) = 15.1 - 2 \, pH$$

Considering a maximum concentration of 10^{-6} molar Fe for most waters, it may be concluded that there is practically no stability field for iron existing in aqueous solution beneath pH = 10.5. "Acid corrosion" is increasingly developed with falling pH and Fe^{2+} ion concentration.

Finally, although corrosion is very strong under acidic conditions (low pH) it becomes less hazardous at neutral pH because the hydrogen ion concentration is then insufficient to maintain the cathodic reaction. As a result, the process of formation of a protective layer can proceed.

It is noteworthy that the "acid corrosion" process leads to the extensive removal of material. From the morphological point of view, hydrogen corrosion is expressed as a relatively uniform corrosion of the metal, this being due to the presence of unlimited (virtually infinite) numbers of coexisting cathodes and anodes.

Oxygen corrosion

In the presence of aerated water, which is generally the case with domestic water supply for consumption, the particular half-cell reaction (gas electrode process) accounting for corrosion involves dissolved oxygen, i.e.:

$$O_2 + 2 H_2O + 4 e^- \rightarrow 4 OH^-$$

This gas electrode is capable of accepting electrons, so it behaves as a cathode (cf. figure 6-3).

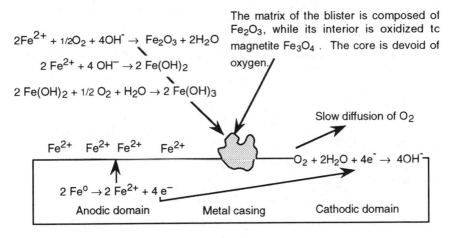

Figure 6-3
Formation of a corrosion blister (lobe) at a point with high oxygen concentration.

For the oxygen half-cell reaction, the equilibrium electrode potential is a function of:

— OH^- ion concentration, hence also the pH.
— Partial pressure of oxygen.

$$E' = E_o - \frac{RT}{F} Ln\left(OH^-\right) + \frac{RT}{4F} Ln\, P_{O_2}$$

If this potential is higher than the metal half-cell potential, then corrosion will result as in the case of iron. The higher the dissolved oxygen concentration, the higher also becomes the gas electrode potential.

In an apparently paradoxical manner, the failure of oxygen to irrigate part of the metal surface will render it anodic, thus corrodable with respect to the rest of the surface, which is itself protected by the presence of oxygen. From this, it is possible to appreciate the undesirable effects of deposits of any kind that hinder the diffusion of oxygen towards underlying zones, thus causing the development of anodic domains.

At the oxygen cathode, the release of OH^- ions will increase the pH of the water, at least in immediate proximity to the metallic surface. Furthermore, Fe^{2+} ions in the presence of oxygen are oxidized to Fe^{3+}. In this context, ferric hydroxide $Fe(OH)_3$ is a sparingly soluble reddish brown compound that forms as a product of corrosion. Therefore, instead of being carried away by the water and leaving a clean surface behind, as is the case with corrosion in a de-aerated medium, the products of corrosion in an aerated medium will accumulate around the anode.

Corrosion products make up the familiar types of scale and blisters that hinder oxygen diffusion to an ever increasing extent and enhance the anodic character of the affected surface. This accounts for the perforations so typical of oxygen corrosion.

In summary, the overall reaction for corrosion in an aerated medium can be deduced from the successive reactions involved (already indicated on figure 6-3), which yield the following:

— At the anode:

$$4\, Fe^o \rightarrow 4\, Fe^{2+} + 8\, e^-$$

— At the cathode:

$$2\, O_2 + 4\, H_2O + 8\, e^- \rightarrow 8\, OH^-$$

Thus, we can write:

$$4\, Fe^{2+} + 8\, OH^- \rightarrow 4\, Fe(OH)_2$$

$$4\, Fe(OH)_2 + O_2 + 2\, H_2O \rightarrow 4\, Fe(OH)_3$$

$$4\, Fe^o + 3\, O_2 + 6\, H_2O \rightarrow 2\, Fe(OH)_3$$

The ferric hydroxide so formed may then be converted into oxide species according to the following reactions:

— which corresponds to the formation of ferric oxide by a process of partial dehydration, thus giving rise to common rust, a substance normally composed of goethite (Fe_2O_3, H_2O).

$$2\, Fe(OH)_3 \rightarrow Fe_2O_3 + 3\, H_2O$$

— which corresponds to the formation of magnetite (Fe_3O_4) by reduction in contact with hydrogen liberated from the cathode.

$$3\ Fe(OH)_3 + 1/2\ H_2 \rightarrow Fe_3O_4 + H_2O$$

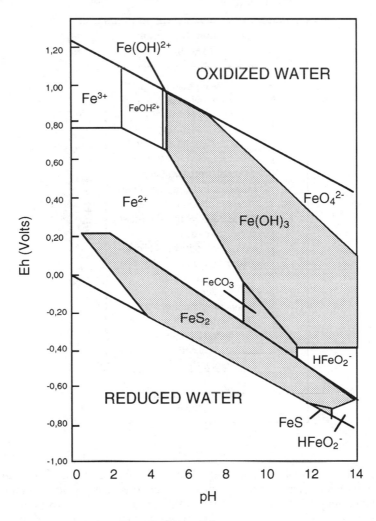

Figure 6-4
Stability fields for solid and dissolved species of iron as a function of Eh and pH at 25°C and 1 atmosphere pressure (Pourbaix diagram).

Once corrosion has started in a well, the oxygen concentration becomes "self-perturbing" (internally buffered). In fact, as soon as a metal oxide scale (blister) forms around a small pit, the oxygen concentration becomes distinctly lower underneath the blister. As a result, the blister becomes a cathode while the casing corresponds to the anode. In most cases, the blister is a highly consolidated structure composed of magnetite (Fe_3O_4) or goethite (FeOOH) or even both together. When the blister contains magnetite, it behaves as an excellent conductor. Corrosive phenomena introduce ferric ions into the blister and bring about the salting out of hydroxides.

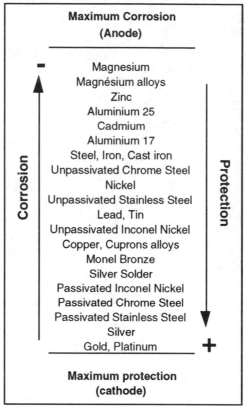

Maximum Corrosion
(Anode)

−

Magnesium
Magnésium alloys
Zinc
Aluminium 25
Cadmium
Aluminium 17
Steel, Iron, Cast iron
Unpassivated Chrome Steel
Nickel
Unpassivated Stainless Steel
Lead, Tin
Unpassivated Inconel Nickel
Copper, Cuprons alloys
Monel Bronze
Silver Solder
Passivated Inconel Nickel
Passivated Chrome Steel
Passivated Stainless Steel
Silver
Gold, Platinum

+

Corrosion

Protection

Maximum protection
(cathode)

Figure 6-5
Electrochemical series.

Corrosion taking place under water is greatly enhanced if the immersed metal is not completely clean, or if it is unequally covered with scale, or made up of parts containing different types of metal that are not insulated from each other. This phenomenon is even more marked when the water is rich in dissolved salts, which increases the electrical conductance.

As mentioned previously, metallic iron is always out of equilibrium with the well water, whatever its quality. Thus, although iron will tend to be dissolved, the actual process of dissolution is a function of the solubility of the hydrolysis product, which is itself chiefly dependent on pH and Eh (cf. figure 6-4). From this diagram, it can be seen that dissolution takes place within the Fe^{2+} field since this ion is soluble, whereas scale formation is favoured within the stability field of $Fe(OH)_3$, which is insoluble.

These phenomena are known by virtue of their overall observed effects, without their underlying causes ever having been determined separately in the field. This is because corrosion phenomena are variably associated and related with the various forms of clogging discussed previously, in particular through the activity of siderophilic and sulphate-reducing bacteria.

The intensity of corrosion is very difficult to assess in advance since too many different phenomena come into play. Various measurement techniques may be envisaged, including flow meter logging (if the survey can be carried out in the course of pumping at normal discharge capacity) and self-potential logging with or without pumping. The overall electrofiltration potential is easily measured between the well head and a reference electrode placed at some distance from the well, by calculating the difference between measurements taken with and without pumping. To avoid perturbation of the electrical measurements, the water well should be of the flowing artesian type or the pump motor should be mechanical rather than electric.

Otherwise, the effects of the natural currents discussed above are often added to stray currents that arise notably in proximity to pipe-lines, metallic conduits and railway lines. In such cases, corrosion can be considerably enhanced.

In this context, the electrochemical series shown in figure 6-5 indicates, for example, that steel has a tendency to corrode more rapidly than copper, with the former acting as an anode and the latter as a cathode. The first metal in the series is the most reactive, hence the most susceptible to corrosion, while the last metal is the most difficult to attack and is thus chemically noble.

Any metal in this list behaves as an anode with respect to all the metals below it and as a cathode with respect to all metals above it. However, it should also be stressed that corrosion is even more pronounced if the two metals are far apart on the electrode potential scale (i.e. well separated in the electrochemical series).

b) Bacterial corrosion

The presence of bacteria in water can bring about the attack of metals, especially iron and manganese. By eating away at metal surfaces, bacteria produce a sort of viscous slime in which they proliferate. This slime contains particles of disaggregated metal, and the bacterial activity continues at depth. While excrescences form in screen slots that are more or less obstructed, the metal beneath these deposits is corroded. Although this phenomenon may superficially ressemble scale formation, after removal of the deposit by water-jet or by increasing the hydraulic flow rate during pumping, the bared metal will show a reduction in its surface area due to the action of microorganisms (bacteria). In such situations, well screens are prone to disintegrate more or less rapidly.

Iron and manganese bacteria are present in most aquifers (cf. figure 6-6). The biological attack of metals is generally due to the following bacteria:

— *Gallionella*: iron oxidation.
— *Leptothrix ochracea*: iron oxidation.
— *Toxothrix trichogenes*: iron oxidation.

— *Leptothrix lopholea*: manganese oxidation.
— *Metallogenium*: manganese oxidation.
— *Hyphomicrobium*: manganese oxidation.
— *Siderocapsa*: manganese oxidation.
— *Siderocystis*: manganese oxidation.

A model for biological corrosion is presented in figure 6-7, while the reactions due to iron bacteria and sulphate-reducing bacteria are summarized in figure 6-8.

— Sulphate-reducing bacteria are widely present in waters, muds and soils; these biotopes in almost all cases contain sulphates which act as final acceptors for electrons. Apart from phosphorylation of the substrate in certain cases, this process represents the only way that bacteria can obtain energy in the form of ATP as a result of electron transfer. Sulphate-reducing bacteria are obligate anaerobes which can only develop in the absence of oxygen and at low redox potential ($< - 100$ mV). The common feature of these species is their ability to reduce sulphates to sulphide by means of a catabolic (dissimilation) reaction.

— Iron bacteria are active in the transformation of iron present in organic complexes and mineral matter, which exists either in the oxidized (insoluble) or reduced (soluble) form:

- The complexation of iron takes place in soil due to the activity of microorganisms. Iron complexed in this manner is taken up into solution and migrates though the soil profile.
- The mineralization of complexed iron is carried out by chemico-organotrophic microbes that make use of the organic part of the iron complex as a source for carbon or nitrogen, thus releasing mineral iron which is then precipitated.
- The reduction of mineral ferric iron can take place through the intermediary of numerous ordinary chemico-organotrophic bacteria. This reduction only occurs under anaerobiosis, and the ferric iron thus serves as an electron acceptor. In the soil, the reduction of ferric iron is a function of the consumption of glucides and organic acids.
- The biological oxidation of mineral ferrous iron requires many different types of bacterium.

The iron bacteria are microorganisms that are even nowadays still poorly known. It is necessary to distinguish several cases as regards their relation to iron and whether they are autotrophic or heterotrophic. Autotrophic iron bacteria, such as *Gallionella* and *Thiobacillus (ferroxidans)*, use iron as a source of energy. In this case, the relation is obligate (strict) and the bacterium cannot grow without ferrous iron; these genera belong to the iron bacteria *sensu stricto*. A second group, comprising the heterotrophic iron bacteria *Leptothrix*, *Sphaerotilus* and *Clonathrix*, do not use ferrous iron as an energy source. Finally, the siderocapsacean bacteria represent a still poorly known group that has not been cultured, encountered near the limit of stability of ferrous iron (in particular, at low partial pressures of oxygen). Under such conditions, it is possible that these bacteria can draw part of their energy from the oxidation of ferrous iron [J.L. GOUY *et al.*, 1984].

The bacteria most frequently mentioned in cases of ferric clogging are *Gallionella, Sphaerotilus, Siderocapsa* and, to a lesser extent, *Toxothrix, Crenothrix, Clonothrix, Siderococcus* and *Naumaniella*. Some of the clogging bacteria (siderocapsaceans, *Gallionella, Thiobacillus*) may be involved in initiating the formation of ferric "nodules", and hence also the genesis of iron ore deposits. Knowing the conditions favoured by these bacteria, and taking account of the pH and Eh as well as the concentrations of iron and organic matter in the water, it is possible to assess the risk of clogging at different aquiferous sites. Four main groups are responsible for the oxidation of ferrous iron in mineral matter:

— Sliding bacteria of the genus *Toxothrix*. These bacteria are widely distributed in ferruginous waters that are relatively cold and poorly oxygenated, but iron does not appear to be indispensable for their growth. They are composed of filamentous cylindrical cells, often U-shaped, which move slowly by sliding over their own excreted mucus.

— Sheathed bacteria:

  • Genus *Leptothrix*, abundantly present in unpolluted waters subject to weak currents, forming considerable deposits of mucilaginous substances impregnated with ferric hydroxide. The waters where they develop show neutral or nearly neutral pH, being rich in carbon dioxide and containing little oxygen or iron (ca. ≈ 2 mg Fe(II) per litre). No growth is observed above concentrations of 12 mg Fe(II) per litre. The bacteria occur in the form of rods and beaded chains which are surrounded by a mucilaginous (or slimy) sheath impregnated with ferric oxide.

  • Genus *Crenothrix*, widely found in water supply systems, drains and wells where the water contains iron, being responsible for the obstruction of numerous catchment works. The bacterial cell is disk-shaped or cylindrical, arranged in beads (tricomes) and surrounded by a sheath encrusted by iron or manganese oxide at the base. The tricomes can attain a length of 1 cm and are attached onto a support; the base of the tricome measures 1.5-5 mm in diameter, while the extremity expands to 6-9 mm.

— Pedunculate (pedicled) bacteria:

  • Genus *Gallionella*, comprising aquatic bacteria which develop in cold waters under conditions identical to those favoured by *Leptothrix*, forming an ochreous and mucilaginous biomass in the water. They cause the obstruction of drains and wells, and also the corrosion of iron conduits. The oxygen concentration is a determining factor in their growth, which is stimulated at 0.1-0.2 mg/l but inhibited at levels above 2.75 mg/l. These bacteria occur in the form of spiralled filaments wound in a double helix, and have an ochreous colour; they form pedicles composed of fibrilla that are impregnated by ferric hydroxide. In liquid media, they develop as yellowish microcolonies incorporating carbon dioxide.

- Genus *Pedomicrobium*, which are soil bacteria occurring as round, oval or rod-like cells with filaments. Both the cells and filaments are impregnated with ferric hydroxide or manganese oxide.

— Gram-negative chemico-lithotrophic bacteria:

- *Thiobacillus ferroxidans*, which is a bacterium found in rocks and acidic waters (pH = 2-5). This species has the property of oxidizing insoluble metallic sulphides such as pyrites or chalcopyrite. It occurs as minute isolated rods or pairs of rods, whose mobility is due to polar ciliae. Bacteria of this type are Gram-negative.

- The family of siderocapsaceans, which may be attached onto aquatic plants or occur free-living in iron oxide biofilms on water surfaces, is also present in wells, drains and in the hypolimnion (lower layer of cooler water) of lakes. As in the case of *Leptothrix* and *Gallionella*, these bacteria form massive deposits of ferric hydroxide. Some species are aerobic, whilst others are microaerophilic or anaerobic. Several genera and species have been described:

  . *Siderocapsa*: spherical or ovoid cells, clustered together and enveloped in a very thick single capsule encrusted with ferric hydroxide.
  . *Naumaniella*: rods enclosed in a clearly defined thin-walled capsule, either isolated or as beaded chains.
  . *Ochrobium*: ellipsoidal or rod-like cells partially enclosed by a thin ferruginous capsule.
  . *Siderococcus*: isolated or clustered spherical cells with filamentous appendices, but lacking a capsule.
  . *Sidercystis*: lives in ochreous-brown colonies in iron-bearing waters.

In summary, it may be considered that any substance or mechanism using cathodic hydrogen will depolarize the system and, as a result, enable the onset of corrosion. An electrolytic cell is thus set up between the anodic and the cathodic areas, and an electric current will flow. In the absence of microoganisms and under anaerobic conditions, cathodic hydrogen will remain on the surface of the cathode, while the ferrous iron on the anode will be oxidized to form a light layer of rust. Therefore, the two electrodes are polarized and the reaction comes to a halt. The reactions which take place are as follows:

— Electrolytic dissociation of water:

$$8 \, H_2O \rightarrow 8 \, H^+ + 8 \, OH^-$$

— At the anode:

$$4 \, Fe \rightarrow 4 \, Fe^{2+} + 8 \, e^-$$

$$4 \, Fe^{2+} + 8 \, OH^- \rightarrow 4 \, Fe(OH)_2$$

$$4 \, Fe(OH)_2 + O_2 + 2 \, H_2O \rightarrow 4 \, Fe(OH)_3$$

— At the cathode:

$$8 H^+ + 8 e^- \rightarrow 4 H_2 \qquad \text{(anaerobic conditions)}$$

$$O_2 + 2 H_2O + 4 e^- \rightarrow 4 OH^- \qquad \text{(aerobic conditions)}$$

The role of microorganisms is to accelerate the electrochemical processes of corrosion. Their action may lead to depolarization of the electrodes, at the anode in the case of iron bacteria and the cathode in the case of sulphate-reducing bacteria.

— At the *anode*, the iron bacteria produce a rust deposit and bring about an anodic and cathodic depolarization while drawing their energy from the conversion of ferrous salts into ferric salts. This process leads to a continuous dissolution of the metal and, eventually, the formation of perforations.

— At the cathode, the sulphate-reducing bacteria mobilize the hydrogen and give rise to a cathodic depolarization.

In general terms, the bacteria do not directly produce corrosion but rather induce its acceleration. Furthermore, bacteria participate in the formation of slime and other deposits which can proceed so far as to cause the complete obstruction of boreholes and conduits. It should be noted that, the metabolism of most bacteria is only sustained by the transformation of mineral salts, from which they draw their energy. Carbon dioxide is only involved insofar as it is the sole source of carbon. Such bacteria are thus capable of developing in the absence of any organic compound.

c) Summary

In certain cases, the different corrosion phenomena observed in wells correspond to precursory signs of the different categories of clogging. Mechanical clogging involves sanding up of the catchment work due to perforation of the casing or the well screen. Electrochemical clogging is caused, for example, by the formation of iron-manganese concretions which reduce the open area of the well screen. Finally, biological clogging is associated with the development of bacteria.

The description of such phenomena clearly illustrates the importance of establishing regular diagnostic reports for catchment structures, even if there are no signs of ageing in their hydraulic behaviour. In this way, a video camera inspection could be carried out at each replacement of the pump, for example, in order to observe the condition of the equipment. Alternatively, preventive treatments could be performed on a regular basis according to the risks that are being run. Certain types of corrosion can be very rapid, leading to the irrevocable abandonment of the catchment structure after an entirely unpredictable lapse of time which may sometimes be of the order of two to three years.

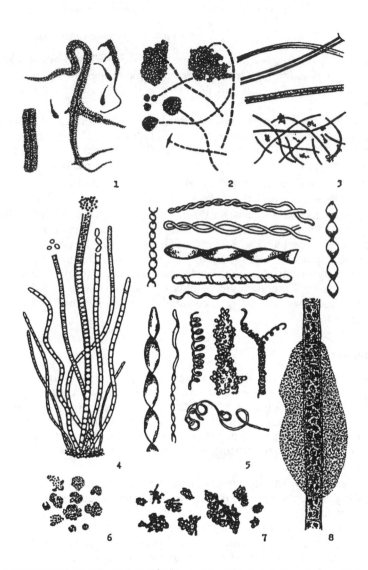

1 Leptothrix crassa, 2 Leptothrix sideropous, 3 Leptothrix ochracea, 4 Crenothrix polyspora, 5 Gallionelle ferruginea, 6 Siderocapsa major, 7 Siderocapsa treubii, 8 Sideromonas confervarum.

Figure 6-6
Various species of iron and manganese bacteria (after B. DUSSART, 1966).

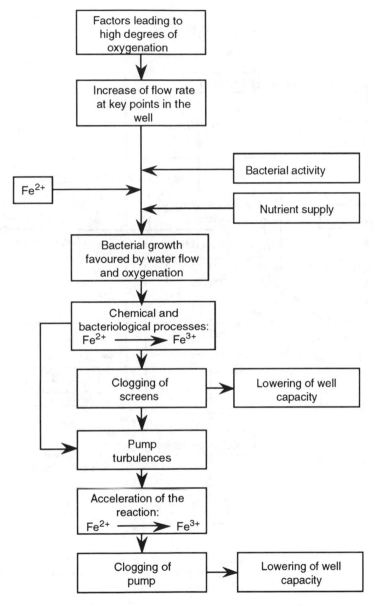

Figure 6-7
Model for biological corrosion process.

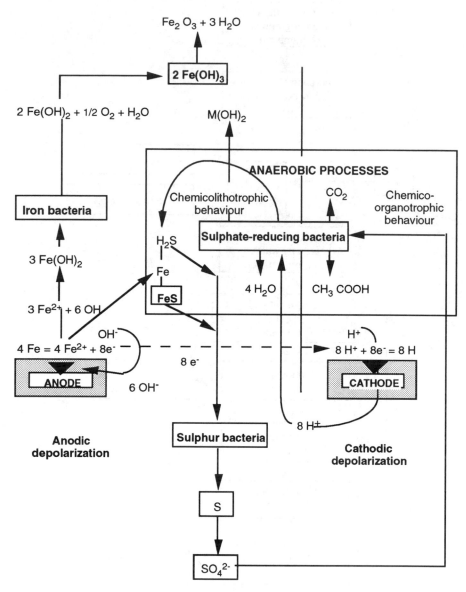

Figure 6-8
Biological corrosion cycle (after J. CHANTEREAU, 1980).

6.3.2. Clogging phenomena

The clogging of water wells is expressed in terms of a progressive lowering of the yield of water catchment structures, which is generally the first characteristic symptom of ageing. This clogging can have diverse origins, the result is always the same, i.e. a decrease in the permeability of the immediate surrounding media (gravel-filter or the formation itself), or an increase in losses of head due to a fall in the open area of the well screen (concretions or scale).

The clogging of water wells, but equally the plugging of river beds and banks in the case of alluvial aquifers, can have grave economic consequences in as much as it necessarily leads to a fall in yield and a resulting rise in cost. The same applies in the case of clogging of artificial recharge systems. Thus, it is essential that the causes of clogging should be determined and the proper treatment measures applied.

In the field, several types of clogging may be observed in a the same catchment structure: these can be mechanical (sanding up or clogging of the pack), chemical (calcareous or ferruginous deposits) or biological.

a) Mechanical clogging

Two types of mechanical clogging are distinguished here, sanding up and clogging of the gravel pack.

Sanding up

There is a risk of sanding up in the following main types of situation:

— When groundwater wells are developed in sandy aquifers through the base of the cribling or the casing, without any measures for retaining the unconsolidated material.

— When the screen slots are either too large or too small. These perforations should be of a definite dimension, being chosen as a function of the grain-size distribution of the formation or the added gravel pack. If the slots are of an inappropriate size, then sanding up of the catchment structure will occur.

— When the packing of gravel around the well screen is unsatisfactory. The emplacement of additional gravel around the well screen is difficult to carry out in a fully homogeneous way.

— Insufficient well development. A suitable characterization of the water catchment system will reduce the duration of development for a given water well.

— When the discharge capacity is higher than the delivery rate. Obtaining clear transparent water is the result of an equilibrium between well discharge rate and the non-entrainment of particulate matter. Any change in water flow rate will perturb the initial equilibrium.

— If the well screen is damaged during its installation or subsequently corroded by aggressive waters.

— If the well screen is incorrectly fitted without using centralizers.

The conditions of water well operation are equally determinant, in particular the running conditions for pumps should be fixed in relation to the characteristics and potential of the catchment structure. Under no condition should the well discharge be greater than the safe yield permitted by the pumping tests, particularly in the case of a start up in the presence of high pumping discharge head. In fact, the variation in load imposed on the pump under such conditions is likely to bring about a change in its speed. As a consequence, there is a risk of introducing a disequilibrium into the aquifer formation and modifying the rate of penetration of waters into the well screen.

In principle, pumping should always be carried out in a steady state regime at a discharge rate slightly lower than the design value for the catchment structure in question. Even though it is possible to envisage, in certain cases, the choice of several pumping cycles whose discharge rates during the said period of start up would be at most equal to the well yield, it would be difficult to impose such an obligation on installations having a large total pressure head.

Several authors have established formulae to simulate the entrainment of particles, basing their studies on the flow velocity of the water passing through the medium while assuming coefficients derived from Stoke's law.

Considering the maximum velocities that can be imposed on pumped water and the resulting velocity field in the formations surrounding the well, it is necessary to remove all aquifer materials having a grain size less than a defined value and to keep in place all coarser materials by means of a well-sorted gravel pack. This is operation is known as development.

The grain-size cutoff value is determined after granulometric analysis, taking into account both the technical constraints, such as the sizes of openings in available well screens, and the grain size of gravel not passing through these openings. The size of retained particles are of the same order of size as the pores, that is to say, about one quarter of the grain-size of the pack material. Although such particles would certainly be retained, grains up to ten times finer than the gravel — according to different authors — will also be retained owing to the bridging effect of grains which cluster upstream from the coarser pores.

If there is no choice of slot size for the well screen, then the duration of development operations will depend on the proportion of material to be removed in the zone where velocities are sufficient to pluck off and carry away rock particles. The entrainment velocity is less than the plucking velocity. Hence, in other terms, under a normal extraction regime following well development at a higher discharge, there should be no plucking of particles provided the rock is not subject to spontaneous disaggregation.

During pumping, it is necessary to avoid bridging of the gravel pack, which is an irreversible process. For example, intermittent pumping without a nonreturn valve will provoke a water hammer in the terrain at each standstill, caused by reinjection of the water contained in the pumping column, which will lead to break-up of the bridging. In such wells, it is common to see influxes of sand during the first few minutes of starting up. This corresponds to a renewal of aquifer development, and sand particles may be so numerous at this stage that it is sometimes necessary to separate the water so obtained from the mains supply.

Some authors [HERZIG *et al.,* 1970] consider coarse particles, with a diameter exceeding 30 μm, as exhibiting volume and mass effects which are predominant over surface area and electric charge effects. These particles are implicitely assumed to be inert in all the material previously included under the term of sand. For grain sizes of the order of one micron, for which physico-chemical surface effects are more important, it is usual to employ the term "fine fraction". For the size class between 3 and 30 μm, surface effects are of equivalent importance to mass- or volume-related effects, while at grain sizes less than 0.1 μm, the particles display the physico-chemical properties of a colloid.

Apart from their purely mechanical effect on particles coarser than the pores, well screens also have some influence on the fine fraction, which undergo numerous interactions with the surfaces they come into contact with.

In fact, this is precisely the effect required of commercially available screens that are expected to retain flocs of colloidal matter during water purification.

For most authors, the concentration of fine particles suspended in the solution decreases logarithmically with the distance traveled through the filter medium, while the efficiency of the filter falls off according to a similar type of logarithmic relation as a function of time.

In later sections, it is shown that these fine or colloidal particles can be mechanically removed by reverse circulation. Chemical methods may also be used, based on the action of acids, surfactants or various surface-tension inhibitors.

Clogging of the gravel pack

Clogging by fine-grained material (sands or clays) is generally reflected by the presence of this material in the pumped water, so particular attention should be paid to the measurement of suspended particulate matter (SPM), especially with regard to its variation in concentration with time.

Filtration puts a fluid containing suspended matter in contact with one or several media that are to be traversed. In a well, the fluid is water and the suspended matter is torn away from the rock. The water passes from the rock towards the gravel pack and then towards the screen. Between these three media, there are two interfaces to cross.

The more gradual is the passage from the fine material (formation) towards the coarse material (gravel pack), the higher the permeability and the greater the tendency for suspended matter to cross over from one medium to the other. With increasing contrast in grain-size distribution, the higher the number of coarse grains enclosing small pores filled with fine material. In this way, the passage of water in the material is progressively restricted to the pores that remain unfilled.

If the pores of the coarse material are larger than certain grains of the fine material, then these latter will penetrate into the coarse medium where they are eventually caught up in constrictions or absorbed against sufaces. This is known as internal clogging, or otherwise internal or deep straining.

If a fluid containing suspended matter penetrates into a medium whose pores are smaller than the particles in suspension, then these latter are blocked on the outside of the medium where they build up a mud cake that tends to oppose the phenomenon responsible for its formation. This is known as external clogging.

b) Chemical clogging

The two phenomena that can trigger chemical clogging are CO_2 release and the supply of O_2. The first brings about the precipitation of carbonates from bicarbonates in solution, as well as a shift of equilibrium from divalent to trivalent iron leading to the precipitation of ferric hydroxides. The second phenomenon gives rise to the formation of insoluble ferric oxides from ferrous iron dissolved in water or from metallic iron in the catchment structure.

Carbonates

The scaling of water wells by carbonates has been thoroughly studied, and it has become apparent that this phenomenon nearly always takes place by forming a mixture of carbonates, sulphates, hydrates and hydroxides that are principally combined with iron and manganese, as well as with accessory amounts of calcium. Scaling phenomena often account for the presence of carbonate precipitates in groundwater. However, this does not take place under the influence of drawdown, as has long been thought, but occurs rather as a result of the increased velocity of water as it enters the well. The formation of a calcium carbonate precipitate occurs according to the following equation:

$$Ca(HCO_3)_2 \xrightarrow{-\Delta P} CaCo_3 \downarrow + CO_2 \uparrow + H_2O$$

The reaction takes place under the effect of a pressure difference (ΔP). The solubility of calcium bicarbonate on the right hand side of the equation is 1,300 mg/l, while the calcium carbonate on the left has a solubility of 13 mg/l.

Flow in the gravel pack and well screen attains a maximum velocity along the pore walls or screen slots. The shear associated with the resulting stream of liquid leads to the release of dissolved gases as bubbles. These gases are chiefly made up of free carbon dioxide and carbon dioxide in equilibrium with the bicarbonates. When the bicarbonates are no longer stable, the water becomes oversaturated in carbonates which are then precipitated little by little in the pumping system and the pressure main.

Finally, the carbon dioxide bubbles will tend to dissolve back into the water once they have traversed the obstacle of the screen face. In most cases, the gas redissolves before it has had the opportunity of reacting with any carbonate, so the water becomes momentarily highly acidic and corrosive with respect to the metallic parts of the well structure. Since the phenomenon is relatively slow, precipitation generally takes place outside the catchment work itself (i.e. in the distribution network and reservoir). Nevertheless, it can occur within the borehole, producing a particularly indurated deposit when other compounds are involved in the crystallization process. Some examples are reported in the literature of scaling on the well screen and the aquifer material. This scale extends for a distance of up to one metre around the borehole, thus reducing the well-yield by more than 60%.

Scale formation is not just restricted to the carbonates of calcium and magnesium. In many cases, precipitates of silica and ferric compounds may also form, sometimes giving a pinkish or red colour to the scale deposited on the metallic parts of the catchment structure. Thus, the behaviour of the water is essentially a function of its composition and the operating conditions of the catchment work (limiting drawdown to the lowest possible value).

It should be pointed out that clogging by scale formation is not necessarily reflected in a progressive loss of the potential yield of the catchment structure. Preferential pathways are sometimes created within the aquifer where flow velocities are very high, thus ensuring a large part of the usual discharge capacity even when the screen is completely blocked. In view of this, there is a clear advantage in carrying out periodic inspections of catchment stuctures.

In this way, scale formation is inseparable from corrosion.

In order to limit scaling as much as corrosion, both of which depend on the release of CO_2 at sites where the flow is speeded up, it has long been considered necessary to reduce the water velocity. However, since water velocity has only an indirect effect on certain types of clogging, some specialists still do not acknowledge its influence.

In the case of chemical clogging, the determining factor is the velocity gradient between adjacent stream threads rather than the velocity itself or the type of flow regime (turbulent or laminar).

Iron-manganese deposits

The release of CO_2 from water as it passes through the gravel pack or the well screen will modify not only the carbonate equilibrium but also the redox potential, notably causing variations in the solubility of oxygen, iron and manganese. In this manner, ferruginous deposits may become intimately mixed with calcareous deposits. By the same token, if there is mixing of waters coming from aquiferous layers that are characterized by different contents of iron and manganese in a soluble form, each of the resulting solutions will be in equilibrium with lower concentrations of the same elements in a sparingly soluble form. The mixing phenomenon changes these equilibria and can lead to the precipitation of hydroxides.

$$Fe(HCO_3)_2 \xrightarrow{-\Delta P} Fe(OH)_2 \downarrow + 2\,CO_2 \uparrow$$

The solubility of ferrous hydroxide on the right hand side of the equation is less than 20 mg/l. If oxygen is supplied to a water containing Fe^{2+} or Mn^{2+} ions, from the surface of an unconfined aquifer or via the borehole in the case of a confined aquifer, then the production of insoluble oxygenated precipitates will be much greater. The most marked clogging will occur at oxygen-rich sites, generally at the top of well screens and in the upper parts of boreholes and horizontal drains. In addition, ferric hydroxide will be precipitated:

$$4\,Fe(OH)_2 + 2\,H_2O + O_2 \rightarrow 4\,Fe(OH)_3 \downarrow$$

which has a solubility of less than 0.01 mg/l. The clogging is severely diminished if the well head is rendered air-tight.

The same principle applies to manganese precipitation:

$$2\,Mn(HCO_3)_2 + O_2 + 2\,H_2O \rightarrow 2\,Mn(OH)_4 \downarrow + 4\,CO_2 \uparrow$$

An oxidation of the iron and manganese hydroxides or and increase in pH gives rise to the formation of hydrated oxides which contain Fe^{2+} and Mn^{2+}. In particular, ferrous ions in solution will react with oxygen to form ferric oxides:

$$2\,Fe^{2+} + 4\,HCO_3^- + H_2O + 1/2\,O_2 \rightarrow Fe_2O_3 \downarrow + 4\,CO_2 \uparrow + 3\,H_2O$$

Even though ferrous and manganous compounds are able to clog catchment works, here again, the composition of the water and its behaviour will lead to different types of effect. It is not uncommon to observe that certain waters containing more than 0.5 mg/l Fe represent no disadvantage as regards the catchment structure, whereas other waters with only 0.10-0.15 mg/l Fe will produce an almost immediate precipitation of ferric hydrate. Otherwise, such effects are easily revealed by tests which also provide a good indication of the possible behaviour of iron with respect to the risk of clogging in catchment works.

pH is a particularly important factor in maintaining iron in solution or controlling its oxidation.

It should be noted that the distinction has only recently been made between ferric clogging of bacterial origin and chemical clogging. In reality, both types are commonly present, which complicates the implementation of efficient solutions in the absence of an accurate diagnosis.

c) Biological clogging

Biological clogging is generally characterized by the presence of filamentous material in the pumped water, in addition to flocs or gelatinous masses, which appear well before any loss in yield of the catchment structure. In most cases, this phenomenon is linked to the activity of iron and manganese bacteria.

Favourable conditions for the development of these bacteria are as follows:
— pH in the range 5.4-7.2.
— Ferrous iron concentration between 1.6 and 12 mg/l.
— Presence of CO_2.
— Redox potential (Eh) should be higher than -10 ± 20 mV.

Iron and manganese bacteria do not live in the presence of high oxygen concentrations, with the exception of *Leptothrix crassa* which develops on the free surfaces of groundwater bodies. These bacteria can grow *en masse* at oxygen levels of less than 5 mg/l, displaying a behaviour known as microaerophilic. It is also possible for such bacteria to benefit from the oxygen produced by nitrate- and sulphate-reducing bacteria.

A study of the literature reveals that iron and manganese bacteria are among the most important causes of ageing in water wells. To summarize the studies carried out to date, including for a large part the work of G. KREMS, the following conclusions can be drawn:

— No material resists clogging.

— The products of clogging are essentially made up of iron and manganese compounds occurring in the form of hardened scale or gels.

— No satisfactory relation has been found concerning the influence of water regime, whether turbulent or laminar, on the process of clogging.

— Purely chemical clogging by iron hydrates is extremely rare.

— Iron and manganese bacteria are the main agents responsible for biological clogging. These bacteria exist in groundwater bodies, but only develop *en masse* when the commissioning of a well brings in additional nutrients and iron due to the increased water circulation velocity. If the water is stagnant, the concentration of dissolved Fe^{2+} should be between 1.6 and 10 mg/l.

— The importance of clogging is greater in wells with high discharge capacity.

— Clogging can be greatly reduced or even eliminated by the monthly treatment of wells with sterilizing reagents, using a dose that leads to a chlorine concentration of 1 g/l in the treated zone.

Otherwise, certain authors have specified the most frequently encountered features associated with obtrusive biological clogging:

— Presence of iron or manganese bacteria.

— Presence of iron and manganese in solution; a minimum concentration of 1.6 mg/l is required in stagnant water, but this can fall to 0.2 mg/l in the case of flowing water.

— Redox potential with respect to the hydrogen electrode should be higher than – 10 mV with a maximum deviation of 20 mV.

— Water flow velocities should be considerably higher than under natural conditions. A study of clogging in Ypresian sands confirms that clogging develops even more rapidly when flow conditions are vigorous. By contrast, clogging is very slight or even negligible when flow velocities are of the order of mm/s.

Biological clogging occurs regardless of the nature of the material involved (e.g. steel, synthetic materials, copper, fibrocement [in French, Fibrociment™ is a tradename for an asbestos cement], etc.), but metals are commonly susceptible to the corrosion that results from clogging. However, it does not systematically affect the top or the base of well screens, but depends essentially on the conditions that are favourable for bacteria. These conditions may differ as a function of the geological formation and the supply of nutrients. The latter frequently exert a dominant influence, particularly when clogging affects the well screen in areas of maximal flow velocity, but not after it attains the gravel pack or the surrounding formations.

Other studies on biological clogging have been carried out, notably in France, which show that clogging arising from bacterial activity does not merely affect the screen or gravel pack, but extends much farther beyond into the aquifer formation. In most cases, clogging is due to the presence of heterotrophic anaerobic and sulphate-reducing bacteria which develop as a result of the nutrient flux brought in by pumping, thus forming a biomass several meters in size which considerably reduces the permeability of the medium.

The role of water circulation velocity in the appearance of this type of clogging has been emphasized in a study of the Ypresian sand aquifer, where more than 60% of the 300 wells are clogged up. BOURGUET et al.,1988 have carried out an experimental physical model that has given rise to the filing of a patent (clogging measurement apparatus). The inventor of this apparatus advocates its use prior to the development of new water well fields, a course of action which would enable an assessment of the clogging hazard specific to a given site. Subsequently, the design of catchment structures and the definition of operating conditions would make it possible to prevent or minimize the risk of clogging.

Studies show that increased malfunctions in catchment structures are due to an enhanced velocity field in the immediate vicinity of the well, which has led certain authors to attribute this phenomenon to overextraction (groundwater overdraft). Bacterial development can thus be explained by:

— The presence of an oligotrophic bacterial population in a slowed down (inactive) life stage.

— An increase in activity of the bacterial population under the effect of pumping. The water movement so produced gives rise to the nutrient flux necessary for bacterial growth.

Although this hypothesis is not unanimously accepted within the scientific community, it nevertheless accounts for the appearance of clogging at various times after the commissioning of a water well. The time to the onset of clogging is highly variable, ranging from some months to several years, and is notably influenced by the occurrence of exceptional climatic events. The hypothesis described here could equally explain the fact that certain catchment structures are affected whilst others are not. In this context, it is known that within a given well field, the distribution of water fluxes is not the same in each catchment work.

Finally, some authors have observed that the bacterial clogging of water wells takes place under certain physico-chemical conditions (pH, redox conditions, etc.). Even though this probably does apply in the reported cases, this not imply that any general conclusion can be drawn. In reality, the phenomenon is clearly far more complex than it appears, in the sense that each tapped aquifer makes up a particular physico-chemical and biological medium that reacts to the environmental constraints specific to each catchment structure.

Experiments on bacterial clogging reveal the role of temperature and drought as triggering factors. The observations were obtained from well sites in France and Belgium (GEOTHERMA, BUYDENS) where the waters contain no iron or manganese (even in trace amounts) and the oxygen concentrations are very low.

The reported characteristics of bacterial development in these areas are as follows:

— "appearance of brown-coloured mucous filaments in the space of four days,

— in a few days, the supply conduits leading to the treatment plant were coated uniformly with a layer more than 1 cm thick that was dislodged in packets by each variation in discharge,

— appearance of ammonia,

— the water is practically devoid of iron (0.04 mg/l). It was not suspected that the water might have contained ferruginous bacteria,

— lowering of the oxygen concentration of the water,

— the phenomenon attains new wells with radial drains. Here again, pumping tests failed to reveal anything".

In the above example, it appeared that the phenomenon was especially sensitive to temperature. Since the groundwater bodies involved here are shallow, the temperature is influenced by external climatic conditions. Bacterial development benefits from the rise in temperature of the groundwater to the extent that, every year, the phenomenon starts in spring (generally in May) and declines in the autumn (October).

The phenomenon then became more complex when waters in the catchment works started to show traces of iron and manganese during the drought of 1976; these contents rose very sharply during the recent droughts of 1989 and 1990 (Central West region of France).

These observations concerning climatic conditions and their influence on the triggering of bacterial phenomena in groundwater wells go towards confirming the situation that has been noticed elsewhere. The case described from Belgium (cited by BUYDENS) refers to water wells that were already 28 years old at the time of the 1959 drought. According to this author, during the drought of 1959, river water participated more than usual to the recharge of the depleted alluvial groundwaters (the water catchment being situated in alluvial aquifers). Thus, the water underwent a less than thorough natural purification, as witnessed by the oxygen deficit, the appearance of ammonia and the rise in organic matter content. BUYDENS concluded from the above that bacteria, which up until then had never manifested themselves, could have found these conditions favourable for their revival.

According to GEOTHERMA as well as BUYDENS, it seems that any aquifer in contact with an organic substrate (plant cover, peat, silt, alluvium, etc.) will run the risk of triggering bacterial problems. When the factors are present in a latent form, it merely requires an exceptional drought or a change in the usual recharge of the catchment structure in order to trigger a number of phenomena which are as rapid as they are unpredictible. This observation is confirmed by the fact that the mineralizing action of iron becomes more pronounced as the overall microbiological activity of the natural environment intensifies. Finally, the study undertaken in 1988 concerning the action of organic matter on alluvial groundwaters in the Rhône-Mediterranean-Corsica basin demonstrates that organic matter is clearly the cause of iron and manganese transformation [GEOTHERMA, 1991].

6.3.3. Failures linked to the resource

The loss of total capacity of a water well may be totally independent of the catchment structure itself but linked to more general factors such as, amongst others, drought and overdevelopment.

a) Rainfall deficit

Groundwater bodies are replenished by rainwater (see Chapter One: the hydrological cycle) which infiltrates into the ground and contributes to the recharging of aquifers. In the absence of precipitation, the equilibrium between inflow and abstraction is broken and a regular lowering is then observed in the level of groundwater bodies. For a given discharge rate, the lowering of the groundwater level brings about a drop in the pumping water-level which can exceed the technical limit of the catchment structure (e.g. dewatering of the pump). From this arises the necessity of adopting a less sustained pumping schedule in order to maintain an acceptable drawdown that is in accord with the catchment structure characteristics.

The cumulative effect of several periods of rainfall deficit, as was the case in France in 1976 and more recently from 1989 to 1992, is the main reason for the lowering of groundwater levels within aquifers. The most strongly affected catchment works are those which are developed in shallow groundwaters, or otherwise tap only the upper part of the water-sheet.

b) Hydraulic perturbations

Under certain conditions, major infrastructure projects such as road and motorway construction, urban development and surface hydraulic works can have a significant impact on the flow of shallow groundwaters. The consequences may be of various types, i.e.: lowering or raising of the usual groundwater level, appearance of chemical or biological pollution.

6.3.4. Failures linked to management of the catchment structure

Failures due to the poor management of water catchment ought not occur. In fact, the optimal conditions for satisfactory management should be established from the outset in view of the results of pumping tests.

Unfortunately, such failures result either from an incorrect interpretation of the pumping tests or non-observance of the operational instructions concerning well operation.

a) Overextraction

Overextraction (or groundwater overdraft) is unfortunately a frequent phenomenon. Its main cause is the extraction of water at a discharge rate greater than the physical yield limit of the well. The consequences of overextraction are summarized in Table VI-II.

TABLE VI-II — *Consequences of overextraction* .

Consequences of overextraction of groundwaters	
CAUSE	- Well discharge greater than maximum water-yield
EFFECTS	- Dewatering of part of the well screen - High water velocity in the water-yielding part of the catchment work - More extensive cone of depression
CONSEQUENCES	- Sanding up - Corrosion and erosion - Scaling - Increase in iron and manganese contents - Destabilization of gravel pack - Bacterial growth - Air binding of the pump - Mediocre discharge capacity

Overextraction can also have other causes. The most commonly encountered cause is linked to pump startup, when the water-level drops generally very rapidly by two-thirds of the column height and then more slowly to the stabilized or pseudo-stabilized pumping water-level. The stopping of the pump leads to precisely the opposite phenomenon, that is, an initial rapid rise of water-level followed by slowing down. Certain pumping schedules are programmed in such a manner that startups and stops are very frequent, thus causing a to-and-fro movement of water-level in the well that is particularly inauspicious in the case of a sandy aquifer formation. Sometimes, there is no need to look any further to explain the sanding up of a water well.

b) Unadapted extraction

There are some water wells that are operated in a particular way, for example, when they tap fractures existing in calcareous, shaly or other types of rock layer. Although the overall pattern of water-yield may eventually allow an important level of discharge, certain precautions need to be taken. In fact, in the case of pumping at an excessively high discharge rate there is a risk of dewatering some of the cracks and fractures in the aquifer formation. Thus, it is necessary to respect the pumping water-level with great accuracy; if not, the water extraction will be ill-adapted to the specific characteristics of the catchment structure, with the risk that sediments will be carried into the well.

Whatever the origin of the failure (catchment structure, pumping, resource), as soon as regular checks show that the discharge capacity of a well has fallen by 10-15% of the initial yield, it is advisable to determine the cause(s) or, in other words, to carry out an ageing diagnosis.

6.3.5. Ageing diagnosis

Water wells evolve slowly with time. In this context, the problem amounts to determining when it is appropriate to apply special treatment for restoring the initial catchment characteristics. Thus, it becomes necessary to perform an ageing diagnosis.

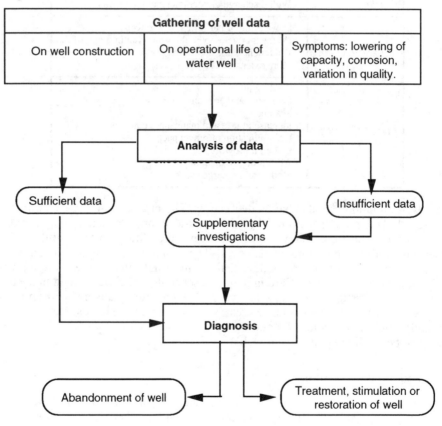

Figure 6-9
Methodology for ageing diagnosis.

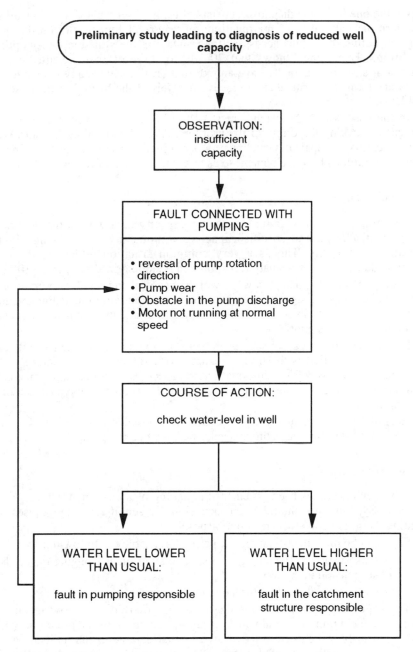

Figure 6-10
Analytical approach for diagnosis of well discharge loss
(after GEOTHERMA document, 1993 - joint Water Authorities series).

On the one hand, the diagnosis consists of using observable signs at a well site to determine the causes of malfunctions in water catchment and, on the other hand, to define the means necessary for remedying the situation under optimal conditions. Experience shows that 20-30% of failures that are attributed to catchment structures actually have an external origin. Before any decision, it is necessary to make a critical and objective analysis of the history of the observed problems.

In the case where the catchment structure itself is involved, the first stage consists of seeking the origin of the problem. After determining the cause, it is then necessary to repair the catchment work. Finally, it is suitable to define the optimal conditions for development so as to avoid any risk of renewed failure.

a) Warning signs

The first symptom of falling discharge capacity of a well is the decline in pumping discharge and specific capacity, without any other change in the operating conditions. This is a very common situation which needs to be confronted efficiently. The analytical approach is summarized in figure 6-10.

During the development of a water well, the careful and regular observation of certain indications (or signs) can prove to be very important in revealing eventual anomalies. A single isolated indication may attract attention, but in most cases does not lead to a diagnosis.

It is necessary to gather several indications, whose convergence will then provide the specialist with strong presumptive evidence on the nature of the problem. On the basis of this information, the specialist will define the means for identifying the precise origin of the problem and determining the type of action to be undertaken.

According to the nature of the indication, it is possible to know which part of the catchment structure has a diminished efficiency due to ageing.

b) Identifying causes

When evidence has been found for the ageing of a catchment structure or a pumping system, this should be an occasion for seeking out the causes and locating them. Several tools are at our disposal:

Physico-chemical investigations, in particular, enable the following:

- Observation of the presence of sand, clay or other clogging materials in suspension in the water.
- Collection of samples for laboratory analysis.

These provide evidence of the degree of aggressivity of the water and the nature of the clogging material which make it possible to form a preliminary idea of the cause. It is then necessary to give further details on the indications and locate accurately the affected zones. Several possibilities for achieving this objective are discussed below in order of increasing complexity.

Pumping tests

First of all, a hydraulic shock test (to be compared, if possible, against previous tests) will give information on the distance to the clogged zone with respect to the well axis; in fact, the form of the experimental curve depends on the position of the clogged layer against the well wall or at some distance away.

In a subsequent stage, stepped pumping tests with increasing discharge rate, when compared with similar previous tests, will show any abnormal increase in head loss due to anomalously high water velocities in openings that remain free. On the contrary, clogging could be such that no openings remain where water velocities are noticeably high.

Pump removal

The following stage consists of removing the pump and the discharge column (rising main), which is then examined with care, especially if the pumping system is thought to be responsible for the lowering of water-yield. The pump strainer can become clogged by corrosion products and the body of the pump can be worn by sand abrasion. The pump housing can be corroded or perforated, as well as the delivery pipe particularly near its connections.

Well logging

After removal of the pump, measurements are carried out by flow meter during water injection or pumping, thus providing information on the velocity field in the catchment zone and the casings. A comparison with previous measurements can reveal the presence of clogged zones and any eventual perforations in the casing.

Important information can also be provided as a result of photographic or video-camera observations. The corrosion and clogging of well screens are clearly visible, while the colour of a deposit may indicate its nature.

Analysis of clogging deposits

It is of fundamental importance to ascertain the nature of clogging deposits. These are occasionally found in the distribution system (water-meters, taps, etc.) or in sediment traps if they are present.

— The presence of sand with a grain size less than the pores of the gravel pack should not be a cause for concern. This is merely the sign of insufficient development at a discharge rate too close to the operating capacity, or for too short a duration, or even that the method was inadequate.

If the presence of sand is accompanied by a lowering of the specific capacity, it is possible that sand has partly filled a portion of the well and thus prevents the screens from playing their role. When the quantity of sand is very large, the height of the gravel pack should also be checked since the sand influx can bring about subsidence of the terrain. This also applies when the grain size is larger than the diameter of the pores in the gravel pack but less than the slots in the screens. If the sand is coarser than the screen slots, their wear or rupture is to be feared.

Sand has a highly aggressive effect on pumps. If sand is abundant, it may settle out in the pumping column and completely block the pump.

— The presence of *calcareous scale*, in grains or flakes, is an indication of an oversaturation in bicarbonate ions which may derive from the degasification of free CO_2.

If carbonate is precipitated in some places, there is every reason to expect CO_2 excess elsewhere, and thus the presence of corrosive waters. Both phenomena may occur simultaneously in well screens, i.e. formation of carbonate scale and perforations caused by corrosion.

It should be noted that waters with neutral pH can become acidic and corrosive simply by virtue of the velocity effect when CO_2 is redissolved after precipitation of the carbonate. This water can attack the body of the pump, the housing and the delivery pipe.

In fact, scale is not exclusively calcareous in composition but contains a mixture of different carbonates, hydroxides and sulphates in combination with calcium, iron and manganese.

— *Iron and oxygen*: the presence of iron in the form of oxides or hydroxides is a characteristic feature of corrosion phenomena.

If there is no decrease in the specific capacity, the most likely origin of the clogging is in the pumping system or in the casing between the screens and the pump intake (suction port).

If the appearance of ferruginous particles is accompanied by a lowering of the specific capacity, it may be assumed that they are the result of the clogging or corrosion of well screens.

In the absence of oxygen, the iron originally in solution in the water or derived from the corrosion of well screens does not exceed a concentration of 0.5 mg/l (expressed as Fe^{2+}). An input of oxygen may come about by contact of the aquifer with unsaturated formations. In such a case, if the screens are too close to the standing water-level oscillation zone where oxygenation can take place, there is a risk of scaling by iron oxides.

— *Bacteria*: the occurrence of bad odours, flocs, gels and slime may reflect the presence of bacteria, whose growth is frequently responsible for the observed falls in capacity of catchment structures.

6.3.6. Summary

Unfortunately, no radical method exists for preventing the ageing of a groundwater catchment structure. Nevertheless, it is possible to increase the service life by respecting some basic rules for the equipment of the well and its mode of operation.

— Choice of well screen. The screens should be in contact with a highly permeable gravel pack and possess the highest possible open areas. The slots should have smooth lips and a wedge-shaped profile which expands in the direction of the water flow. Furthermore, experience shows that the optimal velocity of water flow in the slots is about 3 cm/s.

— Gravel pack. Care should be taken to carry out shrouding correctly within the annular envelope. In fact, any segregation of the gravel may facilitate the process of clogging.

— Well development plays a crucial role in reducing the influx of sand.

— Pumping regime. Wherever possible, it is preferable to reduce the discharge rate and increase the duration of pumping. As a matter of fact, numerous periods of halt are favourable for scale formation. However, pumping time should never exceed 20 hours out of 24.

— Periodic visits for checking and maintenance: in most cases, action is taken at a rather late stage when the discharge falls sharply or when sand is observed in the pumped out water. At this stage, a costly and vigorous treatment, which may already be too late, will not always succeed in improving the discharge capacity of the well. By contrast, provided that it is treated in time, the failure can be readily overcome. Furthermore, it may be added that leaching carried out from the onset of scale formation will often enable the achievement of well operating conditions comparable to those obtained during pumping tests.

If clogging or corrosion cannot be prevented despite the application of these recommendations, the well will have to be restored by various chemical treatments.

Above all, it is necessary to perform a full and accurate diagnosis of the problem under consideration. The following rationale is used in the diagnosis of water well failures:

— Search for the cause(s) of the failure.

— Eliminate its undesirable effects, i.e. restore the well.

— Prevent any renewal of the problem.

6.4 Protection of catchment works

The means of making good the effects of corrosion are discussed in Chapter VI. Nevertheless, from the preventive point of view, there are some simple methods for delaying or attenuating these phenomena; these include the galvanization of casings, cathodic protection and a suitable choice of screen type in relation to the nature of the groundwater.

6.4.1. Passive protection

A reactive metal, ranking high on the electrochemical series, can undergo corrosion which may eventually lead to its dissolution. On the other hand, the same metal may become passivated and thus shielded against attack if a thin film is electrodeposited onto the surface which protects it from electrochemical corrosion.

It should be pointed out that the resistance of a stainless steel surface to corrosion is linked to the quality of the passivated layer

Although galvanization (zinc coating) is a good protection against chemical corrosion, it is completely inoperative in the case of electrochemical corrosion (electrolytic cell). In order to be effective, the content of free chromium in the passive protective film on stainless steels should lie in the range 11-30% (Cr is the dominant element in the metal coating).

However, under certain circumstances, stainless steels are subject to corrosion from halogen salts (particularly the chlorides) which penetrate the passive film and attack the metal. Thus, the process of corrosion can be accelerated in the case of high chloride concentrations.

6.4.2. Cathodic protection

In a medium that is able to behave as an electrolyte, a conducting metal assumes a certain corrosion potential according to the aggressivity of the medium. Two situations may arise:

— The electric current passes from the metal into the electrolyte and metal corrosion is accelerated.

— The electric current passes from the corrosive medium towards the casing (lining tube) and degradation is greatly retarded.

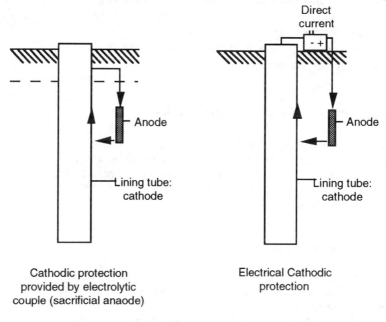

Figure 6-11
Two types of cathodic protection used for groundwater wells

Thus, protection of the casing can be ensured by direct connection of the lining tube (cathode) to a metal acting as the anode, such as zinc, magnesium or aluminium (cf. figure 6-11). The current coming from the anode traverses the corrosive conducting medium and penetrates the metal structure to be protected (cathode). From there, the circuit is completed by a copper wire that leads back to the anode. The metal of the anode is dispersed into the conducting medium, while the metal of the casing pipe remains intact.

This amounts to cathodic protection by means of a sacrificial anode. The anode(s) should be situated beneath the standing water-level, while the mass, dimension and position are dependent on several factors (current, resistivity of corrosive medium, nature of medium, etc.) and should be determined by a specialist.

Another technique consists of using the direct current provided by a generator, with the negative terminal connected to the surface to be protected and the positive terminal to the anode. This is known as electrical cathodic protection.

It is equally possible to place a metal cable that acts as an anode inside the well, connecting it with the casing to be protected. This cable should be replaced regularly.

In the same manner, pumps may be also protected from corrosion.

In fact, pumps are commonly composed of different metals which set up electrolytic cells, thus giving rise to corrosion. Cathodic protection can be installed by placing soluble anodes near the steel shafts (spindles) of the pump motor.

6.4.3. Choice of materials

In a water well, the casing pipe — which is often made of steel — is cemented to the surrounding formations. When the cementation is correctly done, it should ensure a tight seal that affords protection against corrosion under most circumstances.

Well screens, by contrast, are in permanent contact with the groundwater and are consequently far more subject to corrosion. Hence, the choice of material is quite crucial. There has to be a compromise between discharge capacity constraints (e.g. percentage open area) and reliability constraints in view of the pressure and corrosion.

Before selecting the component materials for the screen, it is necessary to carry out a water analysis. The results so obtained enable the choice to be properly adapted to the problem.

In this context, the RYZNAR index characterizes the encrusting or corrosive power of waters. It brings together the pH, the calcium ion concentration, the total alkalinity titre and the content in dissolved salts.

The value of the RYZNAR stability index for a water sample is given by the following equation:

$$I = S - C - pH$$

where:

I : RYZNAR index,

S : value taking account of dissolved salts,

C : value taking account of alkalinity and carbonate hardness.

A knowledge of this index allows a choice of the best adapted material (cf. Table VI-III).

TABLE VI-III — *Choice of material as a function of RYZNAR index*

Value of RYZNAR stability index	Recommended material
7.5 - 18	Plastic
7 - 8	Low-carbon steel
6.5 - 8	ARMCO iron
6 - 8.5	Siliceous red copper
< 9	Super Nickel
< 9.5	Monel 400
< 12	304-type stainless steel
< 15	304-type ELC stainless steel
< 16	316-type stainless steel
< 18	316-type ELC stainless steel

6.5 Artificial recharge of aquifers

Artificial recharge aimed at increasing the resource consists of replenishing an aquifer in order to compensate for the water-level depression linked to intensive aquifer development. Such a procedure of "rational management of natural reservoirs" enables the dynamic storage of resources and their reuse under different conditions of water regime and quality. It allows for changes in water quality, the re-establishment of an equilibrium, the augmentation of the resource and the optimization of the operating regime.

Since 1980, an artificial recharge facility has been in operation at the Aubergenville site (Yvelines department, France - Figure 6-12). The artificial recharging of the aquifer in the Senonian Chalk and alluvial deposits is carried out using waters from the Seine, after clarification, settling and filtering. The waters treated in this way rejoin the aquifer via infiltration basins that are installed in disused gravel-pits. There are a total of seven infiltration basins situated in the catchment zone, allowing a mean recharge rate in 1991 of 850,000 m^3 per month. The total surface area of these basins is of the order of 20 ha. A filter bed composed of sand is emplaced at the bottom of each gravel-pit, using an effective grain size of 0.2-0.3 mm. This layer serves as a mechanical and biochemical substrate for water purification. Although the mean thickness of the filter bed is 0.5 m, this can locally exceed 1 m. The unit infiltration capacity is of the order of 0.4 to 1 m per day, with a water depth of 2-3 m in the basin.

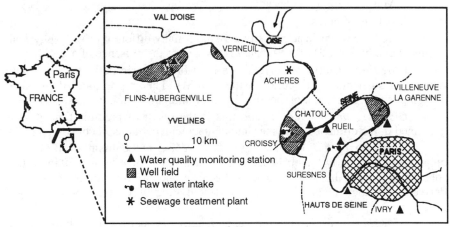

Figure 6-12
Main Lyonnaise des Eaux intake fields in the downstream of Paris (France).

At the Aubergenville site, water percolates across the "natural" filter bed before penetrating into the unsaturated zone. The biological and chemical purification of waters in these gravel-pit basins is due essentially to the action of plankton, atmospheric oxygen and insolation. The bottom of the basin acts as a filter, so the water is cleared of its coarser impurities before completing its purification and rejoining the aquifer.

Using infiltration basins for artificial recharge is a fairly widespread procedure, evidently because of the apparent simplicity of the technique. However, it is now realized that, when considered in detail, its utilization becomes highly complex as soon as any attempt is made to apply optimal operating conditions. In fact, a study of the optimization of artificial recharging carried out by Lyonnaise des Eaux has shown the real complexity of this phenomenon, in particular with respect to clogging. The adaptation of a development project to the hydrogeological conditions of the site is just a preliminary step. It remains to investigate clogging and the means for restricting it, as well as all consequences on maintenance whose financial implications could have an impact on the profitability of water well operation.

The clogging of infiltration basins results mainly from algal proliferation, the settling of suspended matter and bacterial activity in the terrain.

6.5.1 Clogging of recharge basins

A progressive clogging is observed on the beds of all the recharge basins which is reflected by a reduction in the infiltration rate. The clogging process is broadly the result of a combination of two mechanisms:

— Disordering of the soil porosity, as a result of various electrochemical mechanisms:

- Disintegration of aggregates by an excess of ions which disperses the clay or, alternatively, dissolution of their cement in a reducing medium.
- Major swelling of clays.

— Blocking of pores.

This decrease of intrinsic porosity may have many different origins (physical, chemical and biological) or may otherwise be due to the presence of algae.

— *Clogging of physical origin*: the base of the basin acts as a filter with respect to suspended particulate matter (SPM). The importance of physical clogging is thus a function of the SPM concentration in effluent waters.

— *Clogging of chemical origin*: this is the result of precipitation of salts contained in the effluent as it comes into contact with certain soil constituents.

— *Clogging of biological origin*: the precise mechanism of biological clogging is not clearly established, but it is known that bacteria play an important role. In this way, the development of bacteria and the formation of products derived from their metabolism can lead to clogging by obstruction of the soil pores.

— *Clogging by algae*: the presence of nutrients such as phosphorus in the water, combined with sufficient illumination, will enable the development of algae in the basin provided that the temperature is high enough.

It is difficult to specify the importance of chemical and biological clogging. On the other hand, it is quite clear that the role of algae is complex. The presence of algae in a basin is classically thought to offer the following advantages:

— Algal felting favours the filtering of water and the coagulation of particles in suspension.

— The growth of algae extracts nutrients from the medium and can also concentrate harmful substances — particularly the heavy metals — into the vegetal cells.

But these algae also give rise to some drawbacks:

— Release of unpleasant odours.

— Reduction in permeability of the basins by the development of a dense carpet at the ground surface.

Trials are currently in progress to ascertain the overall balance of processes related to the presence of algae. To achieve this, some of the infiltration basins have been stocked with herbivorous fish which eat the algae; at the same time, the quality of water coming from two springs is being studied from the qualitative point of view.

6.5.2 Functioning of aquifer recharge

An analysis of the quantitative data from recharge operations makes it possible to draw up a curve of infiltration rate against time. By operational convention, the infiltration rate is expressed in terms of metres of infiltration per day. The curve of aquifer recharge with time (cf. figure 6-13) exhibits four stages of evolution:

— Phase A: swelling of soil colloids as they first come into contact with water.

— Phase B: progressive dissolution of air bubbles (degasification).

— Phase C: formation of a bacterial coat (useful in biological purification).

— Phase D: Progressive asphyxiation of the bottom of the basin, theoretically leading in the end to complete clogging. This phase is very gradual and can be easily represented as a simple linear regression.

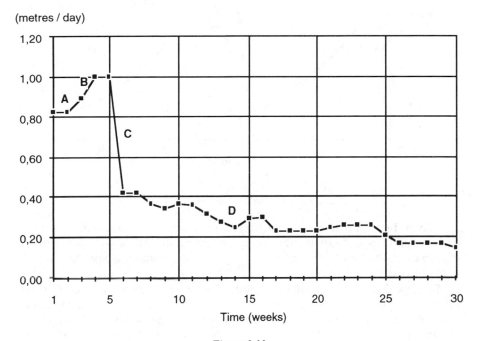

Figure 6-13
Mean recharge of the Aubergenville aquifer as a function of time. Data from infiltration basins n°s 4, 5 and 6 in 1991; infiltration rates in m/day.

An inspection of figure 6-12 reveals that recharge attains its highest values during phases A, B and C. One technique consists of using only the first three phases and then leaving the basin out of operation for a period of time (2-3 weeks) before refilling. The period of stoppage enables the elimination of bacterial activity that develops during phase C. The initial conditions are reestablished following refilling of the basin, and so on with subsequent phases of recharge (figure 6-13).

It is essential to accept a certain degree of clogging in order to preserve flow in the unsaturated medium beneath the basin. This flow probably plays a determining role in the purification of recharge waters passing through the soil. The problem is that clogging is a phenomenon that grows in importance with time, to the point of becoming unacceptable. Therefore, periods of seepage should alternate with periods of draining out, on the one hand to allow aeration of the soil, which promotes the recovery of microbial activity in the soil, and, on the other hand, to eliminate the deposits of suspended matter.

The draining of the seepage basins enables a full recovery of infiltration capacity, as shown by the curves on figure 6-14 and the tests carried out at Aubergenville.

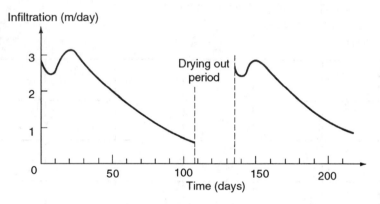

Figure 6-14
Example of optimization of the infiltration cycle.

The problem of managing infiltration systems amounts to determining a cycle of alternating periods of immersion and draining in such a way that optimal efficiency is obtained for the installation. The period of draining also corresponds to maintenance.

Finally, from the quantitative point of view, it is clear that halts in the infiltration cycle lead to a higher degree of recharging than if there are no periods of drying out (cf. figure 6-15). To summarize, these groundwater management techniques appear very attractive from both the qualitative and quantitative standpoints.

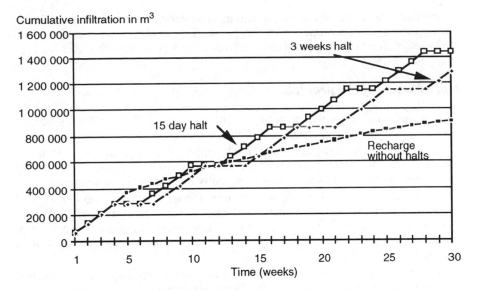

Figure 6-15
Different hypotheses for the temporal management of seepage pits.

6.6 River-bank effects

Since 1989, following the initiative of Lyonnaise des Eaux (LE), a research programme on river-bank effects has been undertaken in collaboration with the Geological Information Science Centre of the Paris School of Mines (CIG in French), the Seine-Normandy Water Authority (AESN in French) and the Institute for Nuclear Protection and Safety (IPSN in French). The objective of this programme is to study in detail the transfer of pollutants between rivers and their adjoining alluvial aquifers, thus answering a major concern of water resource managers. In France, these aquifers provide more than 50% of the drinking water supply to communities in the Seine-Normandy area.

This following section gives a brief presentation of the study site and summarizes the results obtained. The study concerns three categories of pollutants:

— Nitrogenous compounds, due to the significant increase of nitrate contents in watercourses, particularly the Seine, downstream of the Achères waste water treatment plant.

— Radionuclides, derived from accidental and planned releases from nuclear installations (radioisotopes of Co, Cs, Ag, Sb, I, Sr, Ru and Te).

— Pesticides (e.g. Atrazine and Simazine), of which high concentrations have been detected in the aquifers and watercourses of cultivated land during the spring.

Preliminary results show, in particular, considerable biochemical activity in the uppermost few centimetres of the Seine river muds, as well as a marked purification effect taking place within the first metres away from the river bank.

6.6.1 Nature of the problem

The influence of a watercourse on the recharging of an alluvial aquifer depends strongly on the management of the latter. In fact, there are two scenarios:

— Under a natural regime, the alluvial aquifer receives groundwater from the slopes of the river valley and also from the catchment basin. Apart from during short periods of spate, this water is subsequently drained by the watercourse.

— Under a regime disturbed by pumping, the water-level depression created by drawdown in the cone of influence will suck in water from the river and cause local reversals in the flow. The pumped water is derived from the mixing of groundwater *sensu stricto* and river water that has resided for variable periods of time within the aquifer.

In major well fields situated a short distance from watercourses, the proportion of water coming from the river is clearly predominant (of the order of 60-90%). Thus, the water quality in watercourses feeding the aquifer is crucial in determining the water quality of the aquifer to be developed.

Nevertheless, experience shows that waters generally undergo profound modifications when percolating through more or less plugged river beds and banks and then traversing a natural filter made up of several tens or even hundreds of metres of alluvium. These changes, which take place under the influence of complex physical, chemical and biological phenomena, often tend to improve the water quality. This accounts for the fact that, despite the large number of well fields of this type, little is known about the physico-chemical and biological phenomena which condition such modifications.

Confronted with the growth of human activities that are potential sources of pollution, as well as the continuing development of new compounds, it is by no means certain that improvements with respect to conventional pollution will stand the test of time. Thus, it has become indispensable to analyse the phenomena involved and to quantify the action to be taken in the face of pollutants currently considered as the most hazardous, i.e.: nitrogenous compounds, radionuclides and organic micropollutants.

In this way, the LE and AESN considered it important to characterize the geochemical system making up the river and the alluvial aquifer in relation to the different pollutants that are likely to be present in the river water, and to develop models for predicting the quality of pumped waters. These models also need to be checked in the laboratory and *in situ*. Ultimately, the objective is to combine these qualitative models with available hydrodynamic and dispersion models, in order to provide a forecasting tool which takes global phenomena into account.

6.6.2 Choice of study site

At Aubergenville, on the lower reaches of the Seine, the LE has been operating a set of boreholes situated on the southern bank of the river which tap the Senonian Chalk. The shallow groundwaters, which circulate in a terrain made up of chalk and alluvial deposits, are in hydraulic communication with the river. Abstractions carried out by the LE are of the order of 45 Mm^3/year.

Starting in 1976, the LE has undertaken a series of studies aiming to ascertain the hydrogeological mechanisms in operation and evaluate the risks of pollution. Several mathematical models of the aquifer were set up in order to address the problem. After calibrating the hydrodynamic model for steady-state and non-steady state regimes, a pollution risk study was carried out. This was followed by the development of a hydrodispersive model. The large amount of information acquired for this well field has led to its choice as an experimental site.

Within the Aubergenville well field, a preliminary study enabled selection of Rangiport island (well A8) for the installation of a measurement network.

The characteristics of this site are as follows:

— The existence of a strong link between the aquifer and the river.

— Data is available from tracer studies carried out in observation wells and in the Seine.

— Easy access for equipping the site and obtaining field data.

— Well defined domain of study in a restricted space.

— Along with well A8, the eight observation wells (including one in the Seine) situated on Rangiport island form the basic frame of the field measurement network. Apart from a transverse hydraulic profile and differential water-level monitoring, the measurement network was set up in such a way as to enable:

— An initial reconnaissance of the characteristics of the different materials making up the experimental site (sediments, alluvium, Chalk, etc.), including lithology, mineralogy, content in absorbed compounds, fixed bacteria, etc. Otherwise, samples of mud from the Seine were collected to the north of the island.

— The setting up of an analytical monitoring system for waters collected from the Seine and the array of observation wells — on a yearly basis — in order to follow the geochemical modifications of certain molecules during their transfer from river to groundwater. The geochemical context and variations in concentration were also monitored, as well as the presence of certain types of layer which promote purification of the water.

6.6.3 Geology and hydrodynamics of the site

The geology of the site can be summarized as a succession of two major units: present-day alluvial deposits (with pre-Recent deposits at their base) make up a layer about 12 m thick overlying fissured Senonian Chalk. Flow velocity measurements carried out by micro-current meter have shown that the Chalk is water-yielding to a depth of about 10 m beneath its roof.

From the hydrodynamic point of view, the profiles of temperature and total head difference measured in observation wells have enabled the development of a combined flow/thermal transfer model. The permeabilities were approximated by means of an inversion method.

In particular, it emerges from the above that flow takes place mainly in the Chalk and that the banks of the Seine are largely plugged. Seepage from the river towards the aquifer takes place preferentially through the bed of the river rather than through its banks.

6.6.4 Characterization of the biogeochemical system

Waters from the Seine and from observation wells were analysed for the different nitrogenous species, pesticides (atrazine-simazine), the major cations and anions, as well as certain metals. The programme took place over a period of one year, with weekly or monthly sampling according to the substance being analysed.

Sampling in the observation wells was carried out by means of a submerged pump. Temperature, pH, redox potential, resistivity and dissolved oxygen concentration were measured in the field. Otherwise, the pore waters of muds from the Seine were collected and analysed using dialysis cells and core samples taken at different times during the year.

Analytical monitoring in the field

It became apparent that the Seine muds are characterized by strong geochemical gradients within the topmost few centimetres. These gradients are due to biological processes that can be summarized as follows: in the presence of organic matter, the electron-acceptor species O_2, NO_3^-, Fe^{3+}, and SO_4^{2-} are partially or fully consumed or transformed, the oxygen and nitrates disappear, the ferric iron is rendered soluble by conversion to ferrous iron and the sulphates are reduced. Considerable amounts of ammonia make their appearance.

Meanwhile, the aquifer appears as a reduced zone with low redox potential (0-100 mV), very low dissolved oxygen and high concentrations of iron and manganese (up to 8 mg/l of Fe and 700 μg/l of Mn). Nitrates and nitrites are both absent in the waters sampled from observation wells, in contrast to ammonia or organic nitrogen. After a sharp increase across the alluvium, total nitrogen levels then suffer a rapid decline. The same applies to iron and manganese.

In addition to their spatial variations, certain ionic species such as sulphate are also seen to display a seasonal variation.

Laboratory study of nitrogenous compounds

The study was focused in particular on the denitrification potential of the aquifer and muds.

Depth (cm)

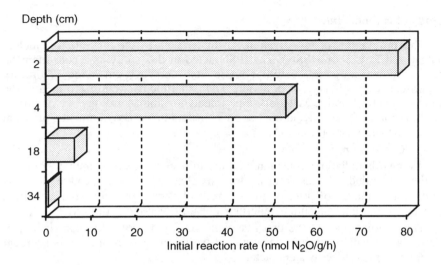

Initial reaction rate (nmol N_2O/g/h)

Figure 6-16
Initial denitrification rate vs. depth in Seine muds.

Depth (m)

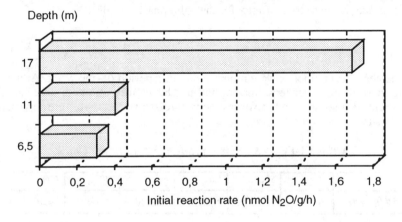

Initial reaction rate (nmol N_2O/g/h)

Figure 6-17
Initial denitrification rate vs. vertical depth within the aquifer.

This approach aims to establish a better definition of the time-variation of nitrogen in different levels of the aquifer and in the Seine river muds. To achieve this, batch trials were carried out under anaerobic conditions, with or without blocking of the denitrification process.

These trials revealed the denitrifying capacity of the aquifer and the Seine muds. In the Seine muds, the reaction is immediate and denitrification rates reach values as high as $1.6.10^{-7}$ mol N/g/h before falling off sharply with depth (cf. figure 6-16). In the aquifer, a relatively long latency time is observed coupled with lower denitrification rates of the order of 10.10^{-9} mol N/g/h (cf. figure 6-17).

Study of radionuclides

This programme of investigation was undertaken by the Institute for Nuclear Safety and Protection (IPSN). The IPSN acted in this study as a consultant on matters concerning radioactivity, while also carrying out some laboratory experiments on this type of pollution. The experiments were performed at the laboratories for "Active Materials Experimentation and Metrology" belonging to the Division of Research and Studies on Transfers in the Environment situated at Cadarache and Orsay (France).

— *Choice of radionuclides:*

As regards radiation protection, in order to respond to an accidental pollution or the controlled releases from nuclear installations, the radioactive elements studied as a matter of priority are, in order of decreasing importance: iodine, caesium, strontium, ruthenium, tellurium, cobalt, silver and antimony.

The radioisotopes of the first five elements listed above are responsible for almost all the public health hazards that would arise from a serious accident involving a pressurized-water reactor.

— *Main results:*

The determination of absorption isotherms at 15°C for the Chalk was carried out in a nitrogen atmosphere. These isotherms lead to a ranking of the studied elements according to their affinity for the solid phase (Chalk):

$$Ag > Cs > Co > Sb > Sr > I$$

The absorption kinetics were monitored over a period of one week. At the end of this period, however, equilibrium had not been fully attained. The partition coefficients *Kd* so obtained (absorbed activity/dissolved activity) are reported in Table VI-IV.

TABLE VI-IV — *Kd of main radioisotopes in the Chalk.*

	^{137}Cs	^{85}Sr	^{125}Sb	^{60}Co	^{110m}Ag	^{131}I
Kd (m³/kg)	68.10^{-3}	$2,9.10^{-3}$	$6,4.10^{-3}$	$57,9.10^{-3}$	∞	4.10^{-5}

Supplementary experiments are currently being undertaken with radioisotopes of iodine and silver, which exhibit a special type of behaviour.

Modelling of transfer mechanisms

The modelling of fluid flow and the transfer of conservative compounds needs to be combined with biochemical phenomena which affect the transfer of nitrogen. The long term objective is to extend this modelling to the whole of the Flins-Aubergenville well field (France).

Modelling of the river-bank effect will make it possible to provide a series of tools for controlling the quality of untreated waters. This system of management starts with water quality observation stations for rivers and aquifers, linking up with pollution transfer models for the fluvial environment and the simulation of river-bank effects. These modelling studies will be soon combined with hydrodynamic and fluid dispersion models for groundwater flow. Such an arrangement for qualitative and quantitative monitoring will lead to a better understanding of physico-chemical, hydraulic and biological phenomena, thus acting in favour of the setting up of an optimized system for water resource management.

6.6.5 Summary

All the results acquired in this study indicate the presence of a river-bank effect between the Seine and the aquifer. The effect can be summarized in terms of an intense biological activity in the uppermost few centimetres of bottom mud as well as a marked degree of purification within a few metres of the river bank.

In the first instance, biological reduction under anaerobic conditions produces large amounts of sulphides, phosphorus, iron, manganese and — above all — ammonia (derived from the transformation of organic nitrogen bound to the suspended particulate matter). Nevertheless, an extremely rapid denitrification also takes place.

Secondly, the purifying capacity of the alluvium is called into operation, thus enabling the removal of large quantities of substances previously taken up in solution.

It is noteworthy that significant quantities of certain pollutants such as the atrazine-simazine pesticides are found in the Chalk, whereas they are absent in the alluvium. This phenomena can be explained by the hydraulic complexity of the aquifer system. The higher transfer velocities revealed by field measurements in this zone restrict the purifying capacity of river-bank materials.

In the light of these results, the predictive model should not only take account of accumulation phenomena occurring in the Seine muds and the purifying capacity of alluvium, but also the rapid transfer mechanisms due to the hydrodynamics of the system.

In other respects, the use of the results obtained from analytical monitoring and laboratory experiments from now onwards provides an approach to the modelling protocol for groundwater quality control at other sites.

6.7 Conclusion

The methods used to preserve water wells, in order to prolong their service life and guarantee the potability of pumped waters, are numerous and in most cases complementary. In this domain, it would appear more advantageous on both technical and economic grounds to favour prevention rather than cure.

A will to preservethe catchment structure against the test of time should be present from the design stage. To achieve this, it is necessary to choose materials (e.g. casings, screens, pumps, etc.) that are best adapted to the physico-chemical constraints of the medium, in other words, the materials that will age under the best conditions.

Subsequently, as described in this chapter, the establishment of protection zones is an essential stage in the implementation of catchment structures. The definition of protected perimeters brings together various constraints of different nature, which may be physical, environmental or economic. These constraints lead ultimately to incompatibilities between, on the one hand, the potential water-yield and also the conservation and rehabilitation of the natural environment and, on the other hand, the nature of modifications affecting the well caused by human activities.

Among the physical constraints, the following may be mentioned:

— Whether the aquifer is confined or unconfined.

— The permeability of formations overlying the aquifer.

— The recharge mechanism of the aquifer and the exchange with other reservoirs.

— The physico-chemical nature of the country rocks and the groundwater.

A final stage consists of monitoring the changes in the yield and capacity parameters of the catchment structure (specific capacity of the well), in addition to the quality of pumped waters.

The thorough and regular monitoring of water wells is a key element in their management, equally from the qualitative as from the quantitative point of view. It enables the early detection of any problems that may arise during the development of the well, thus allowing a rapid and cost effective remedy.

The nature and periodicity of measures taken for checking and monitoring should be defined as precisely as possible in relation to the characteristics of the catchment structure (dimensions, age, condition, etc.), and should then be scrupulously observed.

CHAPTER VII

Restoration of water wells

" In action, be primitive; in foresight, a strategist."

René Char

The restoration of a water well represents a major action both for the operator and the community owning the catchment site. As pointed out in previous chapters, the implementation of a water well is a complex operation. The following serves to show that the task of repairing a deficient well is actually just as long and complicated to carry out. Although there are numerous techniques for restoring or regenerating wells, the main difficulty lies in the correct formulation of the problem and choosing the most suitable method(s). In fact, an analysis of failures shows that, in most cases, the method employed by the drilling company is not necessarily the best adapted to the problem as posed, which may itself have been poorly identified or even completely missed.

For example, it not at all uncommon for a company with a good record in the control of acidification to propose this type of treatment systematically without concern for the actual cause of the blockage nor the possible effects of the acid on clogging phenomena. Faced with this lack of professionalism, as well as the uncertainties and hazards associated with fault diagnosis, certain contracting authorities opt for the construction of new catchment works rather than restoration. On the other hand, it should be recognized that the restoration of certain very old wells is not necessarily justifiable, since such structures become inefficient due their particularly bad state of decay or, above all, their poor initial design.

Otherwise, even if the nature of work is clearly specified, any inadequacy in the operational procedures can be extremely detrimental. In the case of a change of screen in an old water well, for example, the extraction requires the use of a machine that has to operate inside the existing casing pipes. If the casing is corroded and thinned by ageing, it runs the risk of being destroyed during the operation. Because of this, it is necessary to make prior checks that all the technical conditions are satisfied in order to carry out the screen replacement with success. The preliminary diagnosis involves, on the one hand, the general state of repair of all parts directly or indirectly concerned in the operation and, on the other hand, addresses the nature of the measures to be implemented. The fact that these mesures may not be fully adapted can add to the risk of the operation.

It is thus important to carry out a very thorough and full diagnosis, leading not only to an identification of the origin of the problem but also to a definition of the most suitable measures to be adopted before proceeding with well restoration.

Fortunately, much progress has been made over the last decade in our understanding of the phenomena that lie at the origin of catchment structure deterioration, especially in the field of clogging of water wells. However, great difficulty remains in any attempt to make general comparisons and each problem should find its solution in the context of a specific diagnosis.

TABLE VII-I — *Most commonly encountered problems as a function of aquifer type and frequency of maintenance.*

Aquifer rock-type	Most common problems	Maintenance frequency (DWS well)
Alluvium	Clayey-sandy clogging, iron precipitation, presence of scale, biological clogging, drop in discharge capacity, casing rupture	2- 5 years
Sandstone	Clogging of fissures, casing rupture, presence of sand, corrosion	6 -10 years
Limestones	Clogging of crack and fissure network, presence of clay, carbonate precipitation	6 -12 years
Basaltic lavas	Clogging by clays, deposits	6 -12 years
Metamorphic	Clogging of fissure network by clays, mineralization of fissures	12 -15 years
Consolidated sediments	Clogging from Fe and Mn ions in solution, fall in well discharge	6 -8 years
Unconsolidated sediments	Influx of sand and/or clays, scaling, biological clogging, etc.	5 - 8 years

Figure 7-1 indicates the manner in which a preliminary diagnosis is drawn up with respect to the whole range of different repair operations. In the context of the reconditionning of catchment works, two types of repair operation are distinguished: regeneration and restoration.

— *Regeneration* encompasses all the hydraulic and chemical methods employed to combat ageing in water wells, with the aim of increasing their discharge capacity.

— *Restoration* makes use of the same procedures, but applies them to temporarily abandoned wells where the intention is to carry out repairs to enable further development. In such cases, when regular operational data are lacking, the diagnosis is made after establishing a new yield-depression curve and, eventually, also carrying out a video inspection.

In this chapter, the terms regeneration and restoration are taken to mean the same thing and no distinction is drawn between them .

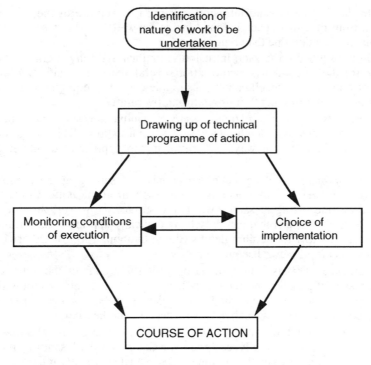

Figure 7-1
Flow chart for preliminary diagnosis

Once a diagnosis of ageing has been made and the type of degradation is defined, the regeneration of a well generally comprises two phases:

— Phase 1: pneumatic and hydraulic procedures.

— Phase 2: chemical treatments.

It is crucial to choose a procedure that is adapted to the type of degradation involved, since an unsuitable treatment will accentuate the lowering of discharge capacity of a well.

7.1 Treatment of sanding up

The sanding up of a water well very often comes about because of corrosion, overdraft or poor design. Unfortunately, there is really no reasonable remedy for the sanding up of a catchment structure once it is up to final equipment specification.

When a well is sanded up, the first priority is to locate the precise source of the sand intrusion. It is nevertheless possible to make a distinction between accidental and permanent sanding up.

— In the first case, sand intrusion is caused by rupture of the well screen (material from the gravel pack is found inside the well) or otherwise some major hydraulic malfunction due to overpumping.

— In the second case, sand intrusion is gradual and may become a serious problem for the catchment structure after several years. This situation is more typical of poor design specification (unsuitable choice of equipment and gravel pack, poor well development) or groundwater overdraft.

Whatever its origin, it is firstly necessary to remove the sand deposit blocking the well (sometimes over depth intervals of several metres). This operation may be performed in different ways according to the depth and diameter of the borehole:

— *Air-lift pumping*. This technique consists of setting up an emulsifying system (cf. Chapter 3.5) by injecting compressed air at the bottom of the well, which allows the lifting of sand towards the surface. However, this procedure requires technical means that can only be realistically applied to small-diameter wells and at shallow to medium depths (down to about 100 m). Otherwise, the flow rates and air pressures become excessive.

— *Scraping*. This is a mechanical operation that enables the extraction of sand by means of a bailer lowered down the well. Although this system is simple, it is very often ruled out due to the frequent damage to equipment caused by the rapid descent of the bailer, which must sink deeply into the sand.

— *Pumping*. Sand removal by pumping may be envisaged for shallow wells, either using a pump specially designed for the purpose (desanding pump or sludger) or a jet-pump when the volumes to be extracted are only minor.

— *Surging or clearance pumping*. A desanding plunger (or surge block) is fitted on the end of the pump inlet and lowered into the well (figure 7-2). These devices are composed of two or three wooden disks joined together by copper or rubber disks. The plunger assembly is manoeuvred by drillpipes attached to the winch.

The spacing of the wooden disks may be varied, thus making it possible to restrict the desanding action to just the intended areas. For a given pump discharge, the rate of input into the basket (retriever) is higher than the value that would have applied if water entered the screen face across its entire surface area. As a consequence, the cleaning out of the formation is more intense and takes place over a shorter time span. The discharge is increased progressively with each zone treated. When this stepped desanding procedure is completed, the basket is removed and the well is pumped at a discharge rate 50% more than the water extraction regime during normal operations.

The movements of the pump serve to agitate the formation and the gravel pack near the screen, thus disrupting any bridging or abnormal void spaces that may have been created. In this type of operation, the pump is started and stopped suddenly. Although the water flows in a defined direction during pumping, the water descends through the piping during stops because the pump has no clack valve at the bottom. This downward impulse produces an inverse movement that agitates the formation. The vigorous nature of the movements means that the screens need to have adequate mechanical characteristics.

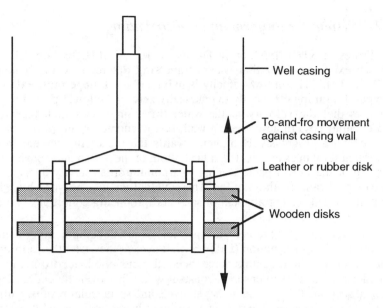

Figure 7-2
Surge block for desanding.

Whatever the method employed, this operation is only provisional in character since it does not eliminate the cause of sanding up. This explains why, in the final analysis, it is necessary to carry out a full examination in order to ascertain the real cause of the deficiency.

— If sand penetrates into the well through perforations in the screen, or if there has been a poor choice of open area dimensions, it may then be possible to fit a new screen on the inside of the old one. This double screening will restrict the intrusion of sand provided that the new screen is chosen appropriately (e.g. a Johnson-type screen with very small slots and high open area). It is also necessary to emplace a new gravel pack between the two screens. Such a procedure has the following drawbacks:

- High cost.
- Reduced internal diameter of the borehole (problem of installing pump).
- Additional loss of head in the water well, thus leading to a drop in yield.

— If clogging is due to influxes of sand linked to excessive speeds of penetration, the most prudent solution is to limit the production yield in order to reduce water velocities vertically beneath the well equipment. This type of solution should be associated with periodic cleaning of the borehole and frequent checking for pump wear.

7.1.1 System for uniform inflow distribution

Professor KIRSCHMER of the Darmstadt Technical Highschool (Germany) has shown experimentally that waters traversing a screen exhibit a non-uniform velocity field. The inflow velocity is maximal at the uppermost extremity (near the pump), falling off rapidly to practically zero in the lower part of the screen. It appears that the velocity of the water traversing the screen face follows an exponential law. In an operating well, this phenomenon gives rise to velocities that become progressively higher towards the top of the screen. In this way, divergent flow lines are created at the bottom of the screen and convergent lines at the top of the gravel pack, leading to a nonuniform velocity distribution in the gravel pack and in the aquifer. These velocities can be very high in places, bringing about the entrainment of sand and producing conditions favourable for the development of bacteria.

To overcome this problem, it is possible to install equipment for maintaining a uniform inflow distribution (UID). Such a system may take the form of a second screen in which the slots have been dimensioned according to the well characteristics so as to ensure a constant water flux over the entire height of the tapped section. This procedure has shown some spectacular results, particularly in sanded up wells where the homogenization of water flow has led to a considerable decrease in the intrusion of sand (up to 99%).

UID systems can also be used with success in the case of wells that show some risk of scale formation and/or bacterial proliferation. A patent has be filed for the UID system manufactured by the company EUCASTREM (see list of suppliers in Appendix). The cost of such a system is of the order of 5% of the total cost of the catchment structure, being evidently a function of the hydrodynamic properties of the well and especially its completed specification.

7.2 Treatment of mechanical clogging

Mechanical clogging is due to the physical entrainment into suspension of materials such as mud, silt, clays and the very fine fraction of pulverized sandstones or sands. It takes place generally under the influence of a destabilization of the water-bearing formation in the near-field of the catchment site, affecting particularly those wells which have undergone excessive or uncontrolled pumping, even on a temporary basis. Due to the hydraulic perturbations caused by pumping, the gravel pack surrounding the screen is destabilized leading to additional losses of head and a consequent lowering of the water-yield.

As an example, let us consider the clogging brought about by a water containing 100-150 ppm (1-1.5 mg/l) of suspended particulate matter (SPM). In a 12"-diameter borehole, an SPM load of 1 mg/l at a yield of 120 m^3/h produces a deposit of 3 kg every 24 hours. All the void spaces can be filled over a distance of 15 cm from the well wall in a period of 7-8 months.

7.2.1 Physical treatment

Mechanical clogging is due to fine particulates which accumulate within the gravel pack (internal clogging) or on its outer face (external clogging), or otherwise block up the screen slots. This phenomenon leads to a reduction in the permeability and, as a result, a lowering of the operating yield of the well.

Several techniques may be employed that are based essentially on destabilization of the gravel pack (breaking up the grain-size sorting that has developed).

As in the case with desanding, it is firstly necessary to clean out the fines, including part of the clay fraction. Here again, the use of a "bailer" should be avoided since this frequently causes deterioration of the well screen.

a) Air-lift pumping

The technique of air-lift pumping described previously can be carried out in an efficient manner by adding desanding plungers which thus enable a treatment of the formations by depth interval.

b) Controlled overpumping

This method consists of pumping while progressively increasing the yield. The starting yield is one fifth of the final yield. By successive rises, a final value is attained which is twice the capacity of the well under steady-state operation. During the procedure, the pumped water is loaded with sand and clay, but becomes progressively clearer at each step. The pumping time can be relatively long (several days at a rhythm of 24 hours pumping per day).

In the case of highly heterogeneous formations, where there are domains of variable permeability, the zones containing fine-grained material are liable to resist desanding. This is the main disadvantage of overpumping, which actuates development of the entire block corresponding to the tapped zone. To remedy this, it is necessary to carry out stepped pumping by treating each zone in turn.

c) Compressed-air treatment

The introduction of compressed air into a well beneath the standing water-level creates an overpressure which forces the water back through the screen towards the aquifer formation. It is evidently necessary to take care to prevent any connection with the outside environment so operations can be carried out in a closed system. Subsequently, either a slow or a sudden decompresion is used to trigger an inflow of water into the catchment structure. The repetition of these movements agitates the gravel pack and the aquifer formation itself, thus leading to disaggregation of the terrain which chiefly affects the loose and weakly bound fine particles of sand and clay attached to the screen face or the surface of the gravel grains.

The amplitude of the resulting drawdown is a function of the pressure of the compressed air. Using a pressure of 7 kg, the groundwater body may be lowered by about 70 m. Following this, the catchment structure is cleaned out by air-lift (emulsifying) and the treatment is completed with pumping.

However, this method is not without some risk to the catchment structure. As a prior step, it should be established that the catchment system can resist the pressures applied and that the screens are in no danger of collapsing.

In the presence of a gravel pack, there is a risk of disturbance (due to the treatment) which might open up preferential channels within the formation.

This procedure may perhaps be employed with the least risk in media that are self-developing. Even though it is excellent for cleaning out water wells, compressed-air treatment does not, however, eliminate the causes of clogging.

d) Water injection

The introduction of water with or without pressure allows the cleaning out of catchment structures that are clogged by unconsolidated and weakly attached deposits. It is advantageous to use water injection in addition to other methods, particularly cleaning by compressed-air or air-lift which remains the most practical treatment to implement.

The simplest method of water injection consists of direct discharge into the well from ground level or from the top of the casings. Water injected in this manner passes from the well into the terrain via the screen slots, while the action of the counter stream detaches the deposits and carries them off into the surrounding formation. The declogging effect becomes even more efficacious as the injected flow rate increases. However, there is always a risk that preferential channels open up and the formation is not uniformly cleaned throughout. Nevertheless, this drawback is partly remedied by following the injection by a series of pumpings which clear out resuspended particles from the catchment structure. The more such pumpings become necessary the more it is advisable to avoid resuspending excessive amounts of detached particles. This is the reason for alternating periods of water injection, pumping and, in some cases, compressed-air treatment, while the duration and flow rate are progressively increased.

It is frequently an advantage to act directly on the sites where deposits are formed, more especially in the depth interval adjacent to the tapped zone. In these cases, horizontal jetting can be applied with success; the method consists of applying a horizontal jet from inside the borehole towards the screen face.

The tool used is of the same type as described in Chapter II (cf. 3.1.4; well development using high-pressure jetting head). It is made up of 1-4 injection nozzles (jets with 4-10 mm internal diameter) each of which is placed horizontally near the closed end of a 50 mm-diameter pipe. The jetting head is lowered into the well and positioned opposite the screens. Arrangements are made for displacing the injectors over the entire depth interval of the tapped zone, and for rotating the injection pipe about its own axis so all the surface area of the screens can be washed. At the surface, the injection pipe is connected to a flushing (or jet) pump.

Several systems exist which are based on a swivel jet or reverse flow cleaning; in this context, we should mention two processes (HD and KW) that are marketed by the firm Carela.

Nevertheless, mechanical declogging (or swabing) is rarely sufficient, notably in the case of very old wells where clogging is due to the presence of fine clayey particles. This is the reason why it is often necessary to plan a complementary phase of chemical treatment.

7.2.2 Chemical treatment

As a complement to the HD and KW processes, the Carela company advocates the use of its products Bio-forte and Plus, which are composed of chemically pure acids, detergents, inhibitors, auxilliary reagents and hydrogen peroxide. It should be noted that this formulation is food-grade approved by European authorities for use in the domain of drinking water supply.

In the case of the Carela HD system, the manufacturer recommends using a mix of 1litre "Plus" in 30 litres "Bio-forte", at a concentration of 10%, for a contact time of 12-24 hours. The product information specifies that the deposits and incrustations should be totally eliminated during the residence time, while, in addition, the borehole should become completely disinfected.

The presence of hydrophilic wetting agents in "Carela Bio-forte" and "Carela Plus" leads to good capillarity of the product in solution, thus enhancing its penetrative in-depth action within the gravel pack and even farther into the aquifer.

In the case of the Carela KW process, the chemical treatment is carried out differently since it is based on the principle of blocking off the screened section of the well at three levels by means of three casing flanges. In this way, two chambers are formed which are isolated from waters located above and below the system being treated. A pump injects water into both of these chambers, thus contributing to the flux traversing the gravel pack. The central flange is fitted with a motorized screw pump which sets up a pressure differential between the two chambers. The overpressure in one of the chambers and the pressure drop in the other drives a flow of chemical product through the screen face and the gravel. By reversing the screw pump, it is also possible to reverse the direction of flow.

The reconditioning (or regenerating) head shown in figure 7-3 comprises stiff or soft brushes, a brush holder with spiral slots, a plunger with three flexible-lipped bearing blocks and four high-pressure nozzles offset at 90°. This arrangement enables washing of the well and/or dosing of the chemical product. A unique feature of this system is the combination of mechanical and hydraulic methods for washing the screened sections. The three plunger blocks give rise to alternating overpressures and pressure drops within the water column. The currents forced through the screen face cause diffusion of the reagents into the gravel pack.

Other types of chemical, such as the sodium phosphates, can be used to disperse the clayey fines; application of these products can be highly efficacious provided that the phosphate compound selected is suitable for the case in question.

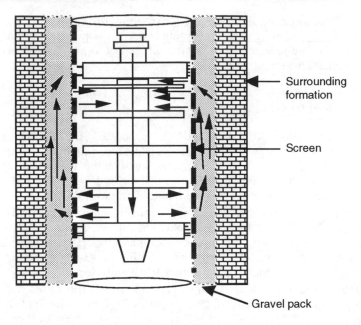

Figure 7-3
Schematic diagram of a well cleaning head manufactured by CARELA©

The choice of products for chemical treatment requires a knowledge of the compounds involved and their properties. The sodium phosphates are derived from a number of different phosphoric acid species.

— *Salts of orthophosphoric acid* (H_3PO_4), which are available commercially in both the anhydrous and the crystallized form:

- monosodium orthophosphate, NaH_2PO_3;
- disodium orthophosphate, Na_2HPO_4;
- trisodium orthophosphate, Na_3PO_4.

— *Polyphosphoric acids* ($H_4P_2O_7n\ HPO_3$) are very numerous. The first member of this series is pyrophosphoric acid ($H_4P_2O_7$), which forms the following well known salts:

- disodium or acid pyrophosphate $Na_2H_2P_2O_7$;
- tetrasodium or neutral pyrophosphate $Na_4P_2O_7$.

The second member of the series is tripolyphosphoric acid ($H_5P_3O_{10}$), whose best known sodium salt is pentasodium tripolyphosphate ($Na_5P_3O_{10}$). The subsequent members form the vitreous polyphosphates, which are commonly characterized by their P_2O_5 content. When this content is lower than 63%, the compound is said to be a short-chain polyphosphate. In their general formula $P_2O_7Na_4n\ NaPO_3$, n is equal to or greater than 2. The following salts are to be found:

- tetrapolyphosphate, with $P_2O_5 = 60.5\%$;
- pentapolyphosphate, with $P_2O_5 = 62\%$.

The long-chain polyphosphates have P_2O_5 contents higher than 63%. Their general formula is $NaPO_3 P_2O_5n$, where n is always a very high number. The extreme member of this series is sodium hexametaphosphate, with a theoretical P_2O_5 content of 69.7% while the commercial product has a content of the order of 68%.

— The metaphosphoric acids (HPO_3) form a series whose first known member is sodium trimetaphosphate or $(NaPO_3)_3$.

The salts of the phosphoric acids exhibit four essential properties:

— *Complexing efficiency or chelating power*, which enables phosphates to fix anions or cations to yield new species known as complexes.

— *Sequestering power* is the capacity to form a complex with a soluble metal salt, the metal species thus loosing its normal cationic properties. It is considered that, in the presence of ferruginous water or ferric precipitates, the use of sodium phosphates promotes the occurrence of sequestering.

— *Solutizing power* characterizes the ability of phosphates to form soluble complex salts by fixing insoluble salts into their molecule. This property appears to be of interest, for example, in the descaling of carbonate or calcium sulphate incrustations.

— *Dispersing* or *peptization power* is a property which enables the dispersal and liquefying of substances. It is a suitable choice in the presence of clayey media.

The above properties vary with the method of preparation of the phosphate, since the complexing efficiency increases with chain length. Thus, pyrophosphate has a low capacity to form complexes while hexametaphosphate exhibits the highest capacity. On the other hand, the dispersing power increases as the chain length decreases. In this way, the dispersing power of pyrophosphates is very high and that of the very-long-chain vitreous polyphosphates is intermediate at their intrinsic pH of 6-8.

With a view to treating sand-clay aquiferous media, it is necessary to achieve deflocculation of the clays. In this context, it appears that the polyphosphates are best suited to yield satisfactory results, while the vitreous polyphosphates in particular offer certain advantages. However, the orthophosphates should be ruled out of account. In reality, obtaining deflocculation is a complex problem and recourse is often made to products that combine dispersing power with chelating properties or, on the contrary, dispersant properties can be associated with a complexing agent. These products belong to the category of special polyphosphates, the most typical of which are Calgon (USA) and Giltex (France).

The properties of sodium phosphates also vary as a function of time. The pyrophosphates, polypyrophoshates and metapyrophosphates are actually produced from orthophosphates and will gradually revert to this form. This degradation is accelerated by high temperatures. It also varies with pH, showing an increase at pH values less than 7 and above 8.5, but the increase is more rapid under acid conditions compared with alkaline conditions.

Some kinetic data are given here for the hydrolysis of polyphosphates into othophosphate as a function of pH in the temperature range 20-40°C.

— Pyrophosphate: at pH 7 and 40°C, 1% converted in 60 hours.

— Pyrophosphate: at pH 9 and 40°C, 1% converted in 400 hours.
— Tripolyphosphate: at pH 7 and 40°C, 1% converted in 30 hours.
— Tripolyphosphate: at pH 9 and 40°C, 1% converted in 100 hours.
— Metaphosphate: at pH 7 and 25°C, 1% converted in 60 hours.
— Metaphosphate: at pH 9 and 25°C, 1% converted in 90 hours.

From this, it can be seen that the properties of sodium phosphates change according to their conditions of use (i.e. temperature, pH, concentration). It is appropriate here to give some details on the conditions of use of the various commercial salts.

As regards the effect of temperature, the dispersing and chelating powers of sodium salts appear to show little variation between 20° and 40°C. The natural environments in which these products are intended to act have temperatures in the range 20-40°C.

The commercial pyrophosphates are available in two forms:
— disodium pyrophosphate $Na_2H_2P_2O_7$, intrinsic pH = 4.2;
— tetrasodium pyrophosphate $Na_4P_2O_7$, intrinsic pH = 10.2.

By associating these two salts, it is possible to cover the whole range of pH from 4.2 to 10.2 and set a precise pH value chosen in advance of the treatment. The use of disodium pyrophosphate allows buffering of the pH in excessively alkaline environments.

Hexametaphosphate-type polyphosphates commonly exhibit an acidic pH of around 6-7, but this can be increased if needed. The intrinsic pH values of special polyphosphates show a wide range of variation (e.g. the pH of Giltex E is 7.1-7.6 and the pH of Giltex N is 8.3-8.8). Otherwise, it should be noted that sodium phosphates offer the possibility of selecting a pH value in advance.

The conditions of application for any given product are dependent on numerous parameters. It is clear that no universal product exists that is suitable for all cases of mechanical clogging. Unfortunately, it has to be acknowledged that chemical treatment is rarely applied in a satisfactory manner.

In the context of polyphosphate treatment, there are generally three main causes of failure.

— 1) As mentioned above concerning the conditions required to obtain deflocculation of clays, it would appear that the medium should be alkaline (with a pH of around 9). By contrast, clays will flocculate at a pH of less than 5.

Thus, one cause of failure may arise from the use of disodium or acid pyrophosphate alone (with a pH of 4-4.3) in an acidic or even alkaline medium, since it is doubtful that a pH compatible with clay deflocculation can be obtained. Moreover, companies using this product are almost unanimous in noting its ineffectiveness. All the same, this does not condemn in any systematic way the use of acid pyrophosphate, provided there is a prior treatment of the medium or the product with an adjuvant that raises its pH (e.g. caustic soda or sodium carbonate).

It is also possible to mix the acid pyrophosphate with neutral or tetrasodium pyrophosphate, the latter product having a pH of between 10.1 and 10.6. This aspect has already been discussed, and it should be borne in mind that, from first principles, it is evidently an advantage to make use of the high dispersing power characteristic of acid pyrophosphate.

However, it should be pointed out that if the pH of the medium is unknown at the time of treatment, then it is subsequently very difficult to explain the results that are obtained.

After treatment with hydrochloric acid, ferric hydroxides may form which entrap the clayey particles and cause development of a gel. In this case, there is no longer just the action of calcium phosphates on the clay, but also an effect on the ferric compounds.

In summary, pH on its own is insufficient to account for the results obtained during treatment, even though it proves necessary to establish alkaline conditions to ensure the deflocculation of clays.

TABLE VII-II — *Properties of some polyphosphates.*

Product name	% P_2O_5	Density	pH of 1% solution	Solubility g/ml water	Complexing efficiency	Dispersing power
Acid or disodium pyrophosphate, $P_2O_7Na_2H_2$	62 - 63	1.15	4 - 4.3	13 g at 20°C	weak	very strong
Neutral or tetrasodium pyrophosphate, $P_2O_7Na_4$	52 - 52.5	1.10	10.1-10.6	5.5 g at 20°C 12.5 g at 40°C 22 g at 60°C	weak	strong
Neutral or sodium metaphosphate, $(NaPO_3)_n$	67.5 - 68.5	1.40	6.30	unlimited	strong	medium
Giltex E (mixture of polyphosphates)	66 - 67.5	1.30	7.1 - 7.6	25 g at 20°C	strong	medium

— 2) A second cause of failure lies in the amounts of product that are employed. An excessively small volume will not produce the desired effect, while too large a quantity will bring about the opposite effect. It should be remembered that an increase in the concentration of tetrasodium pyrophosphate will cause a reversal of its deflocculating effect and give rise to flocculation. It is thus indispensable to know the quantities to be employed in relation to the nature of the product and the role it is intended to play. In fact, amounts vary according to whether the objective is to disperse the clays or form complexes with the calcium or ferric ions.

By convention, the quantities of products to be used are expressed as a percentage of the volume of water contained in the borehole. The following is a compilation of information obtained on dispersant dosages:

- Calgon: 1.7-3.4% (U.S.A)
- The Progil company indicates a value of 2% for products such as Giltex (France). This value is to be taken into account for other dispersant products, which include the pyrophosphates and hexa-metaphosphates.

- For products in its range, Layne USA proposes a value of 1.2% for use in well restoration work. In the "Johnson National Drillers Journal" of July-August 1954, it is stated that, in the treatment of drilling mud, a value of 0.6% should not be exceeded for tetrasodium pyrophosphate, tripolyphosphate, hexametaphosphate or septa- phosphate.

Theoretically, the volume of water considered here is situated adjacent to the screen face over the entire height of the screened section, but augmented in order to take into account the effects of osmosis or local dilution in waters above or below the well screen. This volume should also be increased to compensate for the void space in the formation at the same level as the screen affected by the treatment.

— 3) Another known source of failure is the deterioration in the properties of products during their use. It may be recalled that the deflocculating properties of tetrasodium pyrophosphate, for example, are not permanent and last only for a matter of hours. Nearly all the literature on the subject advises that products should not be left in contact with the terrain for periods longer than 24-36 hours and 48 hours at the most. Some kinetic data are given above for the hydrolysis of polyphosphates into othophosphates as a function of pH.

Finally, experience appears to show that Giltex, for example, should not remain in contact with the terrain for more than 12 hours.

7.3 Treatment of carbonate clogging

Under the term carbonate we include various mixtures of carbonates, bicarbonates, alkaline earth sulphates, hydrates and hydroxides. Due to the influence of different physical factors (flow velocity, aeration, pressure variations, etc.), the groundwater occasionally gives rise to deposits of carbonate or ferruginous scale - and sometimes both at the same time - on parts of the well that are actively tapping the aquifer (i.e. the screens, drains and gravel pack).

This scaling does not necessarily take place in a regular manner. It may affect certain parts of the catchment system, while leaving sufficient channel ways to maintain a certain level of discharge. Nevertheless, a reduction in discharge capacity remains the typical feature of clogging in a well.

The phenomenon of physico-chemical scaling can be influenced by the conditions of well operation, including, for example, the occurrence of frequent start ups and stops in pumping.

Physico-chemical clogging can also result from corrosion. The chemical attack of metals composing certain types of screen may lead to the formation of concretions that obstruct the openings, thus causing localized or widespread clogging deposits or scale that can be observed by video camera inspection.

The study of AVOGADRO and DE MARSILY (1983) brings together the equations proposed by different authors to describe the mechanisms of filtration and retention of colloids. The concentration of colloidal particles in the solution decreases logarithmically with the path length in the filter, while the efficiency of the filter falls off according to a similar relation as a function of time. Within the scientific community, however, there appears to be no consensus on the parameters to be included in these mathematical models.

The ageing of water wells brings about a progressive lowering of discharge capacity which is essentially due to clogging of the screened sections. The origin and composition of such deposits prove to be highly variable. In many cases, an analysis of causes is not carried out and the operator is satisfied with reducing the pump discharge using a closing device until eventual abandonment of the well. Such a procedure entails a lowering of pump efficiency and an increase in operating costs, although methods exist for slowing down the clogging or even restoring the well. However, restoration requires a full understanding of the well characteristics, in addition to the carrying out of supplementary studies from the hydrogeological and chemical points of view:

— Identification of hydrogeological characteristics, particularly their variation in time. This may be done, for example, through the comparison of yield-depression curves.

— Definition of variations in the water quality through a programme of physical and chemical analyses.

— Analysis of clogging deposits and their extent, since the entire height of the screened section need not be systematically affected.

— Survey of well equipment, since certain treatment products are contra-indicated in the presence of materials such as galvanized steel, zinc, etc.

— Choice of most suitable treatments and definition of a restoration strategy comprising a series of physical (mechanical and hydraulic action) and chemical techniques (dissolution of carbonates and sterilization).

The most generally used products for the treatment of carbonate clogging are hydrochloric acid (also known as muriatic acid) and sulphamic acid. The most commonly available commercial grade acid is 33% HCl (with a density of 22 degrees Baumé), which is diluted to a concentration of 10% during the declogging operation. Taking account of the acid pH and the chloride content of the medium, metals such as steel, stainless steel and bronze will undergo chemical attack. Polyphosphates are also used under certain special circumstances (presence of silt or clay deposits). It is evidently necessary to deactivate the acid by adding inhibitors which render the treatment compatible with the metallic components of the well structure.

The most commonly employed inhibitors (deactivators) are as follows:

— Formol, but the frequent presence of a methanol-based antioxidant produces a toxic effect. As a result, its use is not recommended.

— Powdered sulphamic acid, which is less aggressive towards metals than hydrochloric acid but which is more expensive.

— Various inorganic and organic acids are sometimes to be found in specific formulations (e.g. phosphoric acid, ascorbic acid, etc.).

It is noteworthy that, in all cases, whatever the acid used and even in the presence of a deactivator, both zinc and galvanized steel are corroded and should thus be avoided in the equipment of wells that capture calcareous and encrusting waters.

Various combinations of reagents are available as mixtures from suppliers and service contractors that implement acidization treatments. The main products in this category are as follows:

— Herli-rapid-TWB-FCM-1- (liquid), which is distributed in France by the Victor HEINRICH company (based in the department of Bas-Rhin) and manufactured by FELDMAN-Chimie. The principal constituents of this product are given below, with the strength of each component solution expressed as percentage:

- 85% formic acid
- 85% phosphoric acid
- 37% hydrochloric acid
- 99% isopropylic acid
- Inhibitors (FCM IV/1 - FCM IV/2 - FCF IV/8 - FCM IV/10)
- 5% vol. aethix fatty alcohol
- Ascorbic acid DAB 7

In France, use of this product is authorized by the Ministry responsible for health and it has received a favourable notice from the Higher Council for Public Hygiene.

— ID 60 is manufactured by Darc and marketed by Degremont. It is an inhibitor for hydrochloric acid which also acts as a bactericide, being made up of five constituent substances (notably, aminopropionate, a wetting agent and an inhibitor). In principle, each of the constituents is included on a list of food-grade substances (approved in relation to food hygiene). Since this product incorporates a tracer, it offers the advantage of being suitable for use in the monitoring of acidization.

— Carela Bio-forte and Carela Plus, which are distributed by CARELA-France (Bas-Rhin). These products are composed of a mixture of basic constituents (pure hydrochloric acid, citric acid, tartaric acid) as well as corrosion inhibitors containing non-foaming detergents. Hydrogen peroxide is the main component of an additive that is mixed in at the time of treatment.

— Nu-well® acid and Johnson Well Regenerator (JWR) are distributed by Johnson Filtration Systems. The former product is supplied in the form of pellets made up from a sulphamic acid base, and is relatively easy to manipulate. The latter is composed of chemically pure acids - both inorganic and organic - in addition to biodegradable agents.

On of the fundamental problems is to know whether it is reasonable to carry out chemical treatment of clogging while leaving the well pump in place or whether, on the contrary, it is better to withdraw it in view of the risk of damage from declogging products. A recent study undertaken by CIRSEE (CORDONNIER, 1992) shows that there is no simple answer from either the technical or economic standpoints. In particular, attention should be focused on the following points:

- The importance of diagnosis.
- The choice of reagents and the procedure for implementing the treatment.
- Age of the catchment works and the pump.
- Frequency of well operations.

These remarks should evidently be backed up by taking account of the local conditions and also the observations of well operators (number and age of wells, type of deposit, equipment and materials present at well site, restoration systematics, operational experience, etc.), which may be the predominant factors as regards choice of treatment and cost. The above should allow a better approximation of operating expenditure by including the labour costs in comparative budget planning.

Generally speaking, the implementation of declogging treatments usually requires special apparatus which enables brushing of the well walls prior to or simultaneously with introduction of the chemicals under pressure or in steps. Thus, it is advisable to call in specialized firms for such operations.

Finally, some contradictory opinions can be found in the literature concerning the effectiveness of such treatments. Some authors are entirely satisfied while others consider there is no conclusive effect. However, it appears that these judgements are not backed up by a statement of the conditions under which the treatment was carried out. In fact, although various products for the chemical treatment of carbonate clogging are available for purchase, they may be applied under a wide range of different conditions.

In general, the declogging of a well by chemical products is accompanied by mechanical procedures which require specialized measures and apparatus (packer, plunger, etc.). Since it is necessary to evacuate the dissolved matter and then sterilize the well, it is indispensable to define a programme of action suitably prepared in terms of technical means and drawn up by a qualified project manager.

7.3.1 Preventive treatment with polyphosphates

Before discussing the means for treating encrusted wells, a method should be mentioned that prevents scale formation in the first place. In fact, sodium polyphosphates are available that can stop the deposition of scale on screens, pumps or conduits. It is merely sufficient to cause circulation of polyphophates beneath the level of the pump intake.

It is generally acknowledged that 1 g of polyphosphate is required for every cubic metre of water at a temporary hardness of 10 degrees on the French scale. Otherwise, it is also accepted that a dose of 2-5 g/m^3 is suitable in most cases where the waters are circulating.

The two graphs shown on figure 7-4 summarize the action of polyphosphates. On graph no. 1, the rate of $CaCO_3$ precipitation is plotted as a function of quantity of polyphosphate added, assuming water with temporary hardness at 60°C heated to 80°C for one hour. A threshold can be seen above which any further addition is without effect. The value of this threshold concentration is 2 mg/l polyphosphate.

On graph no. 2, the quantity of calcium bicarbonate remaining in solution after injection of 2 mg/l polyphosphate is plotted as a function of temperature. It is noteworthy that, within the normal temperature range of groundwaters (25-40°C), a polyphosphate concentration of 2 mg/l will maintain a minimum of 800 mg/l of calcium bicarbonate in solution.

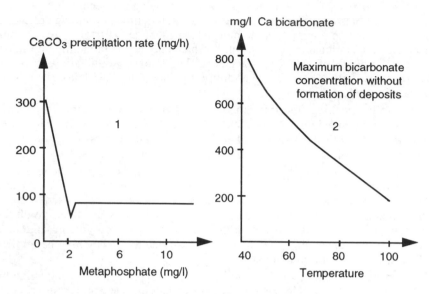

Figure 7-4
Effect of polyphosphate treatment.

7.3.2 Mechanical treatment

Mechanical methods for treating clogging include wall scraping and explosive shots. However, It should be pointed out that very little development of such methods has occurred in practice. The examples given here are cited merely for the sake of the record.

a) Scraping

Broadly speaking, carbonate deposits are very hard so scraping will only produce mediocre results if it is carried out in isolation. Nevertheless, scraping remains as a working method for descaling pipes and screens.

Use can be made of a wall scratcher, which is a pipe brush (or porcupine) composed of flexible wire spokes. The back-and-forth movement of this device in the borehole enables the detachment of scale. However, scraping has no effect on the incrustations that form in the screen slots or within the gravel pack.

This method could be employed with greater success if the incrustations were previously softened or dissolved away.

b) Use of explosives

Applied with the usual precautions, explosive shots will yield results as long as the deposits are fragile and are readily broken up by the shocks.

The operating procedure consists of placing small explosive charges at regularly spaced intervals on the casings and screens. By firing the shots at regular time intervals, it is possible to set up a continuous shock wave which is propagated through the groundwater body and which disrupts the deposits without damaging the casings or the screens.

It is generally acceptable to place a charge of the order of 12-36 g for every metre of well.

7.3.3 Acidization treatment

Acidization involves the injection, either at pressure or under gravity, of hydrochloric acid, sulphamic acid or proprietary chemical products in several passes (batches) and at different levels. Among other uses, this process enables the dissolution of carbonate scale that has been deposited inside the screen face or the slots, on the outer screen face, in the gravel pack or in the water-bearing formation itself.

It should be noted that the volume of acid used in treating an encrusted screen is less than that required for the development of a carbonate formation by acidization. In general, the dose used for descaling is made up of 1.5 to 2 volumes of acid solution for 1 volume of water within the well screen, giving a final concentration equivalent to 18-20 degrees Baumé.

— *Hydrochloric acid* (HCl) at 20 or 22 degrees Baumé is injected at low pressure (maximum 2 bars) or under gravity. It is necessary to carry out several passes by injecting at various levels in the well rather than one.

> As an example, a well 50 m deep with a diameter of 200-300 mm will need a minimum of 3 tonnes of acid injected in three passes.
> .   Residence time: maximum of 1-2 hours.
> .   Agitation: every 15 minutes.
> .   Withdrawal of acid by air-lift pumping.

— *Sulphamic acid* (NH_2SO_3H). The same mode of operation is used as described above. The advantage of sulphamic is that it is supplied as crystals, which allows greater ease of use, whereas the main drawback is its much higher cost.

N.B. - These acids should never be used on equipment made of zinc or galvanized steel.

a) Principle of acidization

Hydrochloric acid (the most widely used in this context) has the property of dissolving the carbonates and bicarbonates of calcium (and magnesium) by transforming them into calcium chlorides (and magnesium chlorides). These compounds are soluble in water, being formed by reactions of the following type:

$$Ca\,(CO_3) + 2HCl = CaCl_2 + H_2O + CO_2$$

$$Ca\,(HCO_3)_2 + 2HCl = CaCl_2 + 2H_2O + 2CO_2$$

1 litre of 15% hydrochloric acid will theoretically react with 221 g of calcium carbonate $(CaCO_3)$ or 203 g of magnesium carbonate (dolomite: (Ca, Mg) $2CO_3$). A concentration of 15% is the optimal strength for the acid. If the HCl concentration is increased, then the viscosity of the dissolution products $(CaCl_2$ and $MgCl_2)$ will also increase. The chlorides become associated with impurities in the terrain, thus forming a type of gel which clogs the catchment structure.

Other experiments have shown that, under normal conditions of pressure and temperature (1 bar at 25°C), 95% of the hydrochloric acid is neutralized after a reaction time of 40 minutes in the presence of limestone and 50 minutes in the presence of dolomite (Ca, Mg) $2CO_3$.

b) Introduction of additives

The tapped water-bearing formations contain iron and aluminium oxides which are dissolved by hydrochloric acid at relatively low pH (2.5-4). On precipitation, these oxides form gelatinous compounds that absorb large quantities of water (up to 40 times their own volume), thus bringing about the clogging of fissures and well screens.

In order to avoid this problem, it is necessary to maintain these oxides in solution during the entire operation by means of citric or lactic acid. Generally speaking, a citric acid solution with a concentration of 10 g/l is sufficient to prevent the precipitation of iron in a limestone formation containing less than 1% Fe_2O_3 by weight, a value that is commonly encountered in rocks of this type. In certain cases, it is possible to make use of a double tartrate of potassium and sodium (also known as Rochelle salt) at a concentration of 4 g/l in 15% HCl.

Otherwise, the tapped formations may also contain calcium sulphate (e.g. gypsum) which needs to be eliminated. Since this compound is only soluble in hydrochloric acid at a pH of 5, ammonium bifluoride (NH_4F_2H) is used to convert the insoluble calcium sulphate into soluble ammonium sulphate. The concentrations commonly employed are 7-8 g of additive per litre of inhibited acid solution.

c) Carrying out of acidization

In most cases, due to the mere effect of gravity, the introduction of acid into a limestone formation does not produce satisfactory results. The acid remains in contact with the well walls, which only leads to an increase in borehole diameter. It is preferable to carry out the acidization under pressure, thus enabling a more extensive diffusion into the fissure network.

In fact, as soon as the pumping pressure becomes significantly higher than the groundwater pressure-head, the acid can penetrate a great distance into the surrounding formations. As a result, there is marked decrease in the local aquifer losses and an increase of the permeability in the vicinity of the well.

By contrast, when the groundwater pressure-head is high it acts against the pumping pressure. In this case, a carrier such as compressed air is used to drive the acid fluid into the formation. This method not only assists the penetration of acid into the water-bearing formation in question, it also enables the rapid return of spent acid and insoluble particles towards the well bore, owing to the sudden decompression during gas release.

There are no precise guidelines regarding the quantities of acid to be used. Nevertheless, it is advisable to perform several successive acidizing treatments with increasing doses at each step. In order to wash the borehole, the first dose is equivalent to the volume of the borehole directly opposite the water-bearing formation. This dose may be doubled or tripled during subsequent treatment steps. The acidization can be prolonged until a notable effect is obtained.

The rate of acid injection also has an important influence on the results achieved. It may be appreciated that the slower the pumping, the easier it becomes to neutralize the acid in the immediate vicinity of the well bore. If, on the contrary, the acid is pumped and injected at a fast rate, it will be still active when it reaches the more distant parts of the formation. Moreover, the injection of acid at depth takes place more readily as the fissures adjacent to the well become increasingly open. This demonstrates that fissures actually need to be chemically attacked during the first acidization step and that pressure should be increased progressively during the pumping.

In practice, the injection of hydrochloric acid is carried out using the mud pump on the drilling rig. Due to the fact that the volume injected is linked to both the pressure and the yield of the pump, higher pump pressures will lead to shorter injection times. The conditions that an injection pump should satisfy can be approximated by the following equation:

$$V = \frac{Q\,P\,t}{60}$$

where:

V : volume of acid to be injected in m^3; it is obtained from the volume of the borehole and calculated as a function of the order of operations.

Q : discharge rate in m^3/h per kg/cm^2 pressure-head of the groundwater body; it is defined from pumping prior to acidization and, in principle, varies with each treatment showing an increase at each step.

P : pump pressure (in kg/cm^2).

t : intended duration of acid injection (in minutes); it is important to bear in mind that this time should ensure that washing the well is complete before the onset of hydroxide formation.

When acidization is finished, pumping should be carried out for a prolonged period (at least 24 hours) with increasing yield in order to eliminate all trace of the treatment products. It can sometimes be an advantage to produce reverse circulation by injecting water. Care is taken to check for the complete elimination of acid by performing analyses on samples of pumped water.

In the Carboniferous of northern France, the discharge of a well treated by acidization was increased by a factor of six. The operation was comprised of four successive acidizing treatments (a total of 14 tonnes of 15% hydrochloric acid) using pressures ranging from 1.5 to 10 bars for durations of 40-60 minutes at each step.

d) An example of acidization

Let us assume we wish to treat a borehole with a volume $0.5 \ m^3$ using a planned operation of four acidization steps. The following volumes of aqueous HCl solution need to be injected after washing:

— V_1 for the first step: $2 \times 0.5 \ m^3 = 1 \ m^3$
— V_2 for the second step: $3 \times 0.5 \ m^3 = 1.5 \ m^3$
— V_3 for the third step: $4 \times 0.5 \ m^3 = 2 \ m^3$
— V_4 for the fourth step: $5 \times 0.5 \ m^3 = 2.5 \ m^3$

It is also assumed that the groundwater discharge before acidization is $4 \ m^3/h$ at a pressure-head of $1 \ kg/cm^2$, while the duration of injection is 10 minutes. The successive values of injection pressure and pump yield are presented in Table VII-III.

TABLE VII-III — *Characteristics of an acidization operation.*

Operational step	Injection pressure (kg/cm^2)	Pump yield (m^3/h)
Well washing	0.75	7.5
1st acidization	1.50	15.0
2nd acidization	2.25	22.5
3rd acidization	3.00	30.0
4th acidization	3.75	37.5

It has been shown in previous sections that water wells change with time. The problem is to determine the suitable moment for applying particular treatments for restoring the original characteristics.

In the case of moderately thick aquifers (20-30 m) susceptible to large variations in groundwater level, the example given above would only be applicable if the variations are less than 10% of the aquifer thickness, which would cause no significant modification in transmissivity. A comparison of specific capacities would make it possible to determine whether the well has deteriorated.

The following example is based on borehole A4 located within the Aubergenville well field; the nearest observation well is ST17. Three sets of mesurements are presented here (cf. Table 7-4):

Situation A.
. Yield measured in February 1977:   405 m^3/h
. Pumping water-level reading: 7.57 m
. Groundwater-level reading in ST17:   13.18 m
. Drawdown A :   5,6 m

Situation B.
. Yield measured in March 1983:   210 m^3/h
. Pumping water-level reading: 8.33 m
. Groundwater-level reading in ST17:   14.99 m
. Drawdown B :   6.66 m

The difference in standing water-level between A and B: 14.99 - 13.18 = 1.80 m.
The mean thickness of the aquifer is 25 m.
The water-level variation/thickness ratio is 1.8/25 = 0.07.
Since this variation is less than 10% of the aquifer thickness, a comparison is valid.

The fall in discharge capacity of the well between February 1977 and March 1983 can be defined by the following relation:

$$\frac{B - B_1}{B_1} = 2{,}35$$

This value corresponds to a reduction of 235% with respect to the original situation.

Situation C.
. Yield measured in May 1983:   363 m^3/h
. Pumping water-level reading: 7.00 m
. Groundwater-level reading in ST17:   15.30
. Drawdown C :   8.30 m

The difference in groundwater level between A and C is 15.3-13.2 = 0.08.
Since the variation is less than 10% of the aquifer thickness, a comparison is valid.

The fall in discharge capacity of the well between February 1977 and May 1983 can be defined by the following relation:

$$\frac{C - C_1}{C_1} = 0{,}84$$

This value corresponds to a reduction of 84% with respect to the original situation.

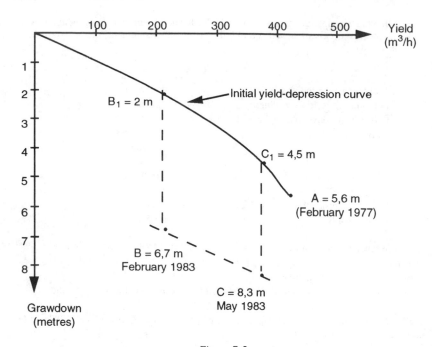

Figure 7-5
Aubergenville well field, borehole A4:
change in specific capacity between February 1977 and May 1983.

Analysis of the situation

Between February 1977 and March 1983, the well deteriorated to such an extent that it was only possible to obtain one half of the original yield. This drop in discharge capacity is reflected in situation B by a lowering of water level in the well that corresponds to 235% of the original drawdown.

After acidization in May 1983, the situation improved but the new sustained yield nevertheless entailed a drawdown 84% greater than the original value.

Conclusion

1   There was clearly excessive delay between A and B, which produced considerable clogging at depth within the well structure.

2   Acidization provides a partial remedy of the problem but does not succeed in restoring the original characteristics of the well.

3   It would have been preferable to take action before the drop in specific capacity had attained 50% in order to have the best chance of restoring the original situation.

7.4 Treatment of iron-manganese clogging

Clogging by ferruginous deposits generally arises from insoluble hydroxides of iron and manganese (formed in the zone of fluctuating groundwater level if the screen is dewatered). It is widely acknowledged that *no treatment is possible* in such cases. However, there are other types of ferriginous deposit that can sometimes be treated by acids or proprietary chemicals.

The origin of iron-manganese clogging lies in the supply of oxygen, usually at the tops of screens or in the upper parts of horizontal drains. The oxygen may come from the surface in the case of an unconfined aquifer, or from the borehole in the case of a confined aquifer.

Three types of deposit may be distinguished:

— Iron and manganese hydroxide compounds.

— Ferruginous deposits mixed with carbonate scale.

— Flocculent, granular and colloidal deposits.

The first category comprises insoluble deposits such as the iron-manganese concretions observed in the zone of fluctuating water level where the screen is drained. *Any attempt to treat this type of clogging will inevitably fail.*

For both of the other categories, acid treatment can be applied in the same way as with carbonates. In this context, mention should also be made of the products "Carela bio plus forte" and "Herli Rapid TWB-FCM1", both of which have proved their worth and are food-grade in quality.

In fact, "Carela bio plus forte" is made up of two complementary components:

— An inorganic acid which allows the solutizing of ferric incrustations.

— Organic acids associated with hydrogen peroxide, which ensure the chemical attack of manganic incrustations.

A balanced mixture of these different products makes it possible to carry out treatment of incrustations at the same time as disinfection of the catchment structure. This is an essential feature, since the majority of iron-manganese deposits have their origin in bacterial activity.

7.5 Treatment of biological clogging

The signs of clogging linked to bacterial activity are nowadays well recognized. This activity is expressed by the presence of masses or aggregates of gelatinous/viscous nature which are fixed indiscrimately over all parts of the catchment structure and the pump. The screens arc progressively blocked up, sometimes to such an extent that the pumped water discharge is reduced by more than 50% of the discharge capacity of the well. This type of clogging appears to be related to nutrient fluxes brought in by pumping which favour the formation of biomass.

In order to combat biological clogging, it is generally advisable to inject sterilizing products either at pressure or under gravity. The products used include chlorine-bearing compounds such as calcium hypochlorite and sodium hypochlorite (Javel water), and well as potassium permanganate and hydrogen peroxide.

Clogging can take place adjacent to the screen face, within the gravel pack or within the aquifer formation. In the first two cases, sterilization treatments (with calcium hypochlorite, sodium hypochlorite, potassium permanganate, etc.) that are carried out at regular intervals or continuously can lead to good results. With continuous treatments, a dosage of 1 g/l active chlorine should be applied adjacent to the affected zone.

In the aquifer formation, and also occasionally in the pack, drastic treatment with sodium hypochlorite would appear promising. However, it remains to be established whether there are any undesirable secondary effects and to specify the effectiveness of this type of treatment as a function of time.

An analysis of the literature shows that many different types of disinfection technique have been tested, ranging from steam injection to gamma-ray and ultrasound treatments. It is evident that the most conventional treatment consists of using various acids or mixtures of acids with inhibitors.

Studies have been carried out for the selection of products having a bactericidal or bacteriostatic action on biological clogging. However, insofar as a favourable action can be demonstrated for any particular product, it is clear that the product in question will not display the same behaviour under all circumstances. This is due to the fact that the physical, chemical and biological conditions are highly variable on the local scale.

According to the Geotherma study of 1991, the most widespread treatment in France, and the one which has also achieved the greatest success, involves brushing of the well followed by intense chlorination. This procedure at lest has the advantage of being an almost standard routine that is harmless to the catchment structure and the aquifer. In summary, the most usual treatment operations make use of the following:

— chlorine dioxide;
— sodium hypochlorite (Javel water);
— sulphamic acid;
— quaternary ammonium compounds;
— potassium permanganate;
— acrolein.

When used on their own, these products have a temporary disinfectant effect. Statistics show that the chlorine-bearing compounds are the most commonly used products. Improved results are obtained, particularly in carbonate aquifers, by combining the treatment with acidization.

To summarize, bacterial clogging appears to be a complex phenomenon which exhibits specific components in each particular case. Prevention is unfortunately difficult to organize, except in the case where factors favourable for the triggering of such clogging (i.e. organic context) are detected at an early stage in the catchment environment. In any chemical treatment, whether it involves curative or a preventive measures, the product used should satisfy two essential conditions:

— Preservation of the potability of waters that are to be extracted shortly after the end of treatment.

— Protection of the catchment site and pumping equipment that necessarily comes into contact with chemicals during treatment. This problem essentially concerns the casing and screens which, in contrast to the pump, cannot be removed.

As regards procedures that can be applied on an industrial scale, three types of treatment are generally distinguished:

— *Curative* treatments are carried out on a well when the threshold of clogging (generally fixed at 30%) is attained. The procedure involves a concentrated dose (several tens of m^3) of one or more chemical substances applied sequentially, injected into the well over a period of several hours so as to limit the effective downtime of catchment operations.

According to statistics, the effective downtime of a water well is of the order of 1-2 weeks; this corresponds to the setting up of work teams, the removal of the pump, various operations associated with the treatments and, finally, reinstallation of the pump.

— *Discontinuous preventive* treatments, which may be carried out periodically (e.g. every six months) on recent wells or on older clogged wells that have previously undergone a curative treatment. The injection of treatment chemicals is performed at lower doses that with curative treatments, even making use of observation wells arranged for this purpose around the borehole (cf. figure 7-6).

At each treatment, it is necessary to halt water extraction for 1-2 days in order to facilitate the dispersal of the treatment solution into the terrain and to withdraw the spent residues. The treatment product should be anti-corrosive and, as far as possible, should be chosen so that removal of the well pump is unnecessary.

— *Continuous preventive* treatments involve the permanent injection of inhibiting solutions at very weak doses through observations wells situated around the borehole (cf. figure 7-6).

The treatment chemicals and the required dose should be chosen in such a way as to inhibit the clogging process without altering the quality of the water supply.

In general, treatment is carried out starting at the clogged borehole. To ensure the effectiveness of the procedure, it is recommended to extend the treatment action to successive annular volumes of aquifer around the screen face. The curative treatment can be applied with one or more products according to the nature of the problem to be resolved.

The use of a single product does not appear as a valid option in practice, since no one substance can combine the action of bactericide with the destruction of organic matter. Nonetheless, an analysis of the situation will indicate the volumes of chemicals that should be used. The successive application of two or three products is the most generally adopted solution.

The development of a biomass in the aquifer is the major problem posed by bacterial clogging. In terms of treatment, it is advisable to determine the volume of contaminated aquifer and then define a strategy which enables the destruction of the clogging biomass as well as its evacuation.

The approach to be adopted is as follows:

— Estimate the volume of contaminated aquifer.

— Define the successive annular volumes to be treated, by taking account of the type of bacteria, the nature of the aquifer, the physico-chemical properties of the water and the nature of the declogging product(s) to be employed.

— Establish a treatment protocol.

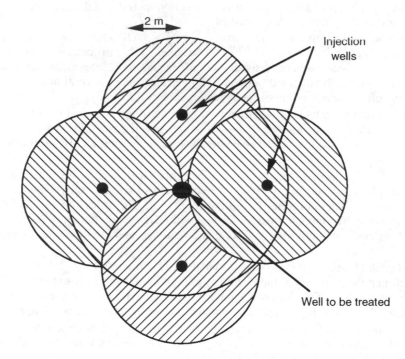

Figure 7-6
Preventive treatment arrangement with four observation wells.

7.5.1 Estimation of contaminated aquifer volume

The flow velocity of water near the catchment site appears to be the main cause of bacterial proliferation insofar as it is a vector for nutrient flux. The problem amounts to determining the distance from the well at which water flow becomes so weak that biomass development no longer occurs.

A study of clogging in boreholes sunk into the Ypresian provides a clear illustration of this problem; 78% of catchment works with an aquifer water flow greater than 1cm/s are clogged, whereas 60% of unclogged wells exhibit lower flow velocities.

From the practical point of view, by knowing the aquifer porosity, the operating yield of the well and the height of its screened section, it is possible to calculate the velocities at which groundwaters will circulate through the different cylindrical volumes of aquifer. It is appropriate to calculate the distance from the well at which water flow within the aquifer becomes sufficiently weak to discourage biomass development.

In the case of the Yypresian well field, it is apparent that the bacteria do not form a clogging biomass when the velocity is less than 1 cm/s. Thus, the problem in this example consists of determining the distance at which water flow in the aquifer is less than 1 cm/s:

Given that $Q = VS$ where $S = 2\pi rhp$ we can find a value for the distance r such that $V \leq 1.10^{-2}$ m/s by using expressions with the following variables:

Q : groundwater discharge in m^3/s

V : flow velocity in m/s

S : cross-sectional area of water-yielding terrain in m^2

P : porosity

r : radial distance in metres from borehole axis

h : height of screened section in metres

We can write:

$$S = \frac{Q}{Vmin}$$

$$r = \frac{Q}{Vmin . 2 \pi h . P}$$

If Q is expressed in m^3/h and V in m/s we obtain: $r = \dfrac{Q}{226 . h . P}$

The velocity of water in the undisturbed aquifer can also be calculated and compared with the velocities obtained near the well. BOURGUET *et al.* [1988] have reported values of 0.2-3.0 cm/s in Ypresian formations near the catchment site, whereas the groundwater flows naturally at velocities of the order of 3 μm/s. It can be seen that the natural flow velocity is multiplied by a factor of 10^3 to 10^4 as a result of pumping.

7.5.2 Definition of successive annular volumes for treatment

Considering the type of bacteria, the nature of the aquifer, the physico-chemical properties of the water and the nature of the declogging products to be employed, it is advisable to ascertain the volumes of solutions that need to be injected during treatment.

The volume Vc of water contained in a borehole is given by the formula:

$$Vc = \pi r_o^2 h$$

where:

r_o : radius of borehole

h : depth of water in borehole

The volume *Vt* of the clogged annular space (cf. figure 7-7) to be treated is obtained from the following equation:

$$Vt = \pi\,(\,r_t^2 - r_0^2\,)\,.\,h = \pi\,(r_0 + r_t)\,e\,h$$

where: *e*: thickness of annulus; *h*: depth of water in well

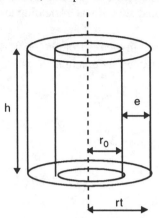

Figure 7-7
Definition of concentric annular volumes to be treated.

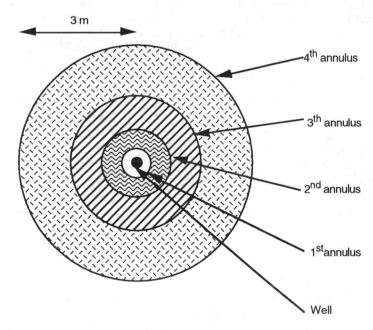

Figure 7-8
Curative treatment: plan view of annular aquifer areas to be treated.

Since the terrain is in principle more highly clogged nearer the borehole, it is more advantageous at first to treat narrow annular zones which are then widened with increasing distance from the borehole axis.

TABLE VII-IV — *Volume of aquifer annulus for each metre depth of well (this value should be multiplied by the aquifer porosity to obtain the true volume to be treated).*

Casing diameter (mm)	Volume depth ratio (m³/m)	0.25 m annulus (m³/m)	0.5 m annulus (m³/m)	1 m annulus (m³/m)	2 m annulus (m³/m)	3 m annulus (m³/m)	4 m annulus (m³/m)	5 m annulus (m³/m)
200	0.03	0.35	1.10	3.77	13.82	30.14	52.75	81.64
300	0.07	0.43	1.26	4.08	14.44	31.09	54.01	83.21
400	0.13	0.51	1.41	4.40	15.07	32.03	55.26	84.78
500	0.20	0.59	1.57	4.71	15.70	32.97	56.52	86.35
600	0.28	0.67	1.73	5.02	16.33	33.91	57.78	87.92
700	0.38	0.75	1.88	5.34	16.96	34.85	59.03	89.49
800	0.50	0.82	2.04	5.65	17.58	35.80	60.29	91.06

7.5.3 Treatment protocol

The treatment protocol is classically divided into three stages:
— First injection of a solution of the treatment chemicals into the terrain, using a volume equal to the void space volume of the innermost aquifer annulus and respecting a residence time that ensures effective treatment of the outermost zones attained by the injected solution.
— First cleanout pumping to evacuate residues and slime produced by treatment.
— Renewed injection of treatment chemicals, using a volume equal to the void space volume of the second aquifer annulus. When the required volume has been injected, water flushing is carried out with a moderate flow rate in order to drive the active product across the first treated annulus. This enables the product to come into contact and completely impregnate the second annulus, thus preparing it for treatment (surge effect).

After a residence time for treating the second annulus, the operation is continued by a second phase of pumping for cleaning out the respective aquifer zone. Following this, the operation may be repeated for the treatment of a third annulus, and so on for any subsequent treatments.

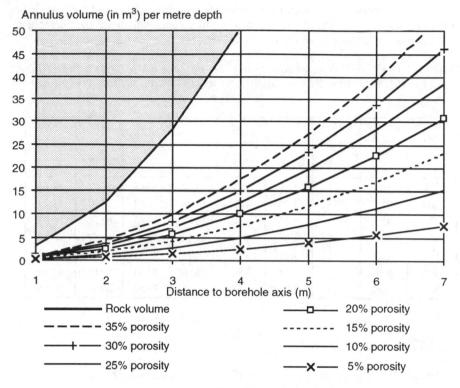

Figure 7-9
Evaluation of volumes involved in declogging treatment.

Finally, it should always be borne in mind that, whatever the precision of the calculations, the determination of porosities for the gravel pack and the developed aquifer zone remains nevertheless rather empirical. Moreover, the aquifer will consume variable amounts of reagents due to the presence of physico-chemical heterogeneities as well as the biomass that it is likely to contain.

7.5.4 Sodium hypochlorite treatment

When bacterial proliferation occurs in operating water wells, treatment with sodium hypochlorite (NaClO, also known as Javel water) represents an effective method that is easy to implement. In particular, this method has been employed to treat bacterial clogging affecting boreholes in the Ypresian of the Paris Basin (Villeneuve-la-Garenne well field in the department of Hauts de Seine, France). The chlorine present in sodium hypochlorite is introduced at the bottom of the borehole, where it has a twofold action:

— as a bactericide;
— to break down organic slime.

The aim of this treatment is not only to eliminate the bacteria and bacterial slime on the well screens but also to obtain a penetrative action within the gravel pack and surrounding formations.

Sodium hypochlorite is an alkaline oxidizing agent which occurs in the form of an unstable solution characterized by its content in active chlorine (chlorimetric titre is quoted in degrees, $1°$ being equivalent to 3.17 g active chlorine per litre).

On this basis:

— 1 litre of sodium hypochlorite solution with a titre of $18°$ contains 57 g of active Cl_2.

— 1 litre of sodium hypochlorite solution with a titre of $48°$ contains 152 g of active Cl_2.

In general, it is recommended to treat at a concentration of 2,000 ppm (i.e. 2 g active chlorine per litre). To implement the treatment, it is necessary to calculate the volume of water contained in the borehole, in the gravel pack and in the surrounding formations. Prior to this, it is advisable to estimate the perimeter around the borehole within which the treatment is intended to act.

Evidently, this treatment should be suited to the degree of clogging and the volume of aquifer to be treated. It is appropriate to calculate the volumes of the annular aquifer zones involved and adapt the treatment according to the extent of bacterial development. This is achieved by considering the velocity of the nutrient flux (water flow in the aquifer) and increasing the volume treated up to the limit of nutrient supply, where velocities become so low that biomass can no longer develop.

From the practical point of view, the introduction of sodium hypochlorite solution is carried out with a PVC hose in order to spread the product throughout the submerged section of the borehole. Since the density of the solution (d = 1.21) is higher than that of water, the product must be introduced at a level above the zone to be treated. So as to enable the product to penetrate into the surrounding terrain, it is necessary to introduce a quantity of water five to ten times greater than the volume contained in the borehole. It is desirable to have the facilities for carrying out pumping in a closed circuit in order to agitate the solution that is left in place for at least 12 hours.

Introduction of the product takes place in several successive passes (generally three to five in number) interspersed with cleanout pumping, so as to attain the annular volumes situated at increasing distance from the borehole.

To eliminate all the material detached during the borehole treatment, it is necessary to carry out prolonged pumping at the end of the operation. It is important to ensure the complete removal of solid particulates so they do not become reattached onto the outer part of the gravel pack.

Theoretical example of treatment

Let us consider a borehole of 20 m depth tapping an aquifer with an effective porosity of 15% over an interval 10 m (this section is screened over its entire extent, and captures all groundwaters discharged from the aquifer). The standing water-level is at 8 m beneath ground level. The borehole is entirely fitted from top to bottom with 600-mm-diameter casings.

Calculation of effect of water volume contained in the borehole:

$$Vc = \pi\, r_o^2\, h = 3.14 \times (0.3)^2 \times 12 = 3.4 \text{ m}^3$$

The capacity-related volume corresponds to the water flush required at each pass to drive the treatment solution into the aquifer. In the case of treatment involving several successive annular zones, the volume of the water flushes is given by:

$$Vc = V_1 + V_2 + \ldots + V_n$$

TABLE VII-V — *Calculation of annular volumes involved in hypochlorite treatment (on the left: initial volumes, on the right: volumes remaining to be treated after successive injections)*

Radius of annulus (m)	Volume of treatment solution (m³)	Initial volume of 48° sodium hypochlorite reagent (l)	Volume of successive flushes (m³)	Volume of successive treated zones (m³)	Residual volume of 48° sodium hypochlorite (l)
0.25 m	1.2	16	3.4	1.2	16
0.5 m	3.1	41	4.6	2	25
1 m	9	119	4.3	6	78
2 m	29.4	387	12.1	20.4	268
3 m	61	803	38.4	31.6	416
4 m	104	1,368	90.4	43	565
5 m	158	2,082	165	54	714
Cumulative volume			318	158	2,082

In the case of natural porous media, it is necessary to take account of the fact that a large fraction of the active chlorine is likely to be inhibited by the constituents of the aquifer. In view of the absorption of chlorine by the fines, it is prudent to boost the volume of sodium hypochlorite used in the treatment.

After the final pass, the inner face of the screen is cleaned with a scratcher (porcupine brush).

The above example shows the magnitude of the volumes involved in treatment. The incorrect calculation of these volumes probably accounts for the large number of declogging operations that result in failure. As a matter of fact, since the high-velocity zone extends for several metres around the borehole axis, considerable volumes of aquifer need to be treated. In most declogging operations, however, the treatment does not attain the annular zone situated between 2 and 5 metres from the borehole. In agreement with BOURGUET *et al.* [1984], the present author considers that the treatment should also cover this zone, where various determinant factors are likely to control the clogging process.

Example of sodium hypochlorite treatment

Borehole 5S' in the Villeneuve-la-Garenne well field (department of Hauts de Seine, France), which is operated by Lyonnaise des Eaux, was constructed in 1974. It progressively showed signs of clogging and was treated with hexametaphosphate in 1982 following an incorrect diagnosis. After this inappropriate treatment, the well capacity fell from 115 m³/h with a drawdown of 30 m in 1980 to 53 m³/h with a drawdown of 44 m in

1987. Confronted with this state of affairs, a new diagnosis was made which enabled the identification of the real cause of clogging, in this case the proliferation of bacteria.

The treatment was carried out by injecting several batches of sodium hypochlorite and led to the restoration of more than 100% of the original characteristics due to the fact that well development had continued with the course of time. The treatment was undertaken in conformity with the methodology described previously.

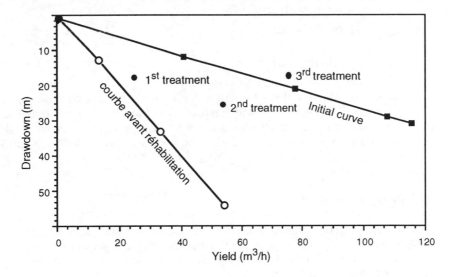

Figure 7-10
Restoration of well 5S located in the Villeneuve-la-Garenne well field.
After a document issued by Lyonnaise des Eaux.

It is noteworthy that the "Burgeap" study, which was based on experiments in test-tubes, advocates a sequence of treatments involving wetting agent/oxidizing bactericide/acid. The optimal treatment would be carried out in two stages:

— An initial oxidizing treatment is carried out using sodium hypochlorite associated with a wetting agent and also possibly an enzyme. This formulation is intended to destroy the biomass.

— A second treatment is composed of a mixture of hydrochloric acid and a bactericidal wetting agent. The role of the hydrochloric acid is to buffer the pH to a value such that the ferric iron precipitated by the hypochlorite is taken up into solution as ferrous ions. A bactericide is used to ensure elimination of the last traces of bacteria that may have resisted the first treatment.

Clearly, this treatment protocol should be adapted according to the degree of clogging in the well to be treated.

7.6 Regeneration after corrosion

It should be remembered that corrosion is caused by water, being a function of its hydrogen ion concentration and/or the presence of the following:
— Dissolved oxygen (even in small quantities).
— Hydrogen sulphide (H_2S).
— Hydrochloric acid (HCl).
— Chlorides.
— Sulphuric acid (H_2SO_4).
— Clays rich in calcium sulphate (gypsum).

These agents influence the pH of the water, which generally becomes acidic (<7), and renders it aggressive with respect to metal. The water circulation velocity and the temperature also have a certain influence. In water wells, the chemical corrosion of metals is much less common than electrochemical corrosion.

7.6.1 Preventive measures

When considering prevention, it is inevitable to come across the same problems addressed in well design. In fact, the most effective preventive measures are those which can be implemented at the design stage. Such arrangements require the following minimum criteria:

— Total compatibility between the metals making up the catchment structure, while ensuring that there is no direct contact between different metals.

— Design for the protection of well casing throughout the non-aquifer section, using water-tight cementation. A special cement should be used for sections traversing formations (e.g. gypsiferous beds) that might attack the cement.

— Installation of casings with adequate thickness, so their oxidation does not lead to general weakening of the whole structure over a short span of time.

— Evaluation of the possible aggressivity of water towards metals as a function of water chemistry. In certain cases, notably in the presence of H_2S,

it is preferable to use PVC equipment.

— Protection of the catchment structure against the effects of telluric currents (cathodic protection) when such phenomena are to be expected, especially in proximity to railway tracks (some boreholes located at 100 m from lines with heavy traffic become perforated in a period of 1-2 years).

— Particular care should be taken if the well taps an aquifer at several levels (alternation of water-yielding and barren intervals in the same formation) due to eventual differences in water chemistry.

— Attention should be paid to avoid excessive water velocities within the well, owing to the risk of erosion and the consequences of embrittlement in certain metal components.

7.6.2 Inspection

The monitoring of catchment structures during their service life is also of very great importance in detecting the onset of corrosion. In this regard, although there are numerous means for carrying out such inspections, the choice should be made by a specialist. For example, video camera examination cannot reveal the extent of corrosion associated with a reduction in casing thickness, whereas other types of investigation may permit such an assessment to be made.

Downhole logging methods which enable the recognition of various types of corrosion should be implemented according to the characteristics of the catchment structure in question. Clearly, in the absence of any information on these characteristics, it is necessary to begin by defining the configuration of the well. Following this, it is possible to specify the complementary logs that need to be carried out.

A range of procedures are currently available which, on the whole, enable satisfactory diagnoses and which are constantly being improved by the development of data processing facilities. Listed below are various types of inspection that are carried out as part of the monitoring of operating wells, where these are also thought to run some risk of corrosion:

— Casing cementation check (detachment, fissures, chemical attack, etc.). It is even possible to detect the formation of dissolution pockets behind the cementation [GRIOLET, 1991].

— Checking the condition of the casing. Various logs allow testing for thickness, deformation, ruptures and perforations. In addition, step pumping tests on certain deep wells can reveal the presence of casing perforations.

— Checking the condition of the screens is undertaken as a supplement to the camera inspection work, enabling the location of individual perforations or ruptures.

— Checking the gravel pack. This last inspection, although not directly concerned with corrosion, makes it possible to assess the condition of the pack in relation to its original position, which may reflect modifications in the hydraulic regime of the catchment structure.

The interpretation of these measurements and checks can only be undertaken by experienced specialists, since any error may have consequences on the choice of restoration that is specified.

7.6.3 Restoration

The restoration of a well damaged by corrosion is generally an awkward and difficult operation. In principle, the water engineer is faced with two types of action, either restoration of the casing (perforations of casings in place, or corroded casings with risk of collapse) or restoration of the screened section (perforated or worn screens). Although these two types of restoration are not systematically linked, in practice it may become necessary to combine them for greater efficiency.

Restoration of the casing consists of re-lining the catchment structure, i.e. placing a new casing inside the existing one and then cementing the annulus. However, this procedure is not always possible if the borehole diameter is too small because the casing placed on the inside may prevent the installation of a pump. This is the disadvantage of wells that are too narrow in relation to their capacity. The risk of such a situation may be evaluated during the design stage.

A technique is currently under study for re-lining boreholes by hot-forming a polyethylene jacket directly onto the existing casing. This process is based on an American technique used for the re-lining of horizontal water conduits (U. Liner process, distributed by Tubafor International). A liner made of low-density food-grade polyethylene is applied to the inside of the well casing, causing very little reduction in diameter with respect to the original pipe and ensuring a fully water-tight seal. This liner also diminishes the well losses and improves the resistance to corrosion. In all probability, the cost of this process should be lower than re-lining with steel.

Restoration of the screen may be carried out in two different ways, either by removal of the existing screen and its replacement by a new one, or the installation of a second smaller screen inside the first and the filling of the annulus with a gravel pack. The first operation is not always possible in practice since a great deal depends on the manner in which the original screen was installed. The second type of operation has the disadvantage of

producing a supplementary loss of head which leads to a significant drop in well capacity.

At this level of restoration, the prior diagnosis should set an objective in terms of results and also establish the cost of the operation. In this way, the project manager can avoid incurrring prohibitive costs with respect to the intended result and in comparison with a newly constructed well.

7.6.4 Treatment of chemical corrosion

First of all, it should be stressed that no remedy exists that can stop chemical corrosion. At the very most, it is possible to prevent its effects by taking protective measures. This problem is discussed in Chapter VI. However, three possible courses of action may be briefly mentioned here:

— Galvanization of the casing.

— The best method for guarding against corrosion is preventive protection, which involves the choice of materials used for equipping the well during implementation.

— In the case of advanced corrosion, the problem may be addressed by an indirect remedy that consists of doubling up (re-lining) the casings and screens. Nevertheless, this system leads to a fall in discharge capacity by increasing the quadratic well losses.

7.6.5 Treatment of electrochemical corrosion

In order for electrochemical corrosion to take place, two conditions need to be satisfied:

— A difference in electrical potential is set up between two different metals, or separate domains of the same metal, even if the spacing between them is infinitesimally small.

— The water contains sufficient concentration of dissolved salts to behave as a conducting fluid (electrolyte).

Various sites may be sensitive to corrosion, including welds on the same metal, slots cut by blowtorch or casing cutter, gashes in the metal adjacent to joints (seals and gaskets) and scuffs on the lining sealer.

When two different metals are connected via an electrolyte, corrosion takes place due to the formation of an electrolytic couple (e.g. a stainless steel screen and a mild steel casing string). To mitigate this inconvenience, two methods can be successfully applied:

— Preventive protection by an appropriate choice of materials for equipping the borehole.

— Cathodic protection.

Both of these methods are discussed in Chapter VI.

7.6.6 Treatment of bacterial corrosion

A number of physical or chemical methods may be distinguished in the context of bacterial corrosion control:

— Physical methods:
- Ultraviolet irradiation.
- "Pasteurization" by raising the temperature.
- Ultrasound treatment.
— Chemical methods:
- Injection of acids.
- Injection of chlorinated solutions.
- Utilization of polyphosphate dispersants.

Although these treatments are efficacious, their action is limited in time, which implies carrying out such operations on a regular basis.

7.7 Cleaning and pumping

After each regeneration operation, whether mechanical or chemical, it is absolutely necessary to clean out the well with air-lift pumping before re-installing the submersible pump. This procedure is carried out by injecting air under pressure - alternately within and beneath the water pipe - and continuing until clean water is obtained without sand or clayey particles.

The operational submersible pump is then re-installed in order to proceed with new step drawdown tests to establish a yield-depression curve for the restored well.

An analysis of this new curve can lead either to the restarting of catchment operations, possibly after a prolonged pumping test, or a further phase of regeneration.

7.8 Role of well age

Well age is responsible for a large part of the deterioration of water wells. In France, the average age of catchment works, taking all types together, is between 10 and 50 years in 70% of cases.

The actual age of a well is clearly a factor in its rate of decay, but a lack of maintenance can worsen the effects of ageing, thus contributing to the appearance of problems that are sometimes difficult to resolve due the condition of the catchment structure.

Old wells commonly suffer from clogging, corrosion and sanding up. Unfortunately, the archaic design of some of these wells means that the only remaining solution in many cases is the pure and simple renewal of the whole catchment structure. In the majority of examples, the effects of ageing can de delayed — and in any case diagnosed — provided that the wells are subject to regular checks.

It is salutary to bear in mind the important contribution of groundwaters to the drinking water supply; according to the latest statistics, 62% of the groundwater abstracted in France is destined for drinking water supplies to the population, with a slight tendency to rise clearly marked in some catchment areas. As regards the total drinking water consumption, groundwaters account for about one half of the total volume of used resources whereas surface waters make up the other half.

Water wells are to be found in all the French departments. The maintenance, upkeep and management of these structures is thus a national concern. More than 30,000 wells in France are used to extract groundwaters, and are operated solely to provide drinking water supplies to the public. The geographical distribution of water wells is evidently highly irregular, the highest concentration being located in sedimentary terrains.

Even if, on the whole, the number of wells should not greatly change in the future, the creation of new catchment works will probably be necessary for several reasons:

— The total inventory of wells in France is decayed and needs to be largely renewed over the next ten or twenty years.

— Isolated wells should be abandoned in preference to the development of major well fields in areas that can be protected.

— Despite eventually requiring treatment at some stage, raw waters from boreholes show variations in quality which render them easier to treat than surface waters.

— The vulnerability of certain resources, including surface waters, gives rise to the need for diversification and the creation of supplementary or back-up catchment structures that are based on new water resources, preferably in the subsurface.

— Certain aquifers are currently underexploited, particularly the karstic aquifers. Their development, although up to now hindered by the difficulties and contingencies of research work, will eventually lead to the creation of new catchment structures.

— Numerous municipalities, particularly rural ones, receive their water supply from a single point source (borehole, well or spring). For reasons of water supply security, a certain number of these sources should be doubled up or redeveloped.

— The technical problems encountered in the exploitation of certain wells will

lead contractors to undertake new catchment projects in preference to restoration or after the failure of treatment.

In such a context, it would be a waste if all these new projects were carried out using the same approach as that adopted in the past. A new approach would take account of the environmental, protection and management aspects, without omitting the contribution of modern techniques to the implementation and monitoring of water catchment operations.

7.9 General course of action prior to preventive maintenance

To ensure the effectiveness of a given treatment, it is necessary to ascertain the causes of incidents that have occurred at the catchment site in question. In order to achieve this objective, the following actions are taken:

— Measurement of pH:
  • At pH < 7, the water is acidic and exhibits corrosive behaviour.
  • At pH > 7, the water is alkaline and has a tendency to form incrustations. Alkaline waters contain carbonates, bicarbonates or hydroxides in solution.

— Sampling of deposits (from pumps, suction pipes, screen, etc.) in order to determine their nature and composition.

— Study of the particulate matter load in pumped waters.

An appropriate treatment is chosen in view of the nature of the sampled deposit:

— Hydrochloric and sulphamic acids (or amino-sulphamic acid) are used for calcium carbonate concretions.

— Polyphosphates are used for oxides of iron and manganese.

— Chlorination (hypochlorite), hydrogen peroxide (H_2O_2) or acetic acid (CH_3COOH) is used for bacterial control.

— Polyphosphates are used for the dispersal of silt and clay.

— Hydrofluoric acid enables the dissolution of silica and is used to treat sandstone formations. This utilization is of particular interest in the oil industry, but the permeability of the matrix should be suitable to tolerate the pressure applied to the rock.

A mixed treatment may be envisaged in cases where several types of deposit are present at the same time. It is necessary to add inhibitors to the acids used in treatment in order to protect the steel components of the well equipment. Apart from chemical treatments, other methods such as scraping, jetting or explosive shocking may be used when this is possible.

The chemical treatment methods comprise:

— Acidization under pressure.

— Open-well acidization.

— Chlorination.

— Jet cleaning for mixed treatment with chlorine and polyphosphate.

7.10 Abandonment of wells

If a catchment structure fails to fulfill the minimum requirements for correct operation, despite all the restoration work that has been undertaken, then it is necessary to proceed with abandonment while observing certain rules. In fact , the structure of the abandoned well will be destroyed by corrosion and all the geological formations traversed by the well will be placed in communication. As a consequence, there is a risk of mixing poor quality levels of the aquifer with more highly water-yielding parts. This may lead to the partial or total destruction of the aquifer.

In addition to this interference between geological levels, surface waters can enter the abandoned well. This inflow includes rainwater, but the case can be aggravated by polluted waters. Thus, the abandonment of a well is a serious problem that should be addressed with caution.

The borehole should be plugged in an impermeable manner, which should evidently rule out its filling by various materials. As a matter fact, filling by an impermeable material would not isolate the groundwater body from other levels - possibly water-bearing - and could thus give rise to a supplementary source of pollution.

The only acceptable way to plug a well is to seal it hermetically with cement. Introduction of the cement by gravity should be avoided since this produces unplanned results. In fact, the cement runs the risk of being diluted before reaching the bottom, a situation which worsens the deeper the well and the larger the bore diameter. The most effective method consists of introducing the clinker cement from the bottom of the well (cf. figure 7-10).

The cement should not be allowed to penetrate into the aquifer in the case of a severely deteriorated screened section or if the tapped formations are highly fissured. A viscous plug made of bentonite or bentonite-polymer mix is placed adjacent to the tapped section. The plug is then capped by cementation.

From the technical standpoint, it can be seen that plugging in accordance with the rules requires certain precautions. The composition of the materials, their density and also their complete harmlessness for the environment should be established as a function of the conditions of well abandonment (whether screens are ruptured or not, nature of the aquifer, etc.).

It should be noted that an enquiry carried out in France in 1991 revealed that more than 60% of abandoned wells were not plugged. As a consequence, such a state of affairs clearly means that, at some stage in the future, all geological formations traversed by a well will be placed in communication (by corrosion, collapse, etc.). Considering the advanced age of the wells in France and the associated high probability of their renewal that should take place over the coming years, this practice of uncontrolled abandonment poses a grave threat to the protection of water resources.

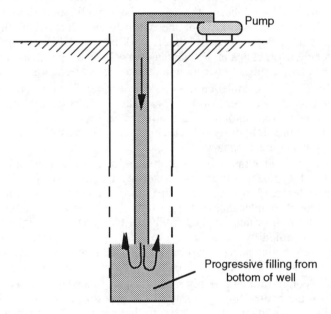

Figure 7-11
Cementation of an abandoned well.

7.11 Conclusion

Knowing the conditions of implementation and the installation of equipment for a given well, combined with regular monitoring during operation, it is possible to reveal the existence of clogging and/or corrosion so a remedy can be applied early enough to avoid a large drop in discharge capacity. On the other hand, when a lowering of the discharge capacity and/or a variation in water quality is observed, the course of action to be undertaken is summarized on figure 7-11. For a well 35 m deep having a diameter of 200 mm with 15 m of screens, a regeneration operation represents 10% of the cost of a completely new well. A recent study has shown that 76% of catchment works in France do not receive regular maintenance from specialist agencies. It appears that upkeep of water wells is not yet customary practice. The most widespread approach consists of making periodic checks.

Nearly all operators point to the chronic lack of technical information concerning the catchment structures they are managing, particularly as regards operational instructions or recommendations.

As specified in the GEOTHERMA study, the origin of this situation is quite often the age of the wells, in cases where their characteristics were not communicated to successive operators. In far too many examples, the only available information concerns the approximate well depth and the casing diameter, while no record exists of the nature of the screens or the original discharging-well tests. Fortunately, this remark is generally only applicable to isolated catchment structures more than ten years old. The operation of well fields can hardly be carried out to satisfaction in the absence of technical information.

An analysis of well completion reports, however, reveals a lack of instructions for the operation of catchment works. This state of affairs generally leads in the course of time to exhaustion of the aquifer, a scenario that is extremely detrimental to the longevity of catchment structures, being sometimes unintentional due to the absence of guidelines.

Experience indicates that there is a problem of communication of technical data between the design office and water well operators. This situation is particularly widespread in rural areas. In France, the unanimous opinion is that there is a failure of maintenance on water wells. The considerable financial outlay in providing drinking water supplies to 95% of the population has unfortunately not been accompanied by the necessary maintenance arrangements. In the vast majority of cases, maintenance measures are dictated solely as a result of losses in yield or serious degradation that lead to difficulties in well operation.

The fault is often far more deeply rooted insofar as a water well is hardly ever considered by the contracting authority as the essential component in its water supply policy, except in the event of a shortfall of capacity. Even when the failure actually occurs, the least expensive solution is sought which, due to lack of information, is often also the most precarious one.

In terms of expenditure, an annual budget of 8-10% of the total cost of the catchment works would enable prevention of most the problems by ensuring proper maintenance.

The indifference of contracting authorities (or developers) towards their water wells in general and maintenance in particular is largely confirmed by the current state of catchment protection zones. The nature of the problem is clearly illustrated in France by the very large proportion of catchment sites where protected perimeters are not in conformity with the regulations.

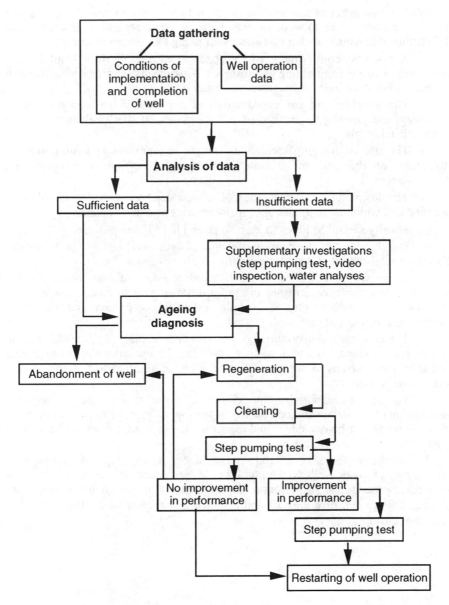

Figure 7-12
Flowchart of action to be taken in the framework of catchment structure restoration.

With the prospect of meeting the needs of future generations, it appears that the time has now come to work towards the active management of groundwaters. It is fitting to construct such a vast endeavour along the following lines:

— A resolute commitment to European scientific facilities and human resources in respect to training and research. This means training and informing all parties involved in the management of water wells.

— The development and coordination of expertise in various disciplines (geology, hydrogeology, geochemistry, geophysics) at the local, national and international levels.

— The use of imaging techniques for the recognition and modelling of structures on different scales (satellite imagery, geophysical surveys, data processing, etc.).

— The design and implementation of all new catchment works according to an approach which incorporates upkeep, monitoring and maintenance.

— Priority should be given to team work and field-based projects.

— Development of the concept of acquired data (well history) management and the establishment of diagnostic reports for all existing wells.

— Water wells should no longer be considered merely as the means for localized extraction of water from an aquifer. Instead, their surface and subsurface environments should be seen as a whole system whose behaviour needs to be studied, controlled and protected.

— The drafting and distribution of contract documents and guidelines for the use of project managers and developers, defining the operational conditions and the rules to be observed during well restoration, in addition to the checks that need to be carried out.

— The use of computerized modelling techniques to simulate and improve assessment of the nature and complexity of hydrogeological problems, both in the field (groundwater body models) and in relation to knowledge engineering (expert systems).

In summary, there is a need to introduce the concept of compulsory monitoring of water wells which involves the application of precise, regular and quantified procedures. At present, computerized catchment control systems are already able to optimize the data and resource management aspects of checking and monitoring.

CHAPTER VIII

Management tools

"The personal computer is a tool that can amplify
a certain part of our inherent intelligence"
Steven Jobs

As described in previous chapters, the implementation of catchment structures, the follow-up of construction work, including their maintenance or even restoration, can all be considered as essential stages in the rational management of water wells. This chapter presents a number of "modern" tools for the management of water wells: research tools, database managers, historical databases, geographical information systems, etc.

The tools presented here depend essentially on computer aids that are used for the rational management of wells.

It has become commonplace to evoke data processing as both an integral part and a consequence of development. The electronic data processing sector shows the highest growth rate among the developing technologies; it represented a market of almost $400 billion in 1986 and this figure is expected to be multiplied threefold by 1995 (cf. Table VIII-I). The impact of computer sciences on industry and on the entire socio-economic environment is so strong that it is impossible today to imagine a development policy which does not call upon information technology. Electronic data processing, because of its wide range of applications in almost all sectors of human activity, is a powerful instrument in the management of technological development and opens up new perspectives in the expansion of industrial, cultural and educational activities. Thus, information technology is becoming an increasingly powerful instrument in every domain: political, economic, technical, social and cultural. Moreover, this is occurring on the national as well as on the world-wide scale as a result of the spread of data communication networks. As such, this technology is of interest to managers and players in the water industry at all levels.

Information technology is playing an unprecedented role in every area of scientific, economic and social activity and is in a continual state of flux. Finally, it must be acknowledged that the information technology culture is essentially English language based. For that reason, in a large proportion of cases it is necessary to describe systems that are of American origin. Nevertheless, it is important to stress the remarkable place occupied by civil engineering software developed in France.

TABLE VIII-I — *World production of computer systems (estimated growth for 1986-95). After OECD document.*

	1986 (billion $)	1990 (billion $)	1995 (billion $)	Mean annual growth rate (%)
Hardware	224	353	621	12
Software	84	174	433	20
Telecommunications and computer services	85	107	143	6
Total	393	634	1.197	13

8.1 Electronic data processing

Given the extremely variable level of market penetration of the computer among production managers, hydrogeologists and field engineers, it would appear reasonable here to recall certain basic concepts.

A computer is an automatic machine for processing information which enables the storage, treatment and restitution of data without human intervention by performing arithmetical and logical operations under the control of recorded programmes.

A computer system is composed of hardware, comprising a set of components and physical devices, and software, which is a set of programmes necessary for its operation. Computers contain a central arithmetical and logical unit surrounded by specialized circuits whose role is to handle the transfer of information coming either from the central memory or from external media (peripherals such as disks, readers, keyboard, printer, etc.) by means of a communications channel termed a "bus". The information is in the form of either executable instructions or data, the whole set making up a programme.

The central processing unit performs three kinds of elementary operation: transfer of information from one place to another in the machine (addressed through the intermediary of the bus), arithmetical operations and logical operations. In simple terms, the macroscopic action of a programme is to control the input (data stream to be processed), to provide instructions for the calculations and choices to be carried out (processing), to pilot the access to memory resources and then produce an output (results).

For a long time, the use of computers was limited to applications of a numerical calculation type. During this period, the prime concern was with the speed of execution of complex calculations involving large amounts of data. Over the last 15 years, two parallel routes have been explored. On the one hand, there has been a race towards numerical power based on specialized processors suitable for vectorial or parallel calculations, and on the other hand, machine languages have been developed to manipulate symbolic data. Machine languages aim to implement in an efficient manner the specific languages used in artificial intelligence (i.e. Lisp or Prolog), with the help of dedicated hardware architectures designed for the high-performance processing of symbols. When machines reached this stage, a whole world of creative feed-back became possible in terms their own design. In fact, the race towards miniaturization, involving an ever increasing integration of the components, now requires the involvement of the machine into its own development cycle in the form of design-aid software such as silicon-chip compilers.

The considerable resources of the computer industry allied to these tools has led to the proliferation of personal computers in which all the constituent components are integrated onto a single card called a mother board. Computing power is now exploding, whether as regards of the memory size, number of registers, speed of the clock cycle, volume of data storage or, finally, in terms of interconnections between all manner of calculators on the global scale.

The consecration of this philosophy was brought home to the author by a Californian start-up five years ago involving a company which has become one of the leaders in the world market of workstations. Their slogan was "the network is the computer", following an approach announced by the world's two largest manufacturers (IBM et DEC).

The above remarks can be simply backed up by inspecting Table VIII-II, which shows the trend in the worldwide sale of computers.

TABLE VIII-II — *Evolution of the number of computers in the world (after: International Data Corporation).*

	Year	Thousands of units sold	$ billion
Large systems	1986	20.8	97.4
	1988	24.3	114.6
Medium systems	1986	247.6	72.6
	1988	330	86.8
Small systems	1986	3,333.2	91.1
	1988	5,219	126.7
Personal computers	1986	52,550	94.2
	1988	95,393	176.6

The applications presented here, which are among the most characteristic of their type, are dedicated to run on minicomputers and microcomputers (Apple, IBM, Compaq, etc.). In the following, emphasis is given to the world of microcomputers, since this sector represents the most widespread type of equipment using the maximum number of applications in the public domain. Although the world of minicomputers also offers numerous applications in the geosciences, they are rarely part of the public domain and this sector is more likely to be dedicated to specific applications developed by research departments for particular needs (aquifer models, warning systems, centralized management, expert systems, etc.).

8.2 General management tools

Until very recently, the boundary between computer design and utilization was characterized by technological differences in the materials which distinguished the domain of the microcomputer from that of the mini-calculator. Today, this boundary is blurred and some overlap exists at the edges of these two domains. Considering the current size of investment budgets for the acquisition of information technology, a generalized trend towards computerization is clearly under way.

The problem of computerization is twofold: on the one hand, the water supply operator is a user who sets out to benefit from the new means put at his disposal in order to optimize his professional activities, and, on the other hand, the hydrogeologist is a design-oriented expert who is going to develop his own tools.

In both cases, the access to tools suitable for solving engineering problems depends on a mastery of the software in question. In this perspective, a simplified approach might consist of separating the calculation software used for numerical modelling from the advanced software incorporating different components (or modules). These modules can make use of databases, image-processing tools, advanced communications functions, artificial intelligence and sophisticated man-machine interfaces. Evidently, such packages include the basic modelling and calculation functions mentioned above which usually still constitute the core of an advanced tool.

— *Engineering software* is generally applied to the solving of specific problems. It is usually written by a professional active in the domain in order to address a particular question. This type of software is found in all branches of the geosciences. Under this heading are included most Earth Science computer applications.

— *Advanced tools* remain poorly known in general. They comprise two categories: specialized software packages and veritable "advanced tools" specific to the geological domain. The programmes are highly elaborate, being the fruit of years of work by teams composed, very often, of engineers and computer scientists. Generally speaking, these high-performance products are reliable and well-documented. However, their release remains restricted in some cases since they may fulfill a strategic function for the company or working group concerned, while training in their use can be time-consuming and reserved for staff who are particularly competent.

A number of software products are presented in the following which belong to both of the categories mentioned above. The intention here is to provide water operators, hydrogeologists, researchers, trainers and water industry employees with an overview of computer developments in their discipline and, eventually, to enable them to find the tools they might require. As usual, it has proved impossible to furnish an exhaustive review, and the software presented here is chosen in relation to its scientific or industrial relevance, its contribution to the resolution of general and clearly expressed problems faced by the profession or, alternatively, with respect to the emergence of a perceptible technological trend.

8.2.1 Engineering research tools

Although belonging to two distinct groups, being either inside or outside the oil industry, geophysicists are making an increasing use of computer methods. There are numerous software products for processing geophysical data, whether applied to seismic surveys, electric logging or even electromagnetic methods. To some extent, each research or engineering design department has developed its own in-house programme(s), even though certain standard programmes are available. For example, the Canadian company GEOSOFT specializes in geophysical cartography and can supply all sorts of programmes destined for this field of application.

In France, COGEMA possesses an internal system of geophysical data processing which comprises a set of programmes for carrying out a range of operations from acquisition to the final editing of documents (data acquisition, transfer, management, editing and utilization). The geophysical department also ensures the support of a tectonic programme using a decentralized unit.

On the international level, it is relevant here to mention AGIS, which is a knowledge-based system for the computerized interpretation of seismic data, developed by I. PITAS from Thessalonica University (Greece), as well as the SEISRISK III programme designed for the evaluation of seismic risk developed by B. BENDER of the U.S. Geological Survey. In addition, EQGEN is an earthquake simulation programme written by N. Y. CHANG of the University of Colorado. Finally, the EMIX34 programme developed by R.S. BELL makes it possible to compute inverse models of electromagnetic conductivity from logging data. Similarly, GREMIX allows the automatic interpretation of seismic refraction data and RESIX enables the interpretation of resistivity data from electric logging acquired by different methods (e.g. Schlumberger, Wenner, pole-dipole, dipole-dipole, dipole-equatorial and dipole-polar). These latter software products are developed by INTERPEX Ltd. There also exist numerous programmes for the migration of scismic reflection data, including a programme written by J. D. UNGER of the US Geological Survey.

Cartography: the development of cartography as an aid to decision-making for planners and managers is beginning to make its appearance in the Earth Sciences. In particular, the BRGM-GEOHYDRAULIQUE group has produced a map of potential groundwater resources in West Africa. In a similar way, thematic cartography derived from satellite imagery provides an exceptional means of evaluation and synthesis of geographical information at a time when the management of natural resources goes hand in hand with development, and where environmental problems are to be considered in their global aspect.

Over the past few years, multispectral remote sensing has become an indispensable supplement to aerial photography in the study and management of natural resources. In the field of satellite imagery, the appearance of new images such as those from the Thematic Mapper of Landsat or the High Resolution Visible sensor of SPOT, has led to completely new developments, given the changes arising from their increased spectral and geometrical resolution. In general, it can be said that satellite information will facilitate both the establishment of new thematic maps in fields not covered until now, and also the revision of already existing maps, with a possible acceleration of the rate of revision and an improvement in the choice of zones to be revised.

Finally, the stereoscopic imagery of the SPOT satellite has led to remarkable applications such as those leading to the automatic generation of numerical models of the terrain from pairs of images. This three-dimensional mode of display opens the way to applications in structural geology, thematic cartography and the use of flight simulation to view inaccessible terrains. These results are the fruit of advances made by French researchers grouped together in the ISTAR company and working in collaboration with INRIA.

8.3 Interdisciplinary tools

After presenting a certain number of specific tools, it is interesting to quote some applications of a multidisciplinary nature which deserve to be better known. These are general programs which concern cartography, statistics, geostatistics, etc. (cf. Figure 8-1).

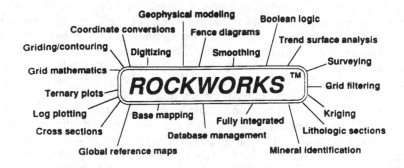

Figure 8-1
An example of an integrated pluridisciplinary tool (after RockWare document).

Some major standard tools should be mentioned here:

— Tools available from the United States Geological Survey (USGS) such as "USGS Microcomputer Programs for Mapping, Statistics, and Geostatistics (Kriging)" which comprise software for applications in the fields of hydrogeology, mining, oil and gas. We also take note of inexpensive software for geologists destined for applications in general geology, the oil or mining sector, geophysical uses, etc., and also STATPAC.

— "Fundamental" tools such as SURFER® or GRAPHER® which make it possible to create maps and obtain a 3-D image or to produce all kinds of graphics which are often indispensable for the illustration of geological and hydrogeological reports. This type of software runs in the Windows® environment and is available notably from Golden Software Inc., 809 14th St., Golden, CO 80401-1866 USA.

— COGS (Computer Oriented Geological Society) sells a catalogue and listing known geological programs, which contains more than 600 programme references grouped by sector of activity and type of computer. It gives the name of the program, a brief description of its functions and the required information for obtaining it: name, address, telephone number of the software developer.

In its monthly journal, at the modest price, COGS also offers floppy disks for MacIntosh, Apple and IBM microcomputers. These programmes implement different applications ranging from stratigraphic symbol frames for laser printers to the table of periodicals accessed with Hypercard. Also included are mathematical and statistical models applied to hydrogeology, cartography or the oil industry.

Tools for scientific calculation, such as geoMATE, enable the optimization of data acquisition and thus have a general interest in geology.

— CONTOUR makes use of ASCII files to produce graphics which display contoured values with any variable. The program is developed by EARTHWARE.

— LITHSEC is designed for the presentation of geological sections and the display of logging data. The program is developed by EARTHWARE.

— GNULEX is a database in the field of stratigraphic nomenclature which includes exhaustive information on formations of the western United States.

— Products marketed by INTERPEX Ltd., P. O. Box 839, Golden, Colorado 80402. USA.

— ROCKWORKS™ (cf. Figure 8-1) distributed by RockWare, Inc., 4251 Kipling St., Suite 595, Wheat Ridge, CO 80033 - USA. Catalogue available upon request.

— Scientific Software Group P.O. Box 23041, Washington, DC 20026-3041, USA, also proposes a catalogue of software and specific hydrogeological publications.

Evidently, this list is not exhaustive. The author has decided to give the present state of the art, while acknowledging the incomplete nature of the task, such is the complexity and continual development of this field. Consequently, the reader is asked to excuse any shortcomings.

8.3.1 Management of historical data

The information involves the use of numerous program, notably ACTIF©, which allow the computerized treatment of the history of drilling and development records. It is a tool which speeds up a number of the routine and tiresome tasks facing those in charge of drilling operations. Furthermore, it contributes to a greater rigour in the data treatment owing to the standardization of the acquired data format. This software makes possible, among other things, the capture and editing of data, the drawing up of characteristic sections and curves, as well as the plotting of Schoeller-Berkaloff diagrams.

The data input comprises 13 screen pages which enable the entry of all the characteristics recorded from the work site: siting, description of the naked hole, its equipment, hydrogeological context, lithological section, logs, well development, hydrodynamic parameters, chemical analyses, etc. A large number of responses can be chosen from the standard lists on display at the relevant moment during entry of the data.

The main procedures available are as follows:

— Establishing a report, consisting of 1-7 A4 pages, which presents all the entered data.

— Graphic plots (on a plotting board) of the technical and lithographic sections, well logs, yield-depression curve and the Schoeller-Berkaloff chemical diagram.

— Export of data in ASCII format, allowing further processing by external software.

The files created by ACTIF© (at least one datafile per catchment structure) can be directly imported by BADGE© into a factual information base on water points. BADGE© is a software package developed to facilitate the management of factual information, that is to say, data independent of time (i.e.: giving characteristics of water extraction points, aquifers, villages, etc.). First and foremost, it is a database management system accompanied by standard processing programs (i.e. extraction and editing of data tables sorted and filtered according to multiple varied criteria, statistics, plotting of points and contouring).

This software is also a convivial interface between the user and the databases to be processed. The databases can be developed according to specification by the user himself. As an example, the following databases have already been developed for or by different BRGM clients:

— wells and boreholes: identification, location, geometrical parameters, hydrogeological context, hydraulic and physico-chemical characteristics;

— sources of pollution: identification, location, nature of the pollutants;

— water-bearing systems: their dimensions, hydrogeological parameters, terms of assessment;

— villages: socio-economic data making it possible to evaluate the sector-based needs, hydrogeological data, audit of resources and requirements;

— irrigated zones: types of crop and irrigation, rotations, water requirements;

— dams: geometrical and technical characteristics are accessed by pulldown menus which make it possible to run all the applications expected of a database manager, i.e.: input and validation of data, importation (dBASE III© or ASCII files), exportation, diagrams, statistical functions, cartography and digitization.

In terms of technical specifications, ACTIF© is written in Quick Basic© language. BADGE© was developed essentially using the CLIPPER© language with databases in the extended dBASE© format. Databases can be installed on any microcomputer operating with DOS© 3.0 or above, possessing 640 kB of memory, a hard disk and a graphics card. The graphic output is implemented on HP-compatible plotters.

8.3.2 Management of time-dependent data

The BRGM has developed a management tool (CHRONO©) for handling time-dependent data, such as variations in groundwater level discharges of springs or rivers, operational yield of water wells, climatological data, etc. This software can manage four different types of temporal database in which the variables are defined by the user:

— bases for the management of data with a daily sampling step (rainfall, temperature, discharge rates, water levels, etc.);

— bases for the management of data with a monthly sampling step (water levels, discharge rates, etc.);

— single-parameter bases for the management of data sampled at random intervals (levels, gaugings, etc.);

— multiparameter bases for the management of data sampled at random intervals (chemical analyses).

The databases are indexed by means of dates and identifiers.

The different software functions are accessible through a multi-windowed display system: choice of working base, data entry, search and selection from a parameter, a station and/or a date, screen display and printing of tables and graphs, elementary statistical parameters, cartography, conversion of data into different units, creation of new parameters, etc.

Furthermore, CHRONO© allows the combination of data imported from different bases by using multicriteria selection with a graphic display. For example, it is possible to display the time variations of two parameters — on the same graphic screen — from one or several different observation sites. Likewise, cartography of the spatial distribution of an element can be displayed from a selection of several sites over a determined time lapse.

8.3.3 Strategic management

The domain of strategic management was probably one of the first domains to benefit from the appearance of geological computer applications, notably in the case of hydrogeological modelling. At the present time, there is a tendency for geologists to become increasingly consulted by project managers. This situation demands that, over and above their purely scientific expertise, hydrogeologists are expected to bring together concepts of management, planning and administration. Once again, it would seem that information technology is the key element in this process, providing a crossing-point that is difficult to avoid.

In the field of hydrogeology, where large amounts of numerical data are treated, one of the major features in the development of the discipline is the use of computers to process information and build mathematical models. In fact, starting from the quantification of water discharges, electronic data processing has opened the way to modelling, which is a fundamental tool in the administration and planning of groundwater development. But modern computer science also enables the acquisition and processing of field data. In this context, a number of advanced tools have been created in France, notably the set of programmes proposed by the BRGM (cf. figure 8-2).

Different administrative tools have been developed using the acquisition of field data as a starting basis. These include a tool produced by the GEOLAB company for the management of the Niger hydrogeological database — known as PROSPER and developed by BURGEAP — which has made possible the management, programming and checking of a village-water supply project.

In the case of basin studies, numerous data are first stored before processing is carried out with the help of groundwater models. The impressive number of programs available is very striking. Many have been developed on large computer systems, while a considerably smaller proportion is available on microcomputers.

Various types of documentation can be found in the computer trade, covering a wide price range and including technical manuals, conversational programs and a multitude of modules devoted to particular applications. Basically two types of publication are available: the great classics of geohydrology (HOUPEURT, de MARSILY, etc.) and books with a computer science orientation that explain how to create or utilise a groundwater model. We can cite, among others, the monograph by J. BARR and A. VERRUIJT "Theory and application of transport in porous media - Modelling groundwater flow and pollution" edited by D. REIDEL. This didactic work demonstrates the construction and utilization of numerical models with the aim of studying groundwater discharges and the transfer of pollutants.

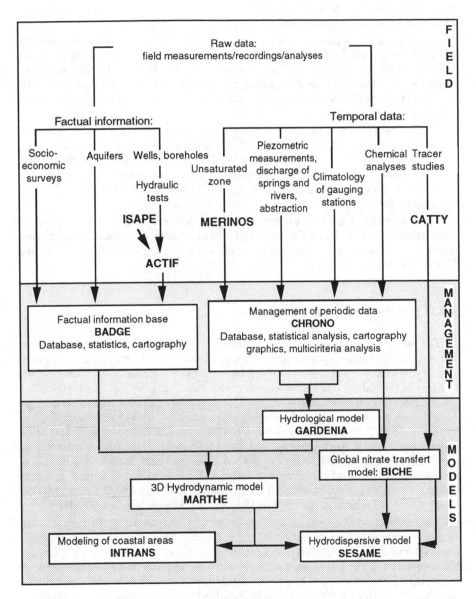

Figure 8-2
BRGM software for the management of groundwater resources
(modified from BRGM document).

It is becoming difficult for system designers to compete with these published works and software products, which are supplied with their source code and which are therefore modifiable by the user at an unbeatably low cost!

There are a multitude of groundwater models on the market, including notably the Icelandic model AQUA, which allows the simulation of discharges and pollutant transfer. French software developed by BRGM, FRANLAB, etc., American software such as TRACER2D and TRACER3D, developed by TENTIME, and the Hydrosoft InterStat software which is based on the method of finite differences.

Finally, different tools provide services in the fields of hydrogeology (interpretation of pumping, migration of pollution, groundwater drawdown, etc.), notably the RIO software, developed by the BRGM, which simulates above- and below-ground discharges and their exchanges. At the same time, it calculates the evolution of water flow lines, water courses, the potential yield of the underlying aquifer and the resulting exchange flow rates.

This domain does not escape from the general trend of evolution in computing, since it has also seen the development of expert system type applications such as HYDROLAB©, created by M. DETAY, P. POYET and E. BRISSON (cf. figure 8-3).

8.4 Trends and perspectives

Ten years ago, a microcomputer could execute less than 100,000 instructions (elementary operations) per second, and it possessed 64 kB (kilo-bytes) of central memory immediately accessible for calculations (a byte being the "volume" in the memory occupied by one character; the prefix "kilo" is used in computer science to introduce a multiplicative factor of 2^{10}, or 1,024). At best, the mass storage device had a capacity of 5 MB (mega-bytes). Nowadays, for example, the Deskpro 386/25 microcomputer of Compaq can execute around 4 million instructions per second (MIPS), has up to 16 MB of central memory and comprises a standard disk storage of 300 MB. In ten years, for a constant cost of computer, the speed of calculation (cf. figure 8-5), capacity of central memory and mass storage have all increased by a factor of 20-50 (a 486 DX at 50 MHz is about 50 times faster than the original IBM PC 8088 operating at 4.77 MHz).

The estimations of experts lead us to believe that these performances will continue to grow at breathtaking rates in the future. We can envisage memories of 256 MB and disks (in all likelihood magneto-optic) of several dozen giga-bytes. Microprocessors are one of the factors in the development of microcomputers: in succeeding generations, the number of transistors assembled on the same silicon chip has risen from 4,000 to 350,000 and the number of bits processed by the registers (or computer word size) has risen from 4 to 8 (a byte), then 16 and finally 32 with the 80386.

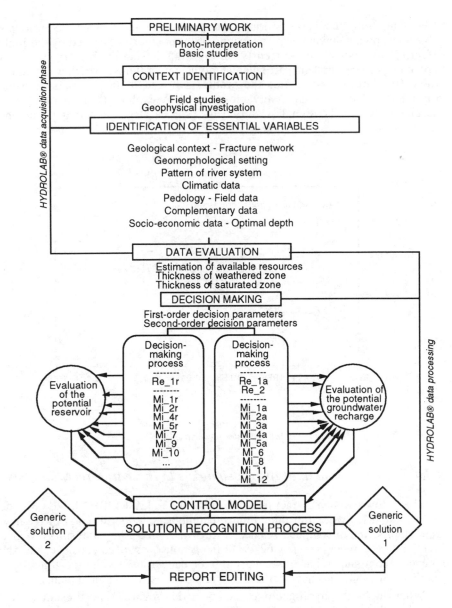

Note: Re = expertise rule; Mi = interpretative model aiming to evaluate the recharge Mi_a and the potential reserve Mi_r.

Figure 8-3
Schematic flow chart of the HYDROLAB® system

However, the very laws of physics give rise to limitations which, from now on, will require the exploration of new approaches. Indeed, the speed constraints of computer circuitry are a limiting factor, all the more so since electrons are subject to slowing down due to the heating of the components caused by the Joule effect. This aspect is currently motivating research in the field of superconductivity, and in particular at high temperatures. Another approach which, in addition, eliminates the problems linked to interactions between components that are excessively close, aims to explore the use of optical calculators where the signal is carried by protons at a speed that cannot be exceeded.

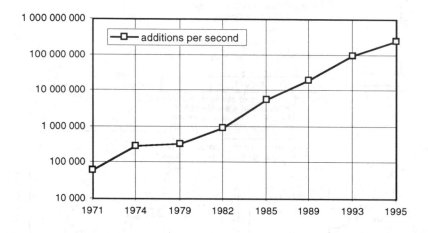

Figure 8-4
Evolution of the computer power from 1971 to 1995.

8.4.1 The commitment of hydrogeologists to these new technologies

The following is an extract from an article by Y. LASFARGUE in "Le Monde Innovation 88". Two out of three French people have a poor opinion of the new technologies, and of computer science in particular", according to results revealed in a CREDOC survey. Precisely 62.7% of the people questioned consider that "the spread of computer science is undesirable or dangerous", while the figure was only 59% in 1984. This new fear is most often expressed in a muted manner.

Where does the hydrogeologist stand in this regard? The present author considers that they may have partially answered this question themselves by introducing information technology into the Geosciences. It would appear that geologists are among the pioneers in electronic data processing, with 63% of geologists regularly using computer-based tools (recent survey published in the review GEOBYTE). Furthermore, certain software products in the geological domain, such as PROSPECTOR of DUDA and HART, have turned out to be precursors of fundamental concepts in the computer sciences. Similarly, the more recent work of R. McCAMMON has contributed to expert systems.

8.4.2 Computer science as a tool in advanced communication

Although this chapter is focused on the conceptual aspects of computer science, its role as a vehicle of information should not be forgotten. The best-known databases are: GEOREF of the American Geological Institute, TULSA (Petroleum Abstract) of the University of Tulsa, GEOARCHIVE of Geosystem, GEOLINE of the Federal Institute for Geosciences and Natural Resources and GEOSCAN of the Geological Survey of Canada. In France, the BRGM administers a subsurface databank known as "GEOBANQUE" (accessed by dialing 36 28 00 03 for the Minitel server in France) which comprises a catalogue of wells and hydrogeological details for approximately 200,000 water extraction points. In addition, bibliographic references are available covering the whole field of Earth Sciences administered jointly with the CNRS (Pascal-Géode and Géode databases); around 80,000 references were included in 1985.

A strong position has already been established in France with the databases GEODE/PASCAL, ARTEMIS, IBISCUS, etc., and it is becoming possible to envisage proper systems of computerized information. These tools are all designed for management purposes; they correspond to the "Management Information Systems" (MIS) which are described in the American literature, and, as such, are appropriate for use in decision-making. Particular mention should be made of MIS that are adapted to the fields of geography (GIS), such as CUSMAP of the US Geological Survey, which enables an improvement in spatial data compilation with a view to optimizing the assessment of mining resources.

8.5 Conclusion

In 1982, Time magazine awarded the title of "Man of the Year" to a machine rather than a human being. That machine was the computer.

Information technology, along with the new forms of computer-based automation that it has engendered, has now become ubiquitous in research departments and is considered indispensable by all industrial enterprises wishing to maintain their profitability and competitiveness. All the indications are that the next revolution in knowledge dissemination and in access to information will take place before the end of this century. It will arise from the technical improvements and falling prices of computer equipment, as well as advances made in the languages associated with artificial intelligence. This mutation will come about from the current use of machines which not only calculate and count but which also reason. The task of such machines will be to assist people in their everyday actions.

The information revolution is just beginning. The information highway will have a significant effect on all our lives in the years to come. Photographs, films and video are currently being converted into digital information. Printed libraries are being scanned and stored as electronic data on disks and CD-ROMs. Newspapers and magazines are now often composed in electronic form.

Moreover, the cost of communications will drop as precipitously as the cost of computing already has.

Personal computers, multimedia CD-ROM software, high-capacity cable television networks, wired and mobile telephone networks — as well as the Internet — are all important precursors of the information highway. E-mail is already easy to use. The concept of the virtual community has infiltrated our real world with the development of computer conferencing systems (e.g. virtual villages that enable people everywhere to carry on public conversations and exchange private e-mail. Virtual communities are social groupings that emerge from the worldwide Net when sufficient numbers of individuals participate in public conversations for long enough — and with positive human feeling — so that webs of personal relationships are formed in cyberspace. A growing number of workers in this info-environment need to be able to absorb, manipulate and market information content. Such an information economy favours small entrepreneurial ventures that can quickly adapt to new technologies. Some hydrogeologists will cross into this brave new world, while others will remain behind.

CHAPTER IX

Conclusion

"To know is not to prove, nor to explain. It is to accede to vision."

Saint-Exupéry

Groundwaters have long remained a poorly understood component of hydrological systems and the global water cycle. These resources are still largely unknown and form the object of misconceptions in the public mind, including a fair number of water professionals also. It is now established that groundwaters make up almost the entire inventory (98 -99%) of liquid water that is stored within or upon the emerged land masses. In addition, underground waters account for a significant proportion of the total fresh water fluxes; nearly a third of all waters circulating in the terrestrial environment are subterranean. As a consequence, groundwaters provide a very widely used resource worldwide, with 600-700 million cubic metres extracted every year. Evidently, the volume of abstractions varies considerably from one country to another according to the climatic regime and the geological setting. In this respect, the countries of Europe — especially the member-states of the EU — can be ranked in an intermediate position. Most of the subsurface of these states contain appreciable groundwater resources, altogether making up 35% of the total water flux on the European land masses. In comparison to surface waters, groundwaters are highly competitive as regards drinking water supply, industrial uses and irrigation.

For example, France draws more than seven billion cubic metres of water every year from her groundwater resources, of which 50% is destined for the water supply to the population, 19.7% for irrigation requirements and 25.8% for industry. The volume abstracted for domestic water supply is secured from some 30,000 wells which, being relatively old, should be largely replaced over the next fifteen years. These old catchment structures, which are often badly designed, otherwise show the effects of clogging and/or corrosion and are also poorly protected.

Faced with this challenge, the author considers it is essential to bring together in the same work the basic principles of water well management in terms of implementation, maintenance and restoration.

— The *implementation* of a water well is a complicated operation that requires feasibility studies (cf. Chapter I) and a choice of suitable materials, in addition to the correct fitting out and development of the well (cf. Chapter II). It follows that the full-time presence of a project manager or a deputy is required on a well construction site (cf. Chapter IV). Finally, the function of a well is not restricted to water catchment alone, but should be part of a long-term vision with the aim of supplying good quality water to the population while providing a correct protection of the resource (cf. Chapter V).

— The *maintenance* of water wells is more than just a concept, since it should become a reality amongst operators. For much the same reason that pumps are serviced, it is also advisable to maintain water wells in good order. Although the means for achieving this are relatively simple (cf. Chapter VI), political will remains one of the key strategic factors in applying the expertise offered by the major water supply companies. Since the ageing of wells is unavoidable, the understanding and recognition of the effects of clogging, corrosion, overextraction and sand intrusion can provide indications that become all the more useful for choosing suitable courses of action or, where appropriate, the carrying out of restoration.

— The *restoration* of wells should become a matter of course (cf. Chapter VII). Taking account of the present economic climate, it is no longer possible to envisage the construction of new wells at great cost while it remains feasible to restore existing ones. Finally, by combining upkeep with restoration, it will become possible to give a new lease of life to catchment works and boost the yield of well fields in many cases (to a much greater extent than if new wells were added). Over the last few years, numerous well management tools have become available (cf. Chapter VIII) which facilitate data archiving and provide help in decision making.

Throughout the present work, the author strives to give a picture of hydrogeology and its importance. In fact, the discipline of geohydrology makes use of approaches from various branches of the exact and natural sciences (mathematics, physics, chemistry, hydraulics, geology, biology, ecology, etc.). Being more than just a sophisticated mathematical instrument, an analysis of local hydrodynamic conditions and a critical analysis of the situation take precedence as regards the understanding of the phenomena. It is at this level that the engineer's shrewdness should make itself felt. The correct analysis of a complex problem in order to set up realistic hypotheses is more difficult than the carring out of some ingenious calculation.

It would appear equally important to provide some formal background to the hydrogeological approach and the procedures of well implementation, maintenance and restoration in the light of European directives emanating from Brussels. Furthermore, since current trends indicate that companies and services of the future will have to comply with various standards (ISO, certification, etc.), a certain degree of formal rigour appears indispensable here.

In addition, the author considers it helpful to lay out guidelines for the interpretation of pumping test data (cf. Chapter III), so that each and every user can attach due weight to understanding the hydrogeological setting before proceeding with mathematical analysis. This temptation to "crunch numbers" becomes even more pernicious if computer programs have a "video games" aspect that tends to mask the physical reality of the phenomenon being studied. In fact, it is far easier to fit a curve than analyse the nature of the observed phenomenon and gain an overall understanding of the processes involved. Bearing this in mind, the author also presents different methods which enable the quantification of quadratic well losses (linked to the type of equipment used). These methods make it possible to assess fairly readily the suitability of the equipment and the degree of well development. Finally, in the interpretation of pumping tests, well development has an impact on the identification of the hydraulic characteristics of groundwater bodies.

All the indications are that geohydrology and its related disciplines are destined to occupy a major place in the management of the environment in the 21st century. Protection of groundwater resources will take on more and more importance especially since we have already identified the problems that will have to be addressed in the future. These problems arise in relation to soil and aquifer pollution, particularly within the unsaturated zone, and will lead to the appearance of major pollution hazards in the coming years. The scale of this phenomenon is just beginning to be evaluated satisfactorily in the case of nitrates and pesticides. However, pollution from very low levels of chlorinated solvents and hydrocarbons will probably rule out the use of certain aquifers due to the persistence of these substances in the environment and the effect of adsorption-desorption processes. Present-day decontamination techniques have difficulty in coping with this type of pollution, so it is advisable to take preventive measures as a matter of priority.

A rational approach to the management of catchment works is indispensable especially since groundwaters are becoming increasingly polluted and also considering the present lack of know-how for the easily controlled decontamination of aquifers. It will probably be necessary to evolve a management policy for the abandonment of poorly protected and isolated wells in preference to the development of major well fields. Since it appears unrealistic to protect hundreds of catchment sites scattered over wide areas of country, it would be better to focus efforts on well identified zones where the resource can be properly protected. Otherwise, it can no longer be assumed that groundwaters are pure, and steps should be taken to integrate the need for their treatment. Because the standards applicable to drinking water are extremely rigorous (even draconian), most raw waters require some form of treatment. Thus, although groundwaters possess certain clear advantages with respect to surface waters (being less vulnerable and of constant quality, etc.), these treatment requirements mean that waters have to be extracted through the operation of well fields. Since water treatment plant are of increasingly complex and sophisticated design, they require a considerable throughput in order to provide an effective return on investment. Furthermore, it is unreasonable to envisage the construction of new treatment plant for small and isolated wells having low capacity and poor levels of protection.

To conclude, it is appropriate to quote the words of the French Prime Minister who, at the close of sittings of the Water Congress in March 1991, stated that is was time to start out on the reconquest of water. The techniques of implementation, maintenance and restoration as applied to water wells will form one of the main pillars of such a challenge, thus providing the main reason for carrying out the synthesis presented in this book. The author trusts that the present work will be favourably received by the different players involved in the water sector, including students, academics, experts, practitioners, technicians and decision makers.

CHAPTER X

Bibliography

"Learning without thought is labour lost;
thought without learning is perilous."

Confucius

The following list is divided into:
— general bibliography, including text books with a strong applications orientation that provide information of direct use to the reader, especially in the regulatory and operational domains;
— basic references, comprising background reading in geohydrology, hydrogeology and related disciplines;
— main scientific publications cited, bringing together an exhaustive compilation of the references consulted by the author.

General bibliography

BOURLET A., GARCIN J-L. — *Code pratique de l'eau. Textes officiels, commentaires, jurisprudence.* Editions du Moniteur, 397 p., (1991).

BRGM — *Méthodes d'études et de recherches des nappes aquifères*, BRGM Ed., 158 p., (1962).

Bulletin Officiel — *Travaux de forage pour la recherche et l'exploitation d'eau potable. Marché publics de travaux, CCTG.* Fascicule n° 76, 76 p. (1987).

CASTANY G. — *Principes et méthodes de l'hydrogéologie.* Dunod Ed., Paris, Dunod Université 236 p., (1982).

CASTANY G., MARGAT J. — *Dictionnaire français d'hydrogéologie.* BRGM Ed., Orléans, 249 p., (1977).

Journal Officiel — *Hygiène alimentaire. Eaux destinées à la consommation humaine.* Brochure n° 1629, 194 p. (1991).

Journal Officiel — *Les servitudes d'utilité publique et les plans d'occupation des sols.* Ministère de l'équipement, 408 p., (1990).

LAUGA R. — *Pratique du forage d'eau.* Seesam Ed., (1990).

LALLEMAND-BARRÈS A., ROUX J-C. — *Guide méthodologique d'établissement des périmètres de protection des captages d'eau souterraine destinées à la consommation humaine.* BRGM Ed., collection Manuels et méthodes n° 19, 221 p., (1989).

MABILLOT A. — *Les forages d'eau - guide pratique.* 2e édition. Technique et Documentation, Paris, 237 p., (1971).

Basic references

AMERICAN WATER WORKS ASSOCIATION — *Water Quality and Treatment. A handbook of Community Water Supply.* McGraw-Hill, 1194 p., (1990).

ASTIER, J-L. — *Géophysique appliquée à l'hydrogéologie.* Masson & Cie Ed., 277 p., (1971).

BARCELONA *et al.* — *Contamination of ground water. Prevention, assessment, restoration.* Noyes Ed., Pollution Technology Review n° 184, 213 p., (1990).

BEAR J. — *Hydraulics of Groundwater.* McGraw-Hill, 569 p., (1979).

BEAR J., VERRUIJT A. — *Modelling groundwater flow and pollution. Theory and applications of transport in porous media.* D. Reidel Publishing Company, 414 p., (1987).

BITTON G., GERBA C. P. — *Groundwater pollution microbiology.* John Wiley & Sons Publishers, 377 p., (1984).

BOWEN R. — *Groundwater.* 2nd Edition. Elsevier Applied Science Publisher, 427 p., (1986).

BOUWER H. — *Groundwater Hydrology.* McGraw-Hill , 480 p., (1978).

BRASSINGTON R., — *Field Hydrogeology.* Geological Society of London Professionnal Handbook Series, 175 p., (1988).

BREMOND R., PERRODON C. — *Paramètres de la qualité des eaux.* Ministère de l'Environnement, 259 p., (1979).

CASSAN M. — *Les essais d'eau dans la reconnaissance des sols.* Eyrolles, 275 p., (1980).

CASSAN M. — *Aide mémoire d'hydraulique souterraine.* Presses des Ponts et Chaussées. 190 p., (1986).

CASTANY G. — *Traité pratique des eaux souterraines.* Dunod, Paris, 657 p., (1963).

CASTANY G. — *Prospection et exploitation des eaux souterraines.* Dunod, Paris, 717 p., (1968).

CHANTEREAU J. — *Corrosion bactérienne - bactéries de la corrosion.* Editions Technique et Documentation, 262 p., (1980).

CLARK L. — *Water wells and boreholes.* Geological Society of London Professionnal Handbook Series, 155 p., (1988).

CORAPCIOGLU Y. — *Advances in porous media.* Vol. 1. Elsevier , 309 p., (1991).

DAVIS S. N., DeWIEST J. M. — *Hydrogeology.* John Wiley & Sons, 463 p., (1966).

DEGREMONT — *Mémento technique de l'eau.* Tomes 1 et 2. Lavoisier, 1459 p., (1989).

DETAY M., POYET P., BRISSON E. — *HYDROLAB$^®$ Le système expert d'aide à l'implantation de forages.* SEG-CIEH 125 p., (1991).

DETAY M., FORKASIEWICZ J., LACHAUD J-C, LEROUX M., POINTET M., GUIRAUD R., TRAVI Y., MARGAT J. — *L'hydrogéologie de l'Afrique de l'Ouest - Synthèse des connaissances : socle cristallin et cristallophyllien et sédimentaire ancien.* Agridoc International, 147 p., (1991).

EAGLESON P.S. — *Dynamic hydrology.* McGraw Hill, 462 p., (1970).

DUSSART B. — Limnologie, l'étude des eaux continentales. Gauthier-Villars, 676 p.,(1966).

DESBRANDES R., — *Théorie et interprétation des diagraphies.* Publication de l'Institut Français du Pétrole. Technip, 545 p., (1968).

DeWIEST J. M. — *Geohydrology.* John Wiley & Sons, 366 p., (1965).

DILUCA C., de REYNIÈS E. — *Forage d'eau. Matériel et techniques mis en œuvre en Afrique Centrale et de l'Ouest.* CIEH-BURGEAP, 273 p., (1983).

DOMENICO P.A., SCHWARTZ F.W. — *Physical and chemical hydrogeology.* John Wiley & Sons, 824 p., (1990).

DOWNING R.A., WILKINSON W.B. — *Applied Groundwater Hydrology.* Oxford Science Publications, Clarendon Press Oxford, 340 p., (1991).

DRISCOLL F.G. — *Groundwater and wells.* Johnson Division, 1108 p., (1986).

ECONOMIDES M., NOLTE K. — *Reservoir stimulation.* Second edition, Schlumberger Educational Services, Prentice Hall, (1989).

FREEZE R., CHERRY J. — *Groundwater.* Prentice Hall, 604 p., (1979).

FRIED, J-J. — *Groundwater pollution : theory, methodology, modelling and practical rules.* Developments in Water Science, vol. 4, Elsevier, (1975).

GOGUEL J. — *Application de la géologie aux travaux de l'ingénieur.* Masson, 373 p., (1967).

HANDA O. P. — *Ground water drilling.* Oxford & IBH Publishing, 236 p., (1984).

HAMILL L., BELL F-G. — *Groundwater ressource development.* Butterworths, 344 p., (1986).

HOUPEURT A. — *Eléments de mécanique des fluides dans les milieux poreux.* Editions Technip, 237 p., (1975).

HOWSAM P. — *Water wells: Monitoring, Maintenance, Rehabilitation.* E. & F.N. Spon, 422 p., (1990).

IFP — *Formulaire du foreur.* Editions Technip, 413 p., (1978).

KHANBILVARDI R.M., FILLOS J. — *Groundwater hydrology, contamination, and remediation in groundwater systems.* Scientific Publications Co., 521 p., (1986).

LETOURNEUR J., MICHEL R. — *Géologie du génie civil.* Armand Colin, 728 p., (1971).

LYONNAISE DES EAUX — *Memento de l'exploitant de l'eau et de l'assainissement.* Lavoisier, (1986).

MARSILY G. de — *Hydrogéologie quantitative.* Masson, Paris, 215 p., (1981).

MAYER DE STADELHOFEN C. — *Applications de la géophysique aux recherches d'eau.* Technique et Documentation - Lavoisier 183 p., (1991).

NIELSEN D. M. — *Practical handbook of ground-water monitoring.* Lewis Publishers, 717 p., (1991).

NYER E. K. — *Groundwater treatment technology.* Van Nostrand Reinhold, 188 p., (1985).

PARIGOT, P. — *Les mesures en cours de forage.* Ouvrage collectif, Technip, 171 p., (1982).

PERSON J. — *Interventions réglementaires du géologue agréé en vue de la protection des eaux destinées à l'alimentation humaine.* Document du BRGM n° 11, BRGM Ed., 193 p., (1979).

PLOTE H. — *Sondage de reconnaissance hydrogéologique, méthode du marteau fond de trou, exécution et surveillance.* Collection Manuels & méthodes n° 12, BRGM, 150 p., (1985).

ROSCOE MOSS COMPANY — *Handbook of groundwater development.* Wiley & Sons, Inc., 493 p., (1990).

SCHOELLER H. — *Les eaux souterraines. Hydrogéologie dynamique.* Masson, Paris, 642 p., (1962).

SCHNEEBELI G. — *Hydraulique souterraine.* Eyrolles, Paris, 362 p., (1966).

TARDY Y. — *Le cycle de l'eau. Climats, paléoclimats et géochimie globale.* Masson, 338 p., (1986.)

TOLMAN C. F. — *Ground Water.* McGraw-Hill, 593 p., (1937).

US DEPARTMENT OF THE INTERIOR — *Ground Water Manual. Water and Power Resources Service*, US Government Printing Office, 480 p., (1981).

WALTON W. — *Groundwater resource evaluation.* McGraw-Hill, 664 p., (1970).

WALTON W. — *Analytical groundwater modelling. Flow and contaminant migration.* Lewis Publishers, 173 p., (1989).

References cited - main published articles

AFB — La protection des captages d'eau, in *Cahiers Techniques de la Direction de l'eau et de la Prévention des Pollutions et des Risques*, n° 24, 101 p., (1989).

AGENCE DE L'EAU RHIN-MEUSE — Contrôle et entretien des captages d'eau souterraine, in *Note technique*, (1989).

AGENCE DE L'EAU RHIN-MEUSE — Etude sur le controle et l'entretien des captages d'eau souterraine, in *rapport Géotherma*, 106 p. (1991).

AGENCE DE L'EAU SEINE-NORMANDIE — Surveillance et entretien des ouvrages de captage d'eau potable. Recommandations aux élus et responsables de la gestion, in *Note technique*, (1988).

AGENCE DE L'EAU SEINE-NORMANDIE — L'alimentation en eau potable, la protection et le traitement des eaux souterraines, in *Cahier technique* n° 12,159 p., (1986).

AHMED F., SMITH P.G. — A simplified method for the detection of Gallionella in water, in *Wat. Res.*, **22**, (3), 395-396 (1988).

ALCALDE R.E., CASTRONOVO D.E., KNOTT E. — Occurrence of iron bacteria in wells in Rio Negro (Argentina), in *Proc. Int. Symposium on Biofouled Aquifers: Prevention and Restoration* (ed. DR. Cullimore), American Water Resources Association, Bethesda, Maryland, U.S.A., 127-136, (1986).

ALFORD G. *et al.* — Contamination of Water Wells by Organisms. Arkansas. *Rap. Water Resources Research Center*, Fayeneville, AR, USA, (1984).

AVOGADRO A., de MARSILY G. — The role of colloïds in nuclear waste disposal. *Mater. Res. Soc. Symp. Proc. 26*, 495-505. *Elsevier, Amsterdam, The Netherlands* (1984).

BANKS D. — A case study of screen failure and borehold rehabilitation in the lower greensand aquifer of southern England in *Proc. of the International Groundwater Engineering.* Conference held at Cranfield Institute of Technology. UK (1990).

BERNARDI A., DETAY M., de GRAMONT H. — Corrélations entre les paramètres géoélectriques et les caractéristiques hydrodynamiques des forages en zone de socle in *Hydrogéologie*, **4**, 245-253 (1988).

BERNARDI A., DETAY M., GAUJOUS D. — Primi risulti di studi e perforazioni eseguita nel camerun settentrionale, in *actes Colloque I.A.H., Rome*, 14 p. (1987).

BESBES M. — Etude des pertes de charge dans les forages. Application à la détermination des transmissivités par essai de pompage de courte durée, in *Publication de la Division des Ressources en eau de Tunisie*, **1**, 15-100 (1971).

BERSILLON J-L, DETAY M., FEUARDENT J-P., VIGNIER V. — Adsorption-désorption de l'atrazine en zone saturée - application aux roches aquifères de la région parisienne : craie sénonienne et calcaire de Champigny, in *Comptes Rendus de l'Académie des Sciences*, Paris, t 318, Série II, 1357-1363 (1994).

BICHARA A.F. — Redevelopment of clogged recharge wells, in *J. Irrig. Drain. Eng.*, **114**, (2), 343-350 (1988).

BICHARA A.F. — Clogging of recharge wells by suspended solids, in *J. Irrig. Drain. Eng.*, **112**, (3) 210-224 (1986).

BIZE J. — Exploitation du complexe alluvions. rivière en bordure de Loire. Etude du colmatage des puits et recherche de remèdes, in *Rapport BURGEAP de 1ᵉ phase* (1984).

BLANC-TAILLEUR M. — Les bactéries ferrugineuses, in *Courrier JOHNSON*, **52**, (1979).

BOLSENKÖTTER H., BRUSSE R., DIELDERICH G., HOLTING B., HOHBERGER K., REGENHARDT H., SEHLOZ W., WILLINGER E., WERNER J. — Hydrogeologische kriterien bei der bemessung von wasserschutzgebieten für grundwasserfassungen, in *Geologisches Jahrbuch Reihe C/ Helf* **36** (1984).

BOUDOU J.P. — Ecologie des bactéries aérobies des eaux souterraines. Fixation du fer et du manganèse. Colmatage des forages d'eau, in *rap. Compagnie Générale des Eaux, Anjou-Recherche*, 85 p., (1984).

BOULY — Etude technico-économique de la présence de fer et de manganèse dans les forages alluviaux du bassin Rhin-Meuse, in *Rapport ENSG* (1980).

BONNET M., UNGEMACH P., SUZANNE P. — Interprétation des essais de pompage en régime transitoire : l'effet de puits et la post-production, in *Chronique d'hydrogéologie* **12**, 113-126 (1967).

BOURGEOIS M. — Corrosion et incrustation dans les forages d'eau. Compilation sur les causes, préventions et remèdes, in *Doc. BRGM* (1976).

BOURGEOIS M. — La corrosion et l'incrustation dans les forages d'eau (choix de l'équipement adapté), in *Doc. BRGM*, rapport 76 (1987).

BOURGUET L., GATELLIER C., HERMIN M-N. — Colmatage microbien des forages et circulation de l'eau - Résultats d'un "colmatomètre" expérimental, in *Revue des Sciences de l'eau* **1-2**, 3-21 (1988).

BOURGUET L., GATELLIER C. — Colmatage des forages, causes et remèdes, in *Hydrogéologie et géologie de l'ingénieur*, **1**, 121-126 (1984).

BOUSSELMI L., TRIKIE E., ENNABLI M. — Comportement électrochimique et analytique de la corrosion des crépines en acier ordinaire et galvanise par les eaux de forage, in *XXIIèmes Journées de l'Hydraulique*, .**2**, 1-8 (1991).

BOWEN G.G. — A case study. Rehabilitation of an iron biofouled public supply borehole, Otter Valley, South Devon. UK. *Proc. of the International Groundwater Engineering Conference held at Cranfield Institute of Technology*, UK, (1990).

BRÉMOND R., CHERET I., PARSY C. — Mesures piézométriques et essais de pompage dans les nappes souterraines - Application à l'estimation de la puissance d'une nappe. *Service de l'Hydraulique de l'AOF*, 78 p., (1960).

BRÉMOND R. — Amélioration par acidification d'un captage d'eau dans une formation calcaire, in *Rapport CIEH*, 24 p., (1962).

BRÉMOND R. — Etude de l'influence des caractéristiques technologiques des puits et forages sur leur rendement et leur longévité. 1ère partie : l'ensablement, in *Doc. CIEH*, (1964).

BRÉMOND R. — Etude de l'influence des caractéristiques technologiques des puits et forages sur leur rendement et leur longévité. 2ème partie : Le colmatage par éléments sablo-argileux et les carbonates, in *Doc. CIEH*, (1965).

BRGM — Hydrogéologie de la craie du bassin de Paris. *Actes du colloque régional de Rouen 25-26 mai. Doc. BRGM* n° 1, 627 p. et n° 3, 155 p., (1978).

BRGM — La maintenance industrielle des installations enterrées. Maintenance des forages de production d'eau, in *Rapport BRGM 89 NPC. 25*, (1989).

BRISSON E., DETAY M., POYET P. — Enseignement assisté par système expert - Application à la gestion des ressources en eau, in *Géologues* **93**, 35-40 (1991).

BURGEAP — Exploitation du complexe alluvions-rivière en bordure de Loire. Etude du colmatage des puits et recherche des remèdes, in *Rapport R.573. E.1383* (1984).

BURGEAP — Etude de colmatage des forages exploitant l'Ypresien. Etude du colmatage chimique et bactériologique a l'injection, pré-modèle physique, rapport de synthèse, in *Rapport AFSN/AFME/SCBPE* (1985).

BURGEAP — L'équipement des villages en puits et forages en fonction des conditions hydrogéologiques dans les Etats ACP d'Afrique, in *Rapport CEE*, 177 p., (1978).

CARELA — Régénération des puits. Analyse préalable par photographie subaquatique, mesure de débit, prise d'échantillons, nettoyage. Exemple de travaux réalises, in *L'Eau, l'Industrie, les Nuisances* (1990).

CARRÉ J.,— Le rôle de l'hydrogéologue agréé, in *Actes du colloque de Saint-Brieux*, 63-65 (1992).

CARTIER Y., JOLLIN K., FEUERSTEIN J., QUINAULT J.M., — Etude expérimentale du transfert de polluants radioactifs entre la Seine et la nappe alluviale d'Aubergenville - Rétention de six radionucléides par la craie, in *Note SERE (CEA)* 91/02(I), 24 p., (1991).

CHERET I. — Colmatage des forages servant a l'alimentation en eau de Cotonou, in *Notes de tournée* (1958).

CLARK L. — Boreholes and wells: their monitoring, maintenance and rehabilitation, in *Proc. of the International Groundwater Engineering Conference held at Cranfiels Institute of Technology, UK* (1990).

CLARK L., RADINI M., BISON M. — Borehole restoration methods and their evaluation by step-drawdown tests: the case history of a detailes study in Northern Italy, in *Quart. Journal of Engineering Geology*, 21, 315-328 (1988).

CLARK L., TURNER P.A. — Experiments to assess the hydraulic efficiency of well screens, in *Groundwater*, 21 (3) 270-286 (1983).

CLARK F.E. — Corrosion and incrustation in water wells, in *FAO irrigation and drainage, paper* 34 (1980).

COMBE M., RICOUR J., RISLER J.J. —Les forages de production d'eau et la gestion des risques, in *Hydrogéologie* 4 (1987).

CORDONNIER J. — Réhabilitation de forage par acidification : risques de corrosion et choix du passivant, in *Rapport CIRSEE* (1992).

CORNACCHIA M., DETAY M., GIORGI L. — Nouvelles données sur l'hydrogéologie Centrafricaine, in *Hydrogéologie*, 3, 165-181, (1990).

COSGROVE T. — The use of borehole CCTV surveys. Proc. of the International Groundwater Engineering, in *Conference held at Cranfield Institute of Technology. UK*, 82-86 (1990).

COSTERTON J.W., LASCHEN E.S. — Influence of biofilms on efficacy of biocides on corrosion causing bacteria, in *Materials Performance*, 23, 34-37 (1984).

CULLIMORE D.R. — *Practical manual for groundwater microbiology*. Lewis Publisher, 250 p. (1992).

CULLIMORE D.R. — Physical ways to control build-up of "rust" in wells, in *Johnson Drillers Journal*, 53 (1), 8, 9, 23-25 (1981).

CULLIMORE D.R. — The study of a pseudornonad infestation in a well at Shilo, Manitoba, in *Ground Water*, 21, 558-563 (1983).

CULLIMORE D.R., MANSUY N. — The control of iron bacterial plugging of a well by tyndallization using hot water recycling, in *Water Poll. Res. Journal. of Canada*, 21 (1), 50-57 (1986).

CULLIMORE D.R., MC CANN A. — *The identification, cultivation and control of iron bacteria in ground water*, in Aquatic microbiology, Academic Press, 219-261 (1977).

CULLIMORE D.R. — Think Tank on Biofilms and Biofouling in Wells and Groundwater Systems, in *University of Regina, Regina Water Research Institute, Regina, SK* (1987).

CULLIMORE D.R. — Looking for iron bacteria in water, in *Canadian Water Well*, 15 (3), 10-12 (1989).

CULLIMORE D.R. — An evaluation of there risk of microbial clogging and corrosion in boreholes, in *Proc. of the International Groundwater Engineering Conference held at Cranfield Institute of Technology. UK*. 25-34 (1990).

CULLIMORE D.R. — Evaluation of the available technology to conveniently monitor and control biological fouling of water wells, in *Proc. Int. Conf. Microbiology in Civil Enginerring. Cranfield, Beds, UK* (1990).

D'ANSELME B., BOUDOU J.P., GOUESBET G. — Description générale du colmatage rapide des forages d'eau en France, in *Water Supply*, **3** (2), 151-155 (1985).

d'ARRAS D., DEMONGEOT D., LOUBEYRE R. — Des outils de sécurité en matière de distribution d'eau potable, in *TSM L'Eau*, **10**, 489-490 (1990).

d'ARRAS D., SUZANNE P. — Protection des ressources en eau, aspects légaux et opérationnels, in *IWEN Annual Symposium*, **4** (1) 4-11 (1991).

DEGALLIER R. — Décolmatage des puits et forages - Manuel pratique, in *Hydrogéologie*, **1**, 3-25 (1987).

DEGALLIER R. — Etude de l'influence des caractéristiques technologiques des puits et forages sur leur rendement et leur longévité : l'ensablement et le colmatage, in *Hydrogéologie*, **1**, 27-50 (1987).

DEGALLIER R. — Interprétation des effets de pompages a débit non constant dans des nappes libres, semi-captives ou captives, non illimitées et pénétrées partiellement par les puits de pompage et les piézomètres de mesure, in *Rapport BRGM n° 78 SGN 316 HYD* (1978).

DEGALLIER R., de MARSILY G. — Détermination des paramètres hydrodynamiques par interprétation de variations brusques de niveau dans des puits, in *Rapport BRGM 78 SGN 028 HYD* (1978).

DEGALLIER R. — Colmatage des puits et forages, in *Manuel pratique. BRGM* (1985).

DEGALLIER R. — La dégradation des ouvrages réduit le débit d'exploitation. Rapport : Colmatage des puits et forages, in *Manuel Pratique. BRGM. AFME* (1985).

DEGALLIER R. — Le décolmatage des puits et forages. L'expérience du forage d'eau, in *Geothermie actualité*, **3**, (2), 42-45 (1986).

DEMASSIEUX L. — Les fondements de l'hydraulique des puits, in *Doc. ENSG - Nancy*, 29 p. (1988).

DETAY M. — Rational ground water reservoir management: the role of Artificial Recharge, in *procedings of Second International Symposium of artificial recharge of ground water* Orlando USA 231-240 (1994).

DETAY M. — Le rôle de la réalimentation artificielle de nappe dans la gestion active des aquifères, in *Hydrogéologie* **1**, 57-65 (1995).

DETAY M. — Crystalline hydrogeology of the west african shield: a numerical approach, in *Proc. 29th Int. Geological Congress in Kyoto*, II-20-5 907, (1992).

DETAY M. — Advanced groundwater resources management: the Lyonnaise des Eaux - Dumez experience, in *Proc. 29th Int. Geological Congress in Kyoto*, II-20-4 899, (1992).

DETAY M. — Techniques modernes d'action sur les eaux souterraines, bilan et perspectives, in *XXIe Journées de l'Hydraulique, Rapport Général de la Question II*. G.G.II.1-13 (1991).

DETAY M. — La prise de conscience de la gestion des ressources en eau : l'investissement formation, in *L'eau souterraine un patrimoine à gérer en commun. Doc. BRGM* **195** (1), 271-281 (1990).

DETAY M. — La formation permanente des ingénieurs hydrauliciens dans les pays en voie de développement : bilan et perspectives, in *La Houille Blanche, revue internationale de l'eau*, **3/4**, 235-241, (1991).

DETAY M. — Analyse statistique des paramètres hydrogéologiques de la première campagne de forages dans le Sud-Ouest Gabonais, in *Bull. du CIEH* **60**, 2-22 (1985).

DETAY M. — *Identification analytique et probabiliste des paramètres numériques et non-numériques et modélisation de la connaissance en hydrogéologie sub-sahélienne.* Thèse de Doctorat d'Etat ès Sciences, 456 p., (1987).

DETAY M., BERSILLON J.-L. — La réalimentation artificielle des nappes profondes : faisabilité et conséquences, in *La Houille Blanche, revue internationale de l'eau,* **4,** 57-61 (1996).

DETAY M., COLLIN J.-C. — Introduction à la gestion active des aquifères : concept et philosophie (Éditorial du numéro spécial consacré au colloque AIH 1995), in *Hydrogéologie* **1,** 3-11 (1995).

DETAY M., HAEFFNER H., BERSILLON J.-L. — The role of artificial recharge in groundwater active management: quality, aquifer storage and recovery, environmental effects, and economical perspectives, in *proceedings of the 7th Symposium on Artificial Recharge of Groundwater* Temple, Arizona, 83-97 (1995).

DETAY M., MAUGENDRE J.-P., CLÉMENT M., LABRE J., RICHARD G. — *PA*ris *C*risis *T*ool: an integrated management system to deal with major crisis in the Paris area (France), in *procedings du congrès AIDE de Durban* (1995).

DETAY M. d'ARRAS D., SUZANNE P. — La gestion des ressources en eau en région parisienne Ouest, in *La Houille Blanche, revue internationale de l'eau,* **4,** 295-308 (1992).

DETAY M., DOUSSAN C., GRENET B., LEDOUX E., POITEVIN G., PICAT P., VIGNIER V. — Etude qualitative et quantitative de l'effet de berge - application à la nappe alluviale de Flins - Aubergenville, in *Proc. XXIIe Journées de l'Hydraulique,* 10 p. (1992).

DETAY M., POYET P. — La place de l'informatique dans les géosciences, évolution et perspectives, in *Géologues,* **91,** 37-49 (1991).

DETAY M., POYET P., CASTANY G., BERNARDI A., CASANOVA R., EMSELLEM Y., BRISSON E., AUBRAC G. — Hydrogéologie de la limite Sud-Ouest du bassin du Lac Tchad au Nord Cameroun - mise en évidence d'un aquifère semi-captif de socle dans les zones de piémont et de "biseau sec", in *C. R. Acad. Sciences, Paris,* t 312, Série II, 1049-1056 (1991).

DETAY M., POYET P. — Design and implementation of a field expert system for village water supply programs, in *Bull. Int. of Engineering Geology,* **41,** 63-75 (1990).

DETAY M., POYET P. — HYDROLAB, an expert system for groundwater exploration and exploitation, in *International Journal of Water Resources Development,* **6,** (3), 187-200 (1990).

DETAY M., POYET P. — Application of remote sensing in the field of hydroengineering geology : the artificial intelligence approach, in *Proc. of the International Symposium Remote Sensing and Water Resources.* 849-858 (1990).

DETAY M., POYET P. — Introduction aux méthodes modernes de maîtrise de l'eau, in *Hydrogéologie,* **1,** 3-25, (1990).

DETAY M., POYET P. — Influence of the development of the saprolite reservoir and of its state of saturation on the hydrodynamic characteristics of drillings in the cristalline basement, in *Selected papers on hydrogeology, from the 28th International Geological Congress,* **1,** 463-471. Verlag Heinz Heise, (1990).

DETAY M., POYET P. — HYDROLAB® : le système expert de l'hydraulique villageoise, in *Bull. du CIEH,* **76,** 25-41, (1989).

DETAY M., POYET P., EMSELLEM Y., BERNARDI A., AUBRAC, G. — Influence du développement du réservoir capacitif d'altérites et de son état de saturation sur les caractéristiques hydrodynamiques des forages en zone de socle cristallin, in *C. R. Acad. Sciences, Paris,* t. 309, Série II, 429-436 (1989).

DETAY M., ALESSANDRELLO E., COME P., GROOM Y. — Groundwater contamination and pollution in Micronesia, in *Journal of Hydrology,* **112** 149-170 (1989).

DETAY M., et DOUTAMBAYE C. — Hydrogéologie statistique du socle précambrien de la République Centrafricaine : Principaux résultats du Programme d'Hydraulique Villageoise en Zone Cotonnière, in *Bull. du CIEH*, **76**, 13-24 (1989).

DETAY M., POYET P. — Environnement et géologie - Application de l'intelligence artificielle en hydrogéologie, in *Géologues*, **85-86**, 41-47, 1988.

DOUSSAN C. — Échanges rivière-nappe, Revue bibliographique, in *rapport interne Ecole des Mines de Paris, Centre d'Informatique Géologique - Lyonnaise des Eaux - Dumez,* 55 p., 1989.

DOUSSAN C., TOMA A., PARIS B., POITEVIN G., LEDOUX E., DETAY M. — Coupled use of thermal and hydraulic head data to characterize river/groundwater exhanges, in *Journal of Hydrology,* **153**, 215-229 (1994).

DOUSSAN C., LEDOUX E., POITEVIN G., DETAY M. — Transfert rivière-nappe et effet filtre des berges - Application aux transferts de l'azote, in *La Houille Blanche, revue internationale de l'eau,* **8**, 16-21 (1995).

DRE — Les périmètres de protection des captages d'alimentation en eau potable en Ile-de-France. Direction Régionale de l'Equipement, document provisoire17 p. (1989).

EICHHOLZ G.G., JONES C.G., HAYNES H.E. — Potential radiation control of biofouling bacteria on intake filters, in *Radiation Physics and Chemistry,* **31** (1/3), 139-147 (1988).

EMSELLEM Y. — Les transferts de pression entre nappes et la drainance dans les ensembles aquifères hétérogènes, in *Chronique hydrogéologique* **11**, 131-152 (1967).

EMSELLEM Y., DETAY M. — L'autoépuration des nitrates dans le sol, synthèse méthodologique, in *Rapport de Recherche du Service de la Recherche des Etudes et du Traitement de l'Information sur l'Environnement (SRETIE). Ministère de l'Environnement* (1989).

EMSELLEM Y., DETAY M., ALLA P., SUZANNE P. — A model for self-purification of nitrates in ground water - Example of the Ansereuilles aquifer (France), in *Proc. 28th International Geological Congress* , **3**, 3-467 (1989).

EMSELLEM Y. — L'interprétation des essais de débit dans les aquifères hétérogènes, in *Annales des Mines* (1964).

EMSELLEM Y. — L'interprétation des essais de débit des nappes souterraines par la méthode d'identification, in *Annales des Mines,* **6**, 375-380 (1965).

EMSELLEM Y., JAIN C., LAGARDE A., CHAUMET P. — Interprétation des essais de pompage dans les aquifères multistrates, in *Chronique d'hydrogéologie,* **12**, 19-29 (1967).

FERRIS J.G., KNOWLES D.B., BROWN R.H. — Hydraulique des nappes d'eau souterraines, théorie des essais sur les nappes aquifères, in *Geol. Surv. Wat Supply Paper, USA, n° 1536-E,* p. 69-174, Traduction BRGM n° 4633 (1962).

FLEURIE C., DETAY M., DEMASSIEUX L. — Influence dc la position géomorphologique sur le comportement hydrodynamique des forages dans le socle plutonique du Burkina Faso, in *C. R. Acad. Sciences, Paris,* t. 315 , Série II, 867-873 (1992).

FOUNTAIN J., HOWSAM P. — The use of high pressure water jetting as a rehabilitation technique, in *Proc. of the International Groundwater Engineering Conference held at Cranfield Institute of Technology. UK,* 180-194 (1990).

FORKASIEWICZ J. — Interprétation des données des pompages d'essai pour l'évaluation des paramètres des aquifères - Aide mémoire, in *Doc. BRGM 69 SGL 293 HYD* (1969).

FORKASIEWICZ J. — Principales méthodes d'interprétation des données de pompages d'essai pour l'évaluation des paramètres des aquifères (milieu poreux), in *Doc. CEFIGRE,* 44 p. (1967).

FORKASIEWICZ J., MARGAT J. — La drainance et les communications entre couches aquifères - notions générales, in *Doc. BRGM DS.66.A110,* 26 p. (1966).

FORKASIEWICZ J., MARGAT J. — Estimation du débit spécifique relatif par essais de puits de courte durée, in *Doc. BRGM DS.66.A120*, 10 p. (1966).

GASS T.E. *et al.* — Manual of Water Well Maintenance and Rehabilitation Technology, in *Doc. National Water Well Association, Worthington, OH, USA*, (1981).

GEHRELS J., ALFORD G. — Application of physico-chemical treatment techniques to a severely biofouled community well in Ontario, Canada, in *Proc. of the International Groundwater Engineering Conference held at Cranfiels Institute of Technology, UK*, 219-235 (1990).

GEOHYDRAULIQUE — Méthode d'étude et de recherche de l'eau souterraine des roches cristallines de l'Afrique de l'ouest, in *Rapport CIEH*, 300 p. (1978).

GEOTHERMA - AE Rhin - Meuse — Guide pratique pour le contrôle et l'entretien des captages d'eau souterraine, in *Doc AERM* 89 p. (1991).

GEOTHERMA - AE Rhin - Meuse — Etude sur le contrôle et l'entretien des captages d'eau souterraine, in *Doc AERM* (1991).

GOTTFREUND E., GOTTFREUND J., GEBER I., SCHMITT G., SCHWEISFURTH R. — Occurence and activities of bacteria in the unsaturated and saturated underground in relation to removal of iron and manganese, in *Wat. Supply*, **3**, 109-115, (1985).

GOTTLIEB O., BLATTERT R.E. —Concepts of well cleaning, in *JAWWA*, 60, (5), 34-39 (1988).

GOUY J.L., LABROUE L. — Sur quelques ferrobacteries isolées dans le sud-ouest de la France : écologie et rôle dans l'environnement, in *Annls Limnol.* **20** (3), 147-156 (1984).

GOUY J.L., LABROUE L., PUIG J. — Le colmatage ferrique des drains, in *Revue Française des Sciences de l'eau*, (1984).

GOUY J.L., LABROUE L., PUIG J. — Le colmatage ferrique des drains. Etude de deux cas typiques d'oxydation autotrophique et de précipitations hétérotropique du fer, in *Revue Française des Sciences de l'Eau* (1984).

GRIOLET C. — Forage de Languimberg (Syndicat des Eaux de la région des Etangs Moselle). Recherche de l'origine de la dégradation de qualité des eaux par essai de pompage, in *raport Ministère de l'Agriculture,SRAEL*, 7 p. (1984).

GRIOLET C., MAIAUX C., RAMON S. — La contamination saline des grès du trias inférieur sous couverture en Lorraine par le biais de forages anciens perfores et abandonnes, in V^e *colloque international sur les eaux souterraines, CEMPA IAH, Taommina, Italie*, 21 p. (1985).

GRIOLET C. — Méthode de diagnostic de l'état du tubage des forages, in *Rapport SRAE Lorraine* (1988).

GRIOLET C. — Méthode de diagnostic de l'état du tubage des forages, par pompage, in *Rapport préliminaire. Ministère de l'Agriculture,SRAEL-DES-1988-288*, 54 p. (1988).

GRIOLET C. — Diagnostic de l'état du tubage des forages, par mesure continue de la résistivité électrique et de la température de l'eau en début de pompage, in *Proc. XXIe Journées de l'hydraulique* (1991).

GUILLEMIN C., ROUX J-C. — Pollution des eaux souterraines en France - Bilan des connaissances, impacts, et moyens de prévention, in *Ouvrage collectif. Collection manuels & méthodes n° 23, BRGM*, 262 p. (1992).

HACKETT G., LEHR J.H. — Iron Bacteria Occurence, Problems and Control Methods, in *Water Wells.* National Water Well Association, Worthington, OH, USA, (1986).

HACKETT G. —Les bactéries du fer responsables des biofilms dans les puits. Revue des différentes espèces et des divers traitements chimiques de désinfection, in *Water Well Journ.*, (1987).

HACKETT G. —A review of chemical treatment strategies for iron bacteria in wells, in *Water Well Journ.* **41** (2), 37-42 (1987).

HAESSELBARTH U., LUDEMANN D. — Die biologische Verockerung von Bnunnen durch massentwicklung, in *Eisen-und Mangan-Bakterien Bohrtechnik Brunnenban Rohrleitungsbau*, 363-406 (1967).

HAESSELBARTH U., LUDEMANN D. — Biological incrustation of well due to mass development of iron and manganese bacteria, in *Water Treatment and Examination*, **21**, 20-29 (1972).

HALLBERG R-O., NALSER C. — Oxidation processes of iron in ground water - causes and measures, in *Proc. of International Symposium on Biofouled Aquifers: prevention and restoration*, 43-147 (1986).

HAMAN Z., CRUSE K., WILKINSON P.D., LAMBLIN J.M. — Well and borehole construction and rehabilitation, in *World Water Supply, Proc. of 14th World Congress of IWSA, Zurich*, SS3, 1-16 (1982).

HANTUSH, M.S. — Analysis of data from pumping tests in leaky aquifers., in *Trans. American Geophysical Union*, **37**, 702-714 (1956).

HANTUSH, M.S., JACOB C.E. — Nonsteady radial flow in an infinite leaky aquifer and nonsteady green's functions for an infinite strip of leaky aquifer, in *Trans. American Geophysical Union*, **36**, 95-112 (1955).

HARKER D. — Effect of acidification on chalk boreholes, in *Proc. of the International Groundwater Engineering Conference held at Cranfiels Institute of Technology. UK*, 158-167 (1990).

HELWEG O.J. — Evaluation des pertes de performance des forages d'eau au cours du temps, in *Groundwater* (1982).

HENDY N.A. — Isolation of thermophilic iron-oxidising bacteria from sulfidic waste rock, in *Journal of Industrial Microbiology*, **1**, 389-392 (1987).

HESSENAUER M. — Mauvaise qualité d'eau et entretien des captages d'eau potable Recherche d'une liaison de cause a effet en Meurthe et Moselle, in *Doc. AFB. Rhin Meuse* (1988).

HOWSAM P. — Biofouling in wells and aquifers, in *Journal of the Inst. of Water and Environmental Management*, **2** (2), 209-215 (1988).

HOWSAM P. — Well performance deterioration: an introduction to cause processes, in *Proc. of the International Groundwater Engineering. Conference held at Cranfiels Institute of Technology, UK*, 19-24 (1990).

KAISER P. — Les bactéries du fer, in *Rapport interne. Institut Pasteur* (1981).

KASSECKER F. — Remise en état d'un puits tubé, in *Gas Wasser Warme* (1948).

KENNETH R., APPLIN, ZHAO — The kinetic of Fe(II) oxidation and well screen encrustation, in *Groundwater*, **27** (2), 168-174 (1989).

KOENING L. — Effet de la régénération sur les frais d'exploitation des puits et les rendements de puits neuf et ancien, in *JAWWA*, (1960).

KOENING L. — Economic aspects of water well stimulation, in *JAWWA*, **52**, 631-637 (1960).

KOENING L. — Survey and analysis of well stimulation performance, in *JAWWA*, **52**, 333-50 (1960).

KOENING L. — Effects of stimulation on well operating costs and its performance on old and new wells, in *JAWWA*, **52**, 1499-1512 (1960).

KOENING L. — Relation between aquifer permeability and improvement achieved by well stimulation, in *JAWWA*, **53**, 652-670 (1961).

KREMS G. — Etude du vieillissement des puits, in *Hydrogéologie*, 1, 51-54 (1987).

KREMS G. — Studie üder die Brunnenalterung, in *Bundesministerium des Inners, Unterabteilung Wasserwirtschaft, Berlin*, (1972).

KRUSEMAN G.P., DE RIDDER N.A. — Analysis and evaluation of pumping test data, in *International Institute for Land Reclamation and Improvement Wageningen, the Netherlands*, Bull. 11, (1970).

KRUSEMAN G.P., DE RIDDER N.A., MEILHAC A. — Interprétation et discussion des pompages d'essai, in *International Institute for Land Reclamation and Improvement Wageningen, the Netherlands*, 213 p. (1974).

KUMARDAW R. — Rehabiliation of problem wells in Orissa Drinking Water Supply Project, in *Proc. of the International Groundwater Engineering Conference held at Cranfiels Institute of Technology, UK*, 321-337 (1990).

KUNETZ G. — *Principles of direct current resistivity prospecting*. Gebrüder Borntraeger Ed., Geoexploration monographs **1** (1), 103 p. (1966).

LAHOUD A., DETAY M., DEMONGEOT D., SUZANNE P. — Outils informatiques de prévention et de gestion des situations de crise en région parisienne, in *Colloque "L'eau dans la ville" HYDROTOP 92*. (1992).

LALLEMAND-BARRES A. — Colmatage - décolmatage, étude documentaire, in *rapport BRGM 85 SGN 030 EAU*, 44 p., (1985).

LALLEMAND-BARRES A., ROUX J-C. — Guide méthodologique d'établissement des périmètres de protection des captages d'eau souterraine destinés à la consommation humaine, in *Collection manuels & méthodes n° 19, BRGM*, 221 p. (1989).

LAMBLIN J.M. — Colmatage des forages captant la nappe captive des sables Sparnaciens de la région parisienne. Observations et premiers résultats d'un traitement expérimental, in *Congres AIDE Zurich*. SS3 14-16 (1982).

LAURENT J. — Entretien et réhabilitation des ouvrages de captage. Les effets du vieillissement, les contrôles de surveillance. L'entretien des ouvrages. entretien des ouvrages de captage. entretien des installations de pompage. diagnostic. la réparation. coût des travaux rentabilité de l'entretien, in *Revue du Génie Rural* , **12**, 69-72 (1988).

LAURENT J., COURSIL J.P. — Les puits d'Eau, in *Revue travaux* (1989).

LAURENT J., LE GUYAFER M., BONNEFOY A. — La réhabilitation des ouvrages de captage, in *Revue Travaux* 647, 64-68 (1989).

LAURENT J. —Le captage des nappes sableuses, in *Courants*, **2**, 26-31 (1990).

LINN C.L. — Le maintien de débit des puits par nettoyage chimique, in *JAWWA* (1954).

LYONNAISE DES EAUX ET DE L'ECLAIRAGE. Captage des eaux souterraines. Anonyme, in *Doc. Interne*, 40 p.

LYONNAISE DES EAUX-DUMEZ — Etude des mécanismes de transfert de pollution entre une rivière et une nappe alluviale : Application au champ captant d'Aubergenville, in *rapport d'avancement Centre d'Informatique Géologique - Lyonnaise des Eaux* (1991).

MAC LAUGHLAN R.G., KNIGHT M.J. — Corrosion and Incrustation in Groundwater Bores: A Critical Review, in *Research Publication 1/89, Center for Groundwater Management and Hydrogeology, University of New South Wales*, 42 p. (1989).

MAIAUX C. — Diagraphies de contrôle de forages captant la nappe profonde du Grès Vosgien en Lorraine et en Alsace. Interprétation des mesures, in *Rapport BRGM 85 SGN 112 LOR* 45 p., (1985).

MAIAUX C. — Travaux d'auscultation par diagraphies du forage de Sarrebourg et visite du forage des tissages de Pambervilliers, in *Rapport BRGM*, (1985).

MAIAUX C., BABOT — Diagraphies de contrôle de forages captant la nappe profonde des grès vogiens en Lorraine et en Alsace, in *Rapport BRGM* (1985).

MAIAUX C., RICOUR J., NEISS B. — Réhabilitation des forages profonds, méthodes appliquées en Lorraine, in *Revue Techniques et sciences municipales* (1986).

MAIAUX C., RICOUR J., NEISS B — Pathologie des forages d'eau, in *Revue Techniques et sciences municipales*, (1986).

MANEVY I. — Identification de l'aide en ligne nécessaire à la création d'un outil d'EAO dans le domaine de la recherche hydrogéologique en Afrique, in *Rapport de Maîtrise Géosciences et Géotechniques - Université de Nice*, 261 p. (1990).

MANSUY N., NUZMAN C., CULLIMORE D.R. — Well problem identification and its importance in well rehabilitation, in *Proc.of the International Groundwater Engineering Conference held at Cranfiels Institute of Technology, UK*, 87-99 (1990).

MARLE C. — Ecoulement monophasiques en milieu poreux, in *Revue de l'IFP*, vol. XXII, **20** (1967).

MARGAT J. — Poids des eaux souterraines comme source d'approvisionnement en eau en France - Données statistiques, in *Doc. BRGM n° 195*, 313-337 (1990).

MATTHESS G., FOSTER S.S.D., SKINNER A.C. — *Theoretical background, hydrogeology and practice of groundwater protection zone.* Heise Ed., International Contribution to Hydrogeology, 6, 204 p. (1985).

MC CORMICK R. — Filter-pack installation and redevelopment techniques for shallow recharge shafts, in *Ground Water*, **13** (5), (1975).

MÉGNIEN C. *et al.* — Atlas des nappes aquifères de la région parisienne, in *du BRGM* (1970).

MÉGNIEN C. — Hydrogéologie du centre du bassin de Paris, in *Mémoire du BRGM* **98**, 530 p., (1979).

MINISTERE DE L'EQUIPEMENT — Cahier des Clauses Techniques Générales Marches publics de travaux. Travaux de forage pour la recherche et l'exploitation d'eau potable. Fascicule n° 76 (1987).

MOODY L.F. — Friction factors for pipe flow, in *Trans. ASME* (1944).

MORAMAN T., DORRIER R.C. — Exploration télévisée des forages, in *Groundwater Monitoring* (1984).

MOUCHET P. — Action de l'eau sur les matériaux. Equilibre calco-carbonique, film protecteur, traitement de correction, in *SETEC Bâtiment, formation continue en traitement d'eau* (1981).

MÜLLER O., DETAY M. — Modélisation du comportement des éléments azotés en aquifère alluvial influencé : importance de l'interface surface-nappe, in *Hydrogéologie*, **1**, 3-19 (1993).

MUDGERIKAR R., PHADTARE P.N. — Réhabilitation of failed wells by hydrofracturing, in *Seminar Volume, Statewise Seminars. Maharashtra, Gujarat, Rajasthan, Madhya Pradesh, India* (1989).

NATIONAL GROUND WATER ASSOCIATION — Selection and installation of well screens and gravel packs: an anthology, in *15 papers on well screen and gravel pack selection and installation. NGWA* (1991)

NOVAKOWSKI, K.S. — A composite analytical model for analysis of pumping tests affected by webore storage and finite thickness skin, in *Water Resour. Res.*, **25**, (9), 1937-1946 (1989).

OBUEKWE C.O., WESTLAKE D.W.S., PLAMBECK J.A. — Bacterial corrosion of mild steel under the condition of simultaneous formation of ferrous and sulphide ions, in *Appl. Microbiol. Biotechnol.*, **26**, 294-298 (1987).

PAUL K.F. — Water well regeneration. new technology, in *Proc. of the International Groundwater Engineering Conference held at Cranfiels Institute of Technology, UK*, 168-179.

PHADTARE P.N., GHOSH G. — Monitoring, maintenance and rehabilitation of water supply wells in India. state of art, in *Proc. of the International Groundwater Engineering Conference held at Cranfiels Institute of Technology, UK*, 303-313 (1990).

PITTENDREIGH L.M. — Injection de produits chimiques dans les puits afin de supprimer l'activite biologique et de stabiliser le fer et le manganese dans les eaux souterraines, in *Journal of The New England Water Works Association* (1963).

PLEGAT R. — Dommages causes aux captages d'eau souterraine par les bacteries sulfo-oxydantes, in *L'Eau, l'Industrie, les Nuisances* (1990).

POITEVIN-SOMMER G., DOUSSAN C. — Etude des transferts de micropolluants Seine-nappe alluviale : Premiers résultats, in *Rapport Ecole des Mines de Paris - Lyonnaise des Eaux-Dumez*, 43 p., (1991).

POITRINAL D., ALLA P. — Optimisation par modèle mathématique des investissements dans le cas de la nappe de Croissy, in *Wat. Supply, IWSA,* **3** (2), 219-226 (1985).

POYET P., DETAY M. — Artificial intelligence tools and techniques for water resources assessment in Africa, in *Use of microcomputer in geology,* Plenium Publishing Corporation, 119-159, (1993).

POYET P., DETAY M. — Hydroexpert® an exemple of a new generation of compact expert system, in *Computers & Geosciences,* **15**, (3), 255-267 (1989).

POYET P., DETAY M. — Compact expert system for water resources assessment in Africa, in *Selected papers on hydrogeology, from the 28th International Geological Congress,* **1**, 417-430. Verlag Heinz Heise (1990).

POYET P., DETAY M. — Hydroexpert® - Un système expert d'aide à l'implantation de forages en hydraulique villageoise - Manuel d'introduction et de référence, in *Rapport de Recherche INRIA* **936**, 38 p. (1988).

POYET P., DETAY M. — Hydrolab® un système expert de poche en hydraulique villageoise, in *Techniques et Science Informatiques,* **8**, (2), 157-167 (1989).

POYET P., DETAY M. — Hydroexpert® - Un système d'aide à l'implantation d'ouvrages, in *Actes des 8ᵉ journées Internationales : Les Systèmes Experts et leurs applications, Avignon,* **2**, 397-410 (1988).

POYET P., DETAY M. — Enjeux sociaux et industriels de l'intelligence artificielle en hydraulique villageoise, in *Convention IA 89, Editions Scientifiques et Techniques Hermès,* .**2**, 621-652 (1989).

PRUI S., FLORES C.V. — Monitoring and diagnosis for planning borehole rehabilitation. the experience from Lima, Peru, in *Proc. of the International Groundwater Engineering Conference held at Cranfiels Institute of Technology, UK,* 377-390 (1990).

REHSE W. — Abbaubare organische Verunreinigungen pathogene Keime und Viren in *Rapport Eidgenössiches Amt für Umweltshutz (Office de l'Environnement de Berne),* n°401 77 (1977).

REHSE W. — Elimination und abbau von organisches fremdstoffen, pathogen keimen und viren in Lockergestein, in *Z. dt geol. ges.* **128**, 319-329 (1977).

REPSON H. — *Well logging in groundwater development.* International contibutions to hydrogeology, IAH **9**, 136 p., (1989).

RICOUR J., DENUD T.H. — Vieillissement du parc de forages et conditions d'abandon des ouvrages de production d'eau dans la région Nord Pas de Calais, in *Doc. BRGM 89 SGN. 243 NPC* (1989).

RICOUR J. — Diagnostic sur l'état des forages profonds sollicitant l'aquifère des grès du Trias inférieur sous couverture, in *Doc. BRGM* (1984).

RICOUR J. — Le coût global des ouvrages souterrains, applications aux forages de production d'eau, in *L'Eau, l'Industrie, les Nuisances* (1990).

RICOUR J. — Quel avenir pour les techniques de réhabilitation des ouvrages de production d'eaux souterraines, in *Doc. BRGM n° 195*, 357-366 (1990).

RIVIERE J., KAISER P. — Le rôle des bactéries sulfato-réductirces et des bactéries du fer dans la corrosion et le colmatage des canalisations, in *L'Eau, l'Industrie, les Nuisances* **82**, 46-51 (1984).

ROOT U. — Physical, chemical and biological aspects of the removal of iron and manganese underground, in *Wat. Supply*, **3**, 143-150 (1985).

ROOT U. — New aspects of microbiological and chemical treatment of drinking water in the aquifer, in *Wat. Supply*, **6**, 229-235 (1988).

ROUX J-C. — La situation des procédures de périmètres de protection en Europe : protection des captages d'eau souterraine destinée à la consommation humaine dans les autres Etats de la Communauté européenne, in *Actes du colloque de Saint-Brieux*, 24-27 (1992).

ROUX J-C. — L'évaluation des périmètres de protection en France, in *Actes du colloque de Saint-Brieux*, 92-95 (1992).

SAFEGE — Fiches de formation SNDE (1991).

SAFEGE — Synthèse documentaire sur la régénération et la réhabilitation des forages d'eau, in *Doc. Lyonnaise des Eaux*, 41 p. (1989).

SAUTY J-P., THIERY D. — Utilisation d'abaques pour la détermination de périmètres de protection, in *Note technique aux géologues agréés en matière d'eau et d'hygiène publique, n°6, Doc. BRGM 75 SGN 430 AME*, 30 p. (1975).

SAUTY J-P. — Utilisation des traceurs pour définir les périmètres de protection, in *Doc. BRGM 87 SGN 287 EAU*, 109 p. (1987).

SEYFRIED C.F., OLTHOFF R. — Underground removal of iron and manganese, in *Wat. Supply*, **3**, 117-142 (1985).

SIRONNEAU J. — La nouvelle loi sur l'eau ou la recherche d'une gestion équilibrée, in *Doc. Ministère de l'Environnement*, 107 p. (1992).

SMITH S. — Case histories of iron bacteria and other biofouling, in *Water Well Journal.*, **41** (2), 30-32 (1987).

SMITH S. — Iron bacteria in water wells: causes, prevention, treatment, in *Agricultural Engineering*, **66** (8), 15-18 (1985).

SMITH S. — An Investigation of Tools and Field Techniques for the Detection of Iron precipitating Bacteria. Groundwater and Wells, in *MS. Thesis, The Ohio State University, Ohio* (1984).

SMITH S. — Detecting iron and sulfur bacteria in wells, in *Water Well Journal.*, March, 58-63 (1984).

SMITH S. — Iron bacteria problems in wells: a new look, in *Water Well Journal.*, January, 76-78 (1984).

SMITH S. — The 'Iron Bacteria' problem: Its causes, prevention and treatment in water wells, in *Winter Meeting of the American Society of Agricultural Engineers* (1983).

SMITH S. — A layman's guide to iron bacteria problems in wells, in *Water Well Journal.*, **34** (6), 40-41 (1980).

SMITH S.A. — Manual of Hydraulic Fracturing for Well Stimulation and Geologic Studies, in *National Water Well Assn, Dublin, OH, USA*, 66 p. (1989).

SMITH S.A. — Well maintenance and rehabilitation in North America: an overview, in *Proc. of the International Groundwater Engineering Conference held at Cranfiels Institute of Technology, UK*, 8-16 (1990).

SMITH S.A., TUOVINEN O.H. — Environmental analysis of iron-precipitating bacteria in ground water and wells, in *G. W. Mon. Rev.*, **5** (4), 45-52 (1985).

SMITH S.A., TUOVINEN O.H. — Biofouling monitoring methods for preventive maintenance of water wells, in *Proc. of the International Groundwater Engineering Conference held at Cranfiels Institute of Technology, UK*, 75-81 (1990).

STREETER V.L., WYLIE E.B. — *Fluid mechanics*. McGraw-Hill (1979).

STÜMPKE H — *Anatomie et biologie des rhinogrades*. Masson, 85 p. (1962).

SUTTON S.E., SUTTON J.S. — Rehabilitate or renew ? An example in Western Zarnbia, in *Proc. of the International Groundwater Engineering Conference held at Cranfiels Institute of Technology, UK*. 371-376 (1990).

THIERY D., VANDENBEUSCH M., VAUBOURG P. — Interprétation des pompages d'essai en milieu fissuré aquifère, in *Doc. BRGM* n° 57, 53 p. (1983).

US EPA — Introduction to Ground-Water investigation. 812 p. (1988).

VAN BEEK, C.G.E.M. — Restoring well yield in The Netherlands, in *JAWWA* **76** (10) 66-72 (1984).

VAN BEEK, C.G.E.M. — Rehabilitation of clogged discharge wells in The Netherlands, in *Quarterly J. Eng. Geology, London*, **22** (1) 75-80 (1989).

VINES K.J. — Ednhaz: a program for analysing step drawdown tests, in *Computers & Geosciences*, **15**, (6), 956-978 (1989).

VUORINEN A., CARLSON L., TUOVINEN O.H. — Groundwater Biogeochemistry of iron and Manganese in relation to well water quality, in *Inter. Syrnp. on Biofouled Aquifers: Prevention and Restoration (Ed DR. CULLIMORE)*, American Water Resources Assoc. *Tech. Public. Series TP5-87-1*, 157-168 (1986).

WALTON W.C. — Leaky artesian aquifer conditions in Illinois, in *Illinois State Water Survey, Report of investigation* **39**, (1960).

WENZEL L.K. — Methods for determining the permeability of water-bearing materials, with special reference to dischargong well methods, in *US geol. survey water-supply paper* 887.

WHITE C.C., HOUSTON J.F.T., BARKER D. — The victoria province drought relief project, geophysical siting af boreholes, in *Ground Water*, **26** (3), 309-316, (1988).

WILLIS R., YEH W.W.G. — *Groundwater systems planning & management*. Prentice-Hall, 416 p. (1987).

WMO — Hydrological aspects of accidental pollution of water bodies, in *Operational Hydrology Report* **37**, 208 p. (1992).

Anonyme — Hydrogéologie, in *édition provisoire EIER*, 100 p. (1972).

Anonyme — Eléments d'hydrologie de l'eau souterraine, 173 p.

Main journals

In English or bilingual English/French

Journal of Hydrology (articles on hydrology and hydrogeology). Usually four issues are published every year. Elsevier Scientific Publication Company Journal Division, P.O. Box 211, Amsterdam (The Netherlands).

Hydrological Sciences Bulletin/Bulletin des Sciences Hydrologiques, official journal of the International Association of Hydrological Sciences (articles on hydrology and hydrogeology, events and news of the Association). Published quarterly. IAHS Treasurer, 1909 K Street N.W., Washington D.C. 20006 (USA).

Water Resources Research (in English only, articles on hydrology and hydrogeology: scientifically, the most prestigious journal in this field). Six issues published every year by the American Geophysical Union 2000 Florida Ave., N.W., Washington D.C. 20009 (USA).

Advances in Water Research (specialized in the numerical processing of water problems). Six issues published every year. Computational Mechanics Publications, Ashurst Lodge, Ashurst, Southampton SO4 5AA (United Kingdom).

Water Research, Journal of the International Association on Water Pollution Research and Control. Edited by Pergamon Press, 12 issues annually. IAWPRC, 1 Queen Anne's Gate, London SW1H 9BT(United Kingdom).

Proceedings of the American Society of Civil Engineers
  - Journal of the Hydraulic Division,
  - Journal of Irrigation and Drainage Division,
  - Journal of Soil Mechanic and Foundation Division.

Groundwater, journal of the National Water Association. 500 W. Wilson Bridge Rd, suite 135, Worthington, OH 43085 (USA).

Journal of Fluids Mechanics

Journal of the Society of Petroleum Engineers.

American Society of Agricultural Engineers, bimensual transactions published by ASAE, 2950 Niles Road, St. Joseph MI 49085 (USA).

American Society of Civil Engineers 345 E. 47th St., New York, NY 10017 (USA).
  - Journal of Irrigation and Drainage Engineering .
  - Journal of Water Resources Planning & Management.

American Waterworks Association Journal, published monthly by AWWA, 666 W. Quincy Ave., Denver CO 80235 USA .

Ground Water Age, published monthly by National Trade Publications, 13 Century Hill Dr., Latham, NY 12110-2197 USA.

International Groundwater, National Trade Publication, Inc. 13 Century Hill Drive, Latham, NY 12110-2197 USA.

Ground Water Monitoring Review, four issues published every year by Water Well Publishing Co. 6375 Riverside Dr., Dublin OH 43017 USA.

Hydrological Sciences Journal, published twice monthly by the International Association of Hydrological Sciences, Journal Subscription Dept., Marston Book Services, P.O. Box 87, Oxford, United Kingdom (Cost of annual subscription: US$ 155 for USA and Canada, £85 outside North America).

Journal of Contaminant Hydrology, four issues per volume, two volumes published every year, published by Elsevier Science Publishers B.V., Journal Dept., P.O. Box 211, 1000 AE Amsterdam, The Netherlands. Customers in the U.S. and Canada contact: Elsevier Scientific Publication Company Inc. Journal Information Center, 655 Avenue of the Americas, New York, NY 10010 USA.

Journal of Hydraulic Research, six issues per volume, published by the International Association for Hydraulic Research, Rotterdamseweg 185, P.O. Box no. 177, 2600 MH Delft, The Netherlands.

Journal of Soil & Water Conservation. Six publications per year, from the Soil and Water Conservation Society, 7515 Ankeny Rd., Ankeny, IA 50021-9764.

Selected Water Resources Abstracts. Monthly publication of the US Geological Survey, MS 425, Reston VA. It is possible to obtain issues on subscription by applying to the National Technical Information Service, US Department of Commerce, 5285 Port Royal Rd., Springfield VA 22161 USA. The SWRA can also be consulted online via two servers:

- Dialog (800-3-DIALOG).
- The SWRAs are also available on CD-ROM from:
  - NISC (National Information Services Corp.) tel: 301-243- 0797, with quarterly updates.
  - Silver Platter, tel: 800-343-0064, with quarterly updates.

Journal of the Soil Science Society of America, bimonthly periodical published by the Soil Science Society of America, 677 S. Segoe Rd. Madison, WI 53711 (USA).

University Microfims International, supplier of microfilm copies. 300 N. Zeeb Rd., Ann Arbor, MI 48106.

Water, Air & Soil Pollution, six volumes (24 issues) published annually by Kluwer Academic Publishers Group, P.O. Box 322, 3300 AH Dordrecht, The Netherlands.

Water Well Journal, monthly periodical published by Water Well Journal Publishing Co., 6375 Riverside Dr., Dublin, OH 43017 (USA). Distributed free of charge to U.S. water well contractors.

Specialist journals in computational hydrogeology

COGSletter is published monthly by the Computer Oriented Geological Society, P.O. Box 1317, Denver, CO 80201-1317 (USA).

GEOBYTE is published every two months by the American Association of Petroleum Geologists, P.O. Box 979, Tulsa, OK 74101-0979 (USA). Price: US$ 35 (for customers in France).

Computers & Geosciences, published by Pergamon Press plc, Headington Hill Hall, Oxford OX3 0BW (United Kingdom).

Computers & Mining, monthly periodical published by Pergamon Press. Comparable to *Directory of Mining Programs*, from Gibbs Associated, P.O. Box 706, Boulder, CO 80306 (USA).

USGS is the catalogue of New Publications of the U.S. Geological Survey, which can be obtained from Department of the Interior, U.S. Geological Survey, 582 National Center, Reston, VA 22092 (USA).

HYDROSOFT, four issues per year at an approximate price of £98, from Computational Mechanics Publications, Ashurst Lodge, Ashurst, Southampton, SO4 2AA (United Kingdom).

Non-periodical series of publications

Advances in Water Sciences, Academic Press / International Association of Hydrogeologists, Proceedings of congresses / International Association of Hydrological Sciences, Proceedings of congresses / Les Cahiers du Centreau, University of Laval, Quebec (Canada) / California Water Resources Center, University of California, Davis, CA 95616 (USA) / Irrigation and Drainage Papers, Food and Agriculture Organization, Rome (Italy), 28 published volumes / Transactions of the Hydraulics congress held in Paris usually every other year by the *Société Hydrotechnique de France* / Natural Resources and Water Series, United Nations Organization, New York (USA) / Operational Hydrology Reports of the World Meteorological Organization, Geneva (Switzerland) / Studies and Reports in Hydrology, UNESCO, Paris, 30 volumes published (papers on hydrology, congress proceedings, etc.) / Technical Documents in Hydrology, UNESCO, Paris / Technical Papers in Hydrology, UNESCO, Paris / United States Geological Surveys, Water Supply Papers,

USGS, Washington D.C. 20242 (USA) / Water Research Centre, information and reports, Medmenham Laboratory, Henley Road, Medmenham, Bucks SL7 2HD (United Kingdom).

Main professional associations

International Association of Hydrogeologists (IAH). Secretary: Dr E.ROMIJN, Provincial Waterboard of Gelderland, Markstraat 1, P.O. Box 9090, NL-6800 GX Arnhem (The Netherlands). Organizes international conferences either every year or every other year.

International Association of Hydrological Sciences (IAHS), 1909 K Street N.W., Washington D.C. 20006 (USA). Edits the *Bulletin of Hydrological Sciences* and organizes international conferences.

American Geophysical Union (AGU), 1909 K Street N.W., Washington D.C. 20006 (USA). Among other titles, the AGU edits *Water Resources Research* (WRR), which is the premier English-language journal in the hydrological sciences. It also organizes two annual meetings (in San Francisco and Washington D.C.) where a most American workers in this discipline present the state of research into water resources.

International Association for Hydraulics Research (IAHR), P.O. Box 177, 2600 MEI Delft (The Netherlands). Periodically organizes international scientific meetings.

International Association of Engineering Geology (IAEG). Secretary General: Dr L. PRIMEL, Laboratoire Central des Ponts et Chaussées, 58 Boulevard Lefèbvre, 75732 Paris Cedex 15 (France).

Appendices

International units	US or UK units
1 millimetre (mm)	0.0394 inch
1 metre (m)	3.2809 feet 1.094 yard
1 cm^2	0.155 square inch
1 m^2	10.76 square feet
1 cm^3	0.061 cubic inch
1 m^3	35.316 gallon
1 litre (l)	0.264 US gallon 0.220 Imperial gallon 0.00629 barrel
1 bar	14.5 pound/sq. inch 2089.0 pound/sq. foot
1 Pascal	0.000145 pound/sq. inch 0.02089 pound/sq. foot
1 kilogram (kg)	2.205 pounds
1 tonne (metric slug)	1.102 short ton or US ton 0.9842 UK ton
degree Baumé (French scale for liquid density)	degree Twaddell

ANNEXE II — *Conversion between and units used in English-speaking countries and internationally recognized units.*

US or UK units	International units
1 inch	25.399 mm
1 foot (= 12 inches)	0.30479 m
1 yard (= 3 feet)	0.9144 m
1 square inch	6.4513 cm^2
1 square foot	0.0929 m^2
1 cubic inch	16.386 cm^3
1 cubic foot	0.028317 m^3
1 US gallon	3.785 litres
1 Imperial gallon	4.546 litres
1 pound	0.4536 kg
1 short ton (US)	0.907 tonnes
1 UK ton	1.016 tonnes
1 pound per square inch (p.s.i)	0.06895 bar
	6895 Pascals
1 pound per square foot (lb. sq. ft)	0.000479 bar
	47.9 Pascals
degree Twaddell	degree Baumé

x	eˣ	Ko(x)	eˣKo(x)	-Ei(x)	-Ei(-x)eˣ	x	eˣ	Ko(x)	eˣKo(x)	-Ei(x)	-Ei(-x)	x	eˣ	Ko(x)	eˣKo(x)	-Ei(x)	-Ei(-x)eˣ
0.01	1.0101	4.7212	4.7687	4.0379	4.0787	0.1	1.1052	2.4271	2.6823	1.8229	2.0147	1	2.7183	0.421	1.1445	0.2194	0.5964
0.011	1.0111	4.626	4.6771	3.9436	3.9874	0.11	1.1163	2.3333	2.6046	1.7371	1.9391	1.1	3.0042	0.3656	1.0983	0.186	0.5588
0.012	1.0121	4.539	4.5938	3.8576	3.9044	0.12	1.1275	2.2479	2.5345	1.6595	1.8711	1.2	3.3201	0.3185	1.0575	0.1584	0.5259
0.013	1.0131	4.459	4.5173	3.7785	3.8282	0.13	1.1388	2.1695	2.4707	1.5889	1.8094	1.3	3.6693	0.2782	1.021	0.1355	0.4972
0.014	1.0141	4.3849	4.4467	3.7054	3.7578	0.14	1.1503	2.0972	2.4123	1.5241	1.7532	1.4	4.0552	0.2437	0.9881	0.1162	0.4712
0.015	1.0151	4.3159	4.3812	3.6374	3.6925	0.15	1.1618	2.03	2.3585	1.4645	1.7015	1.5	4.4817	0.2138	0.9582	0.1	0.4482
0.016	1.0161	4.2514	4.32	3.5739	3.6317	0.16	1.1735	1.9674	2.3088	1.4092	1.6537	1.6	4.953	0.188	0.9309	0.0863	0.4275
0.017	1.0171	4.1908	4.2627	3.5143	3.5746	0.17	1.1853	1.9088	2.2625	1.3578	1.6094	1.7	5.4739	0.1655	0.9059	0.0747	0.4086
0.018	1.0182	4.1337	4.2088	3.4581	3.5209	0.18	1.1972	1.8537	2.2193	1.3098	1.5681	1.8	6.0496	0.1459	0.8828	0.0647	0.3915
0.019	1.0192	4.0797	4.158	3.405	3.4705	0.19	1.2093	1.8018	2.1788	1.2649	1.5295	1.9	6.6859	0.1288	0.8614	0.0562	0.3758
0.02	1.0202	4.0285	4.1098	3.3547	3.4225	0.2	1.2214	1.7527	2.1408	1.2227	1.4934	2	7.3891	0.1139	0.8416	0.0489	0.3613
0.021	1.0212	3.9797	4.0642	3.3069	3.3771	0.21	1.2337	1.7062	2.1049	1.1829	1.4593	2.1	8.1662	0.1008	0.823	0.0426	0.348
0.022	1.0222	3.9332	4.0207	3.2614	3.334	0.22	1.2461	1.662	2.071	1.1454	1.4273	2.2	9.025	0.0893	0.8057	0.0372	0.3356
0.023	1.0233	3.8888	3.9793	3.2179	3.2927	0.23	1.2586	1.6199	2.0389	1.1099	1.3969	2.3	9.9742	0.0791	0.7894	0.0325	0.3242
0.024	1.0243	3.8463	3.9398	3.1765	3.2535	0.24	1.2713	1.5798	2.0084	1.0762	1.3681	2.4	11.0232	0.0702	0.774	0.0284	0.3135
0.025	1.0253	3.8056	3.9019	3.1365	3.2159	0.25	1.284	1.5415	1.9793	1.0443	1.3409	2.5	12.1825	0.0623	0.7596	0.0249	0.3035
0.026	1.0263	3.7664	3.8656	3.0983	3.1799	0.26	1.2969	1.5048	1.9517	1.0139	1.3149	2.6	13.4637	0.0554	0.7459	0.0219	0.2942
0.027	1.0274	3.7287	3.8307	3.0615	3.1452	0.27	1.31	1.4697	1.9253	0.9849	1.2902	2.7	14.8797	0.0493	0.7329	0.0192	0.2854
0.028	1.0284	3.6924	3.7972	3.0261	3.1119	0.28	1.3231	1.436	1.9	0.9573	1.2666	2.8	16.4446	0.0438	0.7206	0.0169	0.2773
0.029	1.0294	3.6574	3.765	2.992	3.08	0.29	1.3364	1.4036	1.8758	0.9309	1.2441	2.9	18.1742	0.039	0.7089	0.0148	0.2693
0.03	1.0305	3.6235	3.7339	2.9591	3.0494	0.3	1.3499	1.3725	1.8526	0.9057	1.2226	3	20.0855	0.0347	0.6978	0.0131	0.2621
0.031	1.0315	3.5908	3.7039	2.9273	3.0196	0.31	1.3634	1.3425	1.8304	0.8815	1.2018	3.1	22.198	0.031	0.6871	0.0115	0.2551
0.032	1.0325	3.5591	3.6749	2.8965	2.9908	0.32	1.3771	1.3136	1.8089	0.8583	1.182	3.2	24.5325	0.0276	0.677	0.0101	0.2485
0.033	1.0336	3.5284	3.6468	2.8668	2.9631	0.33	1.391	1.2857	1.7883	0.8361	1.163	3.3	27.1126	0.0246	0.6673	0.0089	0.2424
0.034	1.0346	3.4986	3.6196	2.8379	2.9362	0.34	1.405	1.2587	1.7685	0.8147	1.1446	3.4	29.9641	0.022	0.658	0.0079	0.2365
0.035	1.0356	3.4697	3.5933	2.8099	2.9101	0.35	1.4191	1.2327	1.7493	0.7942	1.127	3.5	33.1155	0.0196	0.649	0.007	0.2308
0.036	1.0367	3.4416	3.5678	2.7827	2.8848	0.36	1.4333	1.2075	1.7308	0.7745	1.1101	3.6	36.5982	0.0175	0.6405	0.0062	0.2254
0.037	1.0377	3.4143	3.543	2.7563	2.8603	0.37	1.4477	1.1832	1.7129	0.7554	1.0936	3.7	40.4473	0.0156	0.6322	0.0055	0.2204
0.038	1.0387	3.3877	3.5189	2.7306	2.8364	0.38	1.4623	1.1596	1.6956	0.7371	1.0779	3.8	44.7012	0.014	0.6243	0.0048	0.2155
0.039	1.0398	3.3618	3.4955	2.7056	2.8133	0.39	1.477	1.1367	1.6789	0.7194	1.0626	3.9	49.4025	0.0125	0.6166	0.0043	0.2108
0.04	1.0408	3.3365	3.4727	2.6813	2.7907	0.4	1.4918	1.1145	1.6627	0.7024	1.047	4	54.5982	0.0112	0.6093	0.0038	0.2063
0.041	1.0419	3.3119	3.4505	2.6576	2.7688	0.41	1.5068	1.093	1.647	0.6859	1.0335	4.1	60.3403	0.01	0.6022	0.0033	0.2021
0.042	1.0429	3.2879	3.4289	2.6344	2.7474	0.42	1.522	1.0721	1.6317	0.67	1.0197	4.2	66.6863	0.0089	0.5953	0.003	0.198
0.043	1.0439	3.2645	3.4079	2.6119	2.7267	0.43	1.5373	1.0518	1.6169	0.6546	1.0063	4.3	73.6998	0.008	0.5887	0.0026	0.1941
0.044	1.045	3.2415	3.3874	2.5899	2.7064	0.44	1.5527	1.0321	1.6025	0.6397	0.9933	4.4	81.4509	0.0071	0.5823	0.0023	0.1903
0.045	1.046	3.2192	3.3673	2.5684	2.6866	0.45	1.5683	1.0129	1.5886	0.6253	0.9807	4.5	90.0171	0.0064	0.5761	0.0021	0.1866
0.046	1.0471	3.1973	3.3478	2.5474	2.6672	0.46	1.5841	0.9943	1.575	0.6114	0.9685	4.6	99.4843	0.0057	0.5701	0.0018	0.1832
0.047	1.0481	3.1758	3.3287	2.5268	2.6483	0.47	1.6	0.9761	1.5617	0.5979	0.9566	4.7	109.947	0.0051	0.5643	0.0016	0.1798
0.048	1.0492	3.1549	3.31	2.5068	2.63	0.48	1.6161	0.9584	1.5489	0.5848	0.9451	4.8	121.510	0.0046	0.5586	0.0014	0.1766
0.049	1.0502	3.1343	3.2918	2.4871	2.612	0.49	1.6323	0.9412	1.5363	0.5721	0.9338	4.9	134.289	0.0041	0.5531	0.0013	0.1734
0.05	1.0513	3.1142	3.2739	2.4679	2.5945	0.5	1.6487	0.9244	1.5241	0.5598	0.9229	5	148.413	0.0037	0.5478	0.0011	0.1704

Table 1
Values of the functions x, eˣ, Ko(x), eˣKo(x), − Ei(−x), and −Ei(−x)eˣ from M.-S.
HANTUSH (1956).

Glossary

acidizing The process of introducing acid down a well into a water-bearing formation, in order to dissolve part of the carbonate clogging or rock constituents. The general objective of acidization is to increase the discharge capacity by improving the permeability, (also known as acidization or acid treatment).

active chlorine Quantity of chlorine available for a given reaction. (cf. chlorimetic titre).

active management of groundwaters Planned control of aquifers involving actions to improve water quantity and quality.

additive Compounds added during acidization in order to maintain iron and aluminium oxides in solution, (e.g. Rochelle salts, citric acid, lactic acid, ammonium bifluoride).

adjuvant Auxiliary chemical substance serving to assist or contribute to a reaction. *Acid pyrophosphate may be used following treatment of the medium or the product with an adjuvant that raises its pH (e.g. caustic soda or sodium carbonate).*

aerobic Said of an environment in which oxygen (or air) is present, of organisms requiring oxygen for growth or a process that can only take place in the presence of oxygen.

ageing, aging The progressive change in material properties with increased age; in catchment structures, usually marked by a deterioration caused by clogging.

air-lift pumping Syn: air-lift development The most commonly used and most efficient method of well development; air supplied via a line is injected from the base of a tube submerged in the water well. The emulsion created in this way reduces the density of the water contained in the tube.

alkalimetric titre Abbr.: AT, (cf. alkalinity).

alkalinity Sum of the concentrations of hydroxides, carbonates and bicarbonates of the alkali metals and alkaline earths present in a given solution. *The alkalinity of a solution is a measure of its ability to neutralize acidity. On the basis of the alkalimetric titre or the total alkalimetric titre, it is possible to derive the proportions of the three main groups of chemical species that give rise to alkalinity* (c.f. alkalimetric titre, total alkalimetric titre).

alternating pumping Method of well development, which consists of alternating sudden stops and starts in order to create brief and powerful variations of pressure on the water-bearing layer, thus reversing the flow through the screen. This procedure, also known as rawhiding, facilitates the disruption of sand bridging, but there is a risk of wear and tear of the pumping equipment, (cf. controlled overpumping).

ammonium An ion NH_4^+ or radical NH_4 derived from ammonia by combination with a hydrogen ion or atom, occurring in salts having properties that ressemble the alkali metals, and also in quaternary ammonium compounds. *The ammonium ion is very frequently present in groundwaters, resulting in most cases from the anaerobic decomposition of nitrogenous organic matter*, (cf. quaternary ammonia compounds).

anaerobic Said of an environment in which oxygen is absent, of organisms only able to grow in the absence of oxygen or a process that can only take place in the absence of oxygen.

analogue recorder Device for automatic recording of data in analogue form, e.g. water level recorder, pressure transducer system. Measurements from chart recorders need to be digitized prior to data treatment and interpretation, (cf. electronic data recorder).

aquiclude A body of relatively impermeable rock that is capable of absorbing water slowly but does not transmit it rapidly enough to supply a well or spring. *Aquiclude or semi-permeable terrains show very slow circulation* (also known as confining bed, aquitard).

aquifer Syn: aquiferous formation Permeable hydrological formation which allows significant discharge of groundwaters as well as the capture of appreciable amounts of water by economical means. An aquifer can be defined in terms of the reservoir, the hydrodynamic, hydrochemical and hydrobiological mechanisms, the part of the global water cycle involved, as well as the spatial and temporal variability of these characteristics. The aquifer is recharged by effective infiltration, i.e. the amount of water actually reaching the groundwater body, whereas its geometry depends on the characteristics of the geological and hydrodynamic boundaries, (cf. groundwater body).

aquifer development Operation designed to improve well performance by enhancing the transmissive properties of the aquifer; similar to well development, except that the fine fraction of the aquifer itself is removed (c.f. well development, mechanical clogging).

aquifer losses Reduction of hydraulic head in the immediate vicinity of the well due to mechanical clogging, (cf. well losses).

aquifuge 1. (adj.) Said of a geological formation or layer with very low hydraulic conductivity (as in *aquifuge or impermeable terrains*). Impermeable formations serve as boundaries to aquifers; 2. (noun) An impermeable body of rock, with no interconnected openings and thus lacking the ability to absorb and transmit water.

area of pumping depression Syn: contributing region Area within the zone of influence where all flow lines are directed towards the well being pumped. Note: do not confuse with radius of influence.

artesian water Groundwater present in a confined aquifer, which flows into wells without pumping or gushes out onto the ground surface; the piezometric level is situated above the ground surface.

artesian well Syn: overflowing well A well that gushes or flows out onto the surface without pumping.

artificial recharge A man's planned operation of transferring water from the ground surface into aquifers. The quantity, quality, location, and time of artificial recharge are decision variables, the values of which are determined as part of the management policy of a considered groundwater system.

attenuation zone Near-field sanitary protection zone surrounding a water well.

Atterberg limits Series of different behaviours observed with increasing water content of a soil. *Bentonites are characterized by their Atterberg limits*, (e.g. liquid limit, plastic limit, plasticity index).

bacterial corrosion Chemical attack of metals brought about by the proliferation of microorganisms in water wells. This phenomenon is due to the presence bacteria in groundwaters, (cf. biological corrosion).

bacterial proliferation Rapid development of bacteria under favourable microenvironmental conditions. Associated with zones of increased nutrient flux into the well, (cf. biomass).

bactericide Chemical substance capable of destroying bacteria or restricting their proliferation.

barite Symbol: $BaSO_4$ Barium sulphate. *Barite is used for weighting up the drilling fluid*, (also known as barytes).

base exchange Chemical reaction whereby a cationic species in solution replaces another from the solid phase. Secondary phenomena which can bring about modifications in the chemical composition of the groundwater during its residence time in the aquifer.

Benoto process A method for cutting a water well in which the casing penetrates under its own weight or is rammed by hydraulic jacks. *Drilling by cutting is better known as the Benoto process*.

bentonite mud, bentonitic mud Drilling fluid based on natural smectite-rich clay. This type of mud cannot be used in water-wells; it has the characteristic property of reacting and flocculating in the presence of nitrate-rich water.

Bernouilli equation A relation describing the conservation of energy in the laminar flow of an ideal fluid.

biological corrosion The deterioration of metals as a result of metabolic activity of microorganisms, (cf. bacterial corrosion)

biomass Mass of living organisms per unit surface or unit volume in a given environment, (cf. bacterial proliferation).

Boulton method Procedure used for the analysis of pumping tests in unconfined aquifers, in which delayed yield is taken into account. *Boulton has tabulated the limited recharges which,* *for example, can be produced by a secondary aquifer; after an initial yield from the aquifer due to pressure release, water is gradually released from unsaturated zone storage by gravity drainage,* (cf. pumping test analysis).

bridge-slot screen Frequently used type of well screen with raised ribs; it is constructed flat, then rolled and welded. Possesses good mechanical resistance and shows an open area varying between 3 and 27% according to the dimensions of the perforations, (cf. screen).

cementation, cementing The operation whereby cement slurry is pumped into a drill hole and forced in behind the casing for such purposes as sealing the casing to the walls of the hole, preventing unwanted leakage of fluids into the hole or migration of fluids from the hole, closing the hole back to a shallower depth, plugging an abandoned well, etc.

centralizer Cylindrical, cage-like device fitted to a well's casing as it is run to keep the pipe centered in the borehole. Cementing centralizers are made with two bands that fit the pipe tightly with spring steel ribs that arch out to press against the wall of the borehole.

chelating power Syn: complexing efficiency The ability of certain bodies to complex cations to form a so-called ring structure.

chemical attack Chemical weathering of rocks or casings by hydration, hydrolysis and oxidation.

chemical clogging Reduction in the porosity due to processes of precipitation caused by increased oxygenation, (e.g. carbonate clogging, iron-manganese clogging).

chemical corrosion Attack of metals not involving electrochemical processes. In the context of water wells, better termed non-electrochemical corrosion. This type of corrosion corresponds to reactions governed by the fundamental laws of kinetics which occur without being accompanied by an electric current. Note: avoid the term chemical corrosion; non-electrochemical corrosion is correct and standardized

chemical development Declogging procedure involving the use of a chemical treatment. The chemical reagents used comprise acids and polyphosphate, (e.g. acidification, polyphosphate dispersion).

chemical treatment Any procedure for well development making use of chemical methods to loosen and remove material clogging the catchment structure.

chloride In general, a binary compound containing chlorine combined with another element or radical.

chlorides *The dissolved chlorides in water may be derived, among other sources, from salt formations in potassic evaporite basins. Since chlorides are not absorbed in the soil, they can be transported over long distances. In addition, they may originate from excessive pumping* near the sea coast.

chlorimetric titre Quantity of active chlorine present in a unit volume of reagent, expressed in degrees, where $1° = 3.17$ g active chlorine per litre.

cleaning The act of bringing the phenomenon of detergency into effect (detersion and cleaning are terms standardized by ISO).

cleanout pumping Pumping carried out to remove material brought into well during development.

coefficient of permeability Syn: specific permeability The volume of mobile water in m^3 transmitted perpendicularly to the flow direction in unit time (s) through a unit cross-section in m^2, under the effect of a unit hydraulic gradient and within the conditions of validity of Darcy's law.

colloid A suspension of very small particles (a few microns in size) of various insoluble substances in a liquid medium. Flocculation is prevented by the high surface tension and viscosity of water as well as the electric charge on the surface of the particles. Changes in the pH of the medium or the concentration of salts will bring about flocculation and precipitation of the solid phase, (cf. biological clogging).

colloidal deposits Gelatinous or viscous masses of flocculate associated with the biological clogging of water wells, (cf. gel).

conductivity The current transferred across unit area per unit potential gradient. *The conductivity of water is a relatively reliable indicator of its mineralization.* Knowing the hydraulic conductivity of a reservoir, it is possible to calculate the discharge rate of an aquifer; conductivity increases with the concentration of dissolved salts and varies as a function of temperature. It is expressed in microsiemens/cm, and is the reciprocal of volume resistivity.

cone of depression, cone of influence, groundwater hole Concave downward depression in the piezometric (potentiometric) surface of a groundwater body, defining the area of influence of a well.

confined aquifer, artesian aquifer, pressure aquifer Aquifer bounded above and below by impermeable beds, or by beds of distinctly lower permeability than in the aquifer itself. The water-bearing formation is saturated throughout its thickness and is capped by a permeable or semi- permeable layer, (c.f. aquifer, unconfined aquifer, leaky aquifer).

confining pressure Syn: geostatic pressure Sum of hydrostatic and lithostatic pressures, or total pressure exerted on a system. An isotropic pressure resulting from the load of overlying rocks, or hydrostatic pressure resulting from the weight of the water column in a zone of saturation, (cf. lithostatic pressure).

continentality A measure of the degree to which a climate is affected by continental influences (or remoteness from maritime influences).

controlled overpumping Method of well development in which the well is pumped at the physical yield limit for a short period in order to induce surging. The rapid alternation of pumping and non-pumping is known as rawhiding, (c.f. alternating pumping, overpumping).

corrosion Complex set of electrochemical or purely chemical phenomena involving attack of metal casings and screens, the accumulation of carbonate and iron-manganese precipitates, and even bacterial activity; e.g. uniform corrosion, pitted corrosion, cracking corrosion, (c.f. electrolytic corrosion, bacterial corrosion).

critical discharge Symbol: Qc Well discharge corresponding to the velocity threshold between laminar and turbulent flow regimes. *In practice, the pumping discharge should always be lower than the critical discharge,* (cf. critical velocity).

critical velocity Corresponds to the velocity at which critical discharge occurs, (cf. Reynold's number).

Darcy's law States that the rate of movement of water through porous media is proportional to the hydraulic gradient. *The circulation of groundwaters is governed by Darcy's law,* (c.f. hydraulic gradient, discharge).

declogging Removal of encrustations, colloidal deposits or organic material from various parts of the catchment structure.

deflocculation Dispersion of colloidal deposits or mud cake to form a suspension, (cf. polyphosphate treatment).

degrees Baumé Abbr.: Bé Divisions of a scale for measuring the specific gravity of liquids, still used in French-speaking countries to express the concentrations of solutions with densities greater than water. Slightly different from the A.P.I. or Twaddell scale: modulus A.P.I. gravity = [141.5/specific gravity] - 131.5; modulus degree Bé = [140/specific gravity] - 130 i.e. 0° Bé is equivalent to a specific gravity of 1.0 at 15°C, and each division on the scale represents a specific gravity increment of 0.004.

delayed storage coefficient The volume of water released from unsaturated zone storage (under gravity drainage) per unit surface area of aquifer per unit change in head.

denitrification Reduction of nitrate, due to bacterial activity, to form nitrite and then elemental nitrogen, usually in an anaerobic environment.

discharge A measure of water flow at a particular point; in the case of a water well, the discharge rate is usually given in cubic metres per day.

dissolution Process in which a substance looses its cohesiveness, due to the hydrating power of water, leading to a partial or complete breaking of the various electrostatic bonds between the atoms and the molecules of the substance entering solution. Dissolution is said to occur when solvation is complete, (cf. solvation).

domestic (or drinking) water supply well Abbr.: DWS well Catchment structure providing water for household consumer use, not intended for irrigation or industrial uses.

down-the-hole hammer drilling Abbr.: DTH drilling A drilling method utilizing impact accompanied by thrust from the tool, which is itself rotating. A pneumatic hammer fitted with drill bits is attached to the end of the drill string. The percussive action is driven by releasing compressed air into the drill string; the energy used to drive this equipment is high-pressure compressed air. This process is of great interest in hydrogeological work, mainly in hard terrains.

drainage basin Hydrological system bordered by watersheds which delimit the catchment area of a watercourse and its tributaries. The only input of water into a drainage basin, which is assumed to be closed, comes from effective precipitation. Also termed drainage area, catchment, catchment area, catchment basin, gathering ground, feeding ground or hydrographic basin (cf. hydrological system).

drainage factor Characterizes the slow downward percolation of waters from the unsaturated zone into an unconfined aquifer; do not confuse with leakage factor.

drawdown Syn: depletion Lowering or depression of water-table caused in most cases by pumping. Corresponds to the difference between dynamic level and standing water-level for a particular well.

driller's log Syn: instantaneous log Record of daily drilling with description of formations encountered, also containing information on casings, screens and water level in well.

drilling fluid A suspension of finely divided material, such as bentonite, pumped through the drill pipe during rotary drilling. *The rotary drill requires the use of a drilling fluid prepared on site which is injected continuously and under pressure into the hollow drill pipes of the string. The drilling fluid may be composed of clear water, bentonite mud or mud with a synthetic biodegradable polymer base; it has several functions: cooling and lubricating, flushing out of chippings from the geological formations, consolidation of the borehole with cake, protection against water inflow and providing useful information on loss of head.* The density is adjusted by weighting up (barite) or lightening (water), while viscosity is adjusted by thinners or viscosifiers. Can also be called drilling mud, drill fluid, drill mud, mud fluid, mud flush, circulating fluid, circulation fluid, circulation medium, driller's mud, mud water, flushing mud, rotary mud (cf. bentonite mud).

drilling rate Syn: rate of penetration (abbr.: ROP) A measure of the speed with which the bit drills into formations; usually expressed in terms of feet per hour. The rate of penetration in soft terrains is high and can reach 100-150 m a day.

drill pipe string, string of rods, drilling string Lengths of pipe, casing or other downhole drilling equipment coupled together and lowered into a borehole. The whole assembly is made up of the following elements, from top to bottom: a swivel, a square drill pipe, ordinary drill pipes, drill collars and a tool. The drill pipes are hollow so that mud can be injected into the bottom of the borehole.

drill reamer, reamer, drilling reamer, reaming shell, core shell, reamer shell A short tubular piece designed to couple a bit to a core barrel. The outside surface of the reaming shell is provided with inset diamonds or other cutting media set to a diameter to cut a specific clearance for the core barrel

drinking water supply Syn: domestic water supply (abbr.: DWS) Distribution network supplying water that is fit for human consumption.

dynamic level Syn: pumping water-level Height of groundwater level when disturbed by pumping.

earthing Special type of electrical potential method, in which one of the current input electrodes is placed at a point on the conductor while the other electrode is placed at infinity. In this context, earthing may be employed as a strategy in support of prospection. Since the conductor is everywhere at roughly the same potential, the ore body or underground river is delimited by equipotential lines.

effective porosity 1: The volume of mobile water, Ve, that a saturated reservoir can contain and subsequently release when fully drained, divided by the total volume of the reservoir, Vt; i.e.: ne = Ve/Vt. Since a reservoir is never completely drained of its water content, effective porosity is more commonly used in hydrogeology than the more theoretical concept of absolute porosity; 2: the property of a soil or rock containing interconnecting interstices expressed as a percentage of the bulk volume occupied by these spaces, (cf. void space volume).

effective precipitation Amount of water from precipitation that remains available at the ground surface after subtraction of losses due to true evapotranspiration. Fraction of precipitation that can infiltrate into the soil or contribute to run off, (cf. run off, evapotranspiration).

effective transfer capacity, effective velocity, average interstitial velocity Carrying capacity of waters infiltrating through a medium, (cf. purifying capacity)

electric log, electrical log Geophysical survey technique based on resistivity or self-potential; provides information on the thickness and nature of water-bearing formations with a view to

groundwater extraction. An electric log typically consists of the spontaneous-potential (SP) curve and one or more resistivity or induction curves.

electric panel array Survey method consisting of measuring apparent resistivity between two electrodes, for different positions of a third electrode. *Panel-type arrays are commonly employed to reveal or confirm the presence of major structural features, in which case the array is operated at right angles to the inferred structural trace.*

electric survey Electrical prospection based on the conductivity of subsurface formations, or their capacity to conduct an electric current, whether natural or artificial; they do not involve measuring the magnetic field. *Electric survey methods make use of: natural (telluric) currents: spontaneous (or self) polarization and artificial currents: electrical potential (mapping of potential difference, with earthing, etc.), resistivity method (electric logging, resistivity rectangles).*

electrochemical corrosion Form of chemical attack on metals involving galvanic couples or gas electrode processes, (e.g. hydrogen corrosion, oxygen corrosion).

electrochemical series, electromotive series List of half-cell reactions arranged in order of decreasing equilibrium potential with respect to the standard hydrogen electrode; also reflects increasing susceptibility of metals to electrolytic corrosion, (cf. standard electrode potential).

electrofiltration Phenomenon controlling the self-potential measured during an electric log survey, resulting from the migration of waters towards the middle of a porous layer.

electrolytic cell Syn: galvanic couple An electrode system in which a potential is set up between two metals having different tendencies to go into solution.

electronic data logger, automatic data acquisition system Apparatus allowing the acquisition of large quantities of data from a measuring device in real time. Used to store very precise and closely spaced measurements of parameters such as water level, pressure, temperature, conductivity cf. analogue recorder.

electroosmosis electro-osmosis Electromotive force proportional to the logarithm of the resistivity ratio between two different electrolytes. It is an important parameter controlling self-potential in an electric log survey, caused by the transport of water from one electrode to another, (also known as electroendosmosis in the biological sciences).

entrainment Carrying off of solid particles into a flowing medium.

entrance velocity Flow rate at which groundwaters pass through the well screen openings, (cf. uniform inflow distribution).

equilibrium pH Symbol: pHS Precise value of pH beneath which waters become aggressive and above which the waters are encrusting.

evapotranspiration Process whereby water is lost by the soil to the atmosphere though the action of evaporation from wet surfaces combined with transpiration, the exhalation of water vapour by plants, largely from their leaves. *Evaporation takes place from the surface of free bodies of water (oceans, seas, lakes and rivers) as well as from vegetation. Both of these phenomena - evaporation and transpiration - are brought together under the term evapotranspiration.*

extractable storage Syn: mineable storage The maximum volume of water that can be economically extracted from the total storage of an aquifer, (cf. reserve).

ferruginous deposits Scale or colloidal deposits formed from the accumulation of insoluble iron hydroxides, usually due to oxygen corrosion or bacterial activity in the zone of temporary saturation, (cf. iron-manganese hydroxides).

fictitious radius Symbol: Rf The distance at which the drawdown, as calculated by the Jacob expression, falls to zero. It is a function of the transmissivity and the storage coefficient, (also known as the critical radius, "imaginary" radius).

filtrate strength Solid matter load present in a drilling mud sample, evaluated by measuring filtrate volume on a Baroid balance, (c.f. mud filtrate, filtrate volume).

fishing, fish job, grappling Searching or attempting to recover a piece of equipment fallen into a well.

float shoe method Cementation method using a special device which is placed at the lowermost end of the casing string. The float shoe consists of a plastic ball which acts as a plug preventing the passage of fluids from the bottom upwards. It is also in contact with the surface by means of a pipe screwed to the cementation shoe, thus allowing the cement slurry to pass into the annular space.

flocculation Aggregation of clay minerals and other fine particles to form flocs. *Influx from gypsiferous terrains can lead to flocculation.*

floccule Abbr.: floc A loose fluffy or foamy mass formed by the aggregation of fine suspended particles.

flow meter, micro-current meter An instrument used to measure velocity of currents flowing in a water well, (fluid meter, flow indicator, flow-measuring device).

flow net A graphical representation of flow lines and equipotential (piezometric) lines used in the study of seepage phenomena, (also known as hydraulic flow net).

flow superposition The respective and simultaneous effects of several different causes of change in groundwater level at a given point. In porous media, these different water flows can be added together algebraically.

food-grade approved Said of materials suitable for use in food packaging or drinking water supply applications.

fracture permeability Ability to transmit water due to the presence of fracture porosity. Within limestone massifs, the fractures are commonly open, thus providing channels which allow the very rapid circulation of groundwaters; a non-porous rock can behave as a reservoir if sufficiently fractured, (cf. karst aquifer).

fracturing Development technique making it possible to widen existing fractures or create new ones in order to improve the specific capacity of a well. There are two types of artificial fracturing: hydraulic fracturing (or hydrofracturing) and fracturing by explosives.

free plug method A destructible cementation plug is introduced into the casing string to be sealed before or after injection of the cement slurry.

gamma-gamma log Downhole profile of induced radioactivity showing the bulk density of rocks and their contained fluids. A porosity log of the wall-contact type indicating formation density by recording the backscatter of gamma rays, (also known as density log, scattered gamma-ray log, densilog).

gas electrode process Half-cell reaction involving a gaseous species (e.g.: Pt, H_2/H^+).

gases Gaseous compounds held in solution in water (e.g. oxygen, carbon dioxide, methane, hydrogen sulphide). The concentration of dissolved gases is governed by Henry's law, and is an important indication of water quality, (cf. Henry's law).

gel Gelatinous material formed from the coagulation of a colloid, (cf. colloidal deposits).

geophysical log Downhole recording of properties in a well or in adjacent geological formations (e.g. self-potential, resistivity, gamma ray, gamma-gamma, neutron). The three main categories are: formation, structural and fluid logs, (cf. electric log).

gravel pack A mass of very fine gravel that is placed around a well screen. Gravel packing is a method of well completion in which a slotted or perforated liner is placed in the well and surrounded by a very fine-mesh gravel. Different methods such as gravel packs, sand consolidation and well screens are used to control the influx of sand.

gravimetry Geophysical survey technique based on the force of attraction between masses, giving rise to an acceleration g - due to gravity - which affects all bodies placed near the surface of the globe.

groundwater Syn: underground water Subsurface waters contained in aquifers beneath the soil-water (or unsaturated) zone.

groundwater basin That part of a drainage basin situated beneath the ground surface. *A groundwater basin can be made up of one or more aquifers, whose boundaries are constrained by geological structures; its recharge takes place through the infiltration of effective precipitation,* (cf. hydrological system).

groundwater body, water-sheet Mass of subsurface water contained within permeable geological formations, (cf. aquifer).

groundwater level The upper surface of groundwater, or the level below which an unconfined aquifer is permanently saturated with water, (also known as water-table, piezometric level).

groundwater level contour Line of equal piezometric level (also: line of equal hydraulic head, hydraulic head contour, groundwater contour, groundwater table contour).

half-cell reaction An electrode process forming part of a galvanic couple, in which a potential is set up between a given metal species and the electrolyte, e.g.: $Zn - 2e = Zn^{++}$ (at the anode), $Cu + 2e = Cu$ (at the cathode).

Hantush method Technique for analysing pumping tests in a non-steady state regime, in which a semi-log relationship between drawdown and time is established for a given observation well. *Hantush has tabulated the high recharges which bring about permanent stabilization of the observation wells after the initial drawdown*, (cf. pumping test analysis).

Henry's law The amount of a gas dissolved by a given amount of liquid at a given temperature is proportional to the temperature, the volume dissolved being independent of pressure.

historical data Records of technical information acquired during testing, operation and development of a well.

homogeneous aquifer One of the two main types of groundwater reservoir, showing pore-space permeability and made up of sands, gravels and sandstones. *Homogeneous aquifers are associated with alluvial deposits occupying valley floors and account for part of the groundwaters of major sedimentary basins; the groundwater discharge rates are generally low*, (cf. heterogeneous aquifer).

hydration Chemical combination of water with another substance. The weathering of rocks and minerals involves penetration of water into the lattice of crystalline solids; do not confuse with hydrolysis.

hydraulic conductivity A measure of permeability with dimensions of length per unit time, denoted as K in Darcy's law. It corresponds to the coefficient of permeability only when the fluid is water at moderate temperatures; hydraulic conductivity has the dimensions of velocity and is expressed in m/s. Also known as coefficient of hydraulic conductivity or hydric conductivity, (cf. Darcy's law, coefficient of permeability).

hydraulic gradient Change in head through unit length of flow path; the h/l ratio is denoted as i in Darcy's law, where h is the head (height of the water column) in metres, proportional to the weight of the water column and l is the height of the cylinder in metres, (cf. Darcy's law).

hydraulic head Syn: total head The difference in piezometric level between the recharge and discharge areas of a hydrological system, calculated as the sum of elevation potential energy possessed by a given mass of water, its pressure head and velocity head. Also termed potentiometric head, total static head. Note: not to be confused with pressure head, (cf. static head).

hydraulic vibrator Device used to facilitate the lowering or extraction of casings in a water well.

hydrochloric acid Symbol: HCl A solution of hydrogen chloride gas in water, widely used as a reagent, (muriatic acid and spirit of salt are obsolete).

hydrogen sulphide, hydrogen sulfide; symbol: H_2S Gaseous compound giving rise to aggressive and corrosive properties when dissolved in water. Waters containing H_2S will attack steel and form iron sulphide encrustations. It is characterized by a smell of rotten eggs, (also known as hydrosulphuric acid, sulphuretted hydrogen).

hydrogeology Study of the chemistry, physics and environmental aspects of water at the Earth's surface, including the mechanisms of storage and flow in groundwaters. Also used in the more restricted sense of "ground-water geology" only (geohydrology).

hydrographic basin, catchment area, drainage area, drainage basin A region or area bounded by a drainage divide and occupied by a drainage system. Otherwise termed a catchment basin, gathering ground, feeding ground, (cf. river basin).

hydrological cycle Syn: water cycle Natural cycle in which water evaporating at the Earth's surface - mostly from the oceans - passes into the atmosphere and falls back as precipitation.

hydrological system Dynamic system which corresponds to a resource - variable in space and time - forming part of the global water cycle, (e.g. drainage basin, groundwater basin, aquifer).

hydrolysis Decomposition of mineral salts brought about by the action of water, owing to the dissociation into H^+ and OH^- ions which leads to exchange reactions with crystalline solids. Plays an important role in the destructive action of water on silicates and the formation of hydroxyl-bearing minerals such as micas and chlorite. Also known as hydrolytic alteration or hydrolytic decomposition, (cf. hydration).

hydrostatic pressure Pressure exerted by the overlying water, (cf. confining pressure).

impervious fault A structural trap forming a lateral boundary to a groundwater body.

infiltration The natural or artificial introduction (recharge) of water into the ground. Infiltration involves flow into a substance, in contradistinction to percolation, which implies passage through a porous substance (e.g. a bed of solid absorbant). In general, the process consists of the slow laminar movement of water through the pore spaces of soil and rock, while flow in large openings such as caves is not included.

infiltration basin Syn: seepage basin Engineered structure used for the artificial recharge of an aquifer.

infiltration rate Flow rate of water infiltrating through a medium. The infiltration capacity is evaluated by cumulating the rates over a period of time.

iron bacteria Microorganisms capable of metabolizing iron present as minerals or organic complexes (e.g. *Gallionella, Toxothrix*). Siderophile bacteria are present in most aquifers and use iron as a source of energy; some iron bacteria are chemicolithotrophic (e.g. sulphate-reducing bacteria, siderocapsaceans) and do not use ferrous iron at all, while others metabolize manganese instead (e.g. *Metallogenium*), (cf. sulphate-reducing bacteria).

iron-manganese clogging Precipitation of iron and manganese compounds, forming encrustations on various parts of the well structure.

iron-manganese hydroxides Compounds generally responsible for iron-manganese clogging, formed in the zone of temporary saturation if the screen is dewatered, (cf. ferruginous deposits).

Jacob method Method for the evaluation of pumping test data under constant-discharge conditions, (cf. pumping test analysis).

jet cleaning Method in which a tool with pressurized water jets is rotated as it is moved past the section of the screen to be treated. The fine particles penetrate into the screen where they are recovered by pumping or with a bailer; the efficiency of this procedure depends on the type of screen, being optimal for screens of the Johnson type, (also known as high-pressure jetting).

jet cleaning head Syn: jetting head Downhole tool fixed to the drill stem of a rotary rig, used for high- pressure jetting of the screen face.

Johnson-type screen Screen with a continuous horizontal opening running the whole length of the screen. The main advantages of such a screen are: the regularity and precision of the opening, the very low risk of clogging and the highest open area coefficient of all the screen types, (cf. screen).

karst Type of topography developed in limestone terrains, formed as a result of variable degrees of dissolution by meteoric waters, (cf. karst aquifer).

karst aquifer Water-bearing formation where most of the storage is in solution channels and fracture porosity, (cf. fracture permeability, karst).

laminar flow Fluid flow characterized by the gliding of fluid layers (laminae) past one another in an orderly fashion. As a preliminary hypothesis, it is necessary to assume that groundwaters exhibit laminar flow over most of their transport path, (cf. Reynold's number).

Landsat D Recent satellite in a series carrying multispectral scanners and thematic mappers, which provides a better resolution at ground level (30, 20 or 10 m). As a result, such images can reveal just as must detail as aerial photographs at a scale of 1:50,000, (cf. remote sensing).

Langelier's index The value obtained by subtracting the saturation pH (pHs) from the measured pH of a water sample. The Langelier index can be used to determine whether a water is scale-forming or aggressive.

leakage Syn: interformational flow Draining or passage of fluids from one formation to another.

leakage coefficient Syn: leakance factor Parameter with the dimensions of reciprocal time, equivalent to the hydraulic conductivity divided by the thickness of the semi-pervious interlayer. It is a measure of the capacity of a leaky interlayer to transmit water vertically

leakage factor Parameter with the dimensions of length used to characterize the effects of leakage in a semi-confined aquifer. Not to be confused with leakage coefficient.

leaky aquifer Syn: semi-confined aquifer Aquifer overlain and/or underlain by a relatively thin semi-pervious layer, through which flow into or out of the aquifer can take place. *The confining bed and/or substratum of an aquifer is often made up of a semi-permeable hydrological formation. Under certain favourable hydrological conditions (difference of pressure head), the semi-permeable layer allows the exchange of water with the overlying or the underlying* *aquifer. This phenomenon, which is known as leakage, implies the presence of an aquifer containing semi-confined groundwaters.* (c.f. aquifer, confined aquifer, unconfined aquifer).

linear loss of head Symbol: BQ Reduction of hydraulic head generated by laminar flow of water into the aquifer close to the well. It is constrained by the hydrodynamic parameters of the aquifer and increases with the duration of pumping, (also known as linear well loss, linear head loss).

lithostatic pressure Syn: lithostatic load Pressure exerted by overlying ground or strata, (cf. confining pressure).

louvre-slot screen Screen with horizontal rectangular perforations in the form of a hood. Mechanically resistant but with a low open area, (cf. screen).

Lugeon test Type of absorption test providing a value for hydraulic conductivity at a given point in the well. The hydraulic conductivity so obtained arises from fracture porosity; 1 Lugeon permeability unit corresponds to the absorption of 1 litre water per metre depth in the well per minute at a pressure of 1 MPa, (cf. absorption test).

manometer An instrument for measuring pressure, but which can be adapted to measure flow.

manometer tube A waterproof piezometric pipe which goes down into the well and which is fixed along the pump column.

maximum acceptable drawdown Criterion determined by physical and technical constraints of the aquifer complex/catchment structure, expressed by the critical discharge, Q_c, and the corresponding critical drawdown, s_c, as measured from well tests.

mechanical clogging Movement of fine particulate matter into the near-field well environment and gravel pack, leading to aquifer losses. Particularly important in overpumped wells, (cf. aquifer development).

mechanical declogging Syn: swabing Physical method of development based on surging water backwards and forwards by means of a plunger. Used to clean screens, gravel pack and aquifer matrix.

mechanical weathering, disintegration, crumbling Disaggregation of rock masses under the effect of weathering, leading to the release of debris of various different grain sizes.

mud filtrate Fluid remaining after deposition of mud cake. If the drilling fluid is too thin, filtrate may be driven into the surrounding permeable formations. Measurements are taken with a Baroid press, in which is placed a sample of the mud to be strained, (cf. filtrate strength).

mud loss Leakage of drilling fluid through well walls into the surrounding terrain. *A fairly frequently encountered situation arises from total loss of mud which places the well structure in real danger* .

natural polymers Organic products derived from Guar gum, used to enhance viscosity during drilling operations (e.g. Revert, Foragum). For the same amount of matter with the same viscosity, their molecular structure enables them to produce ten times more gel than a bentonitic mud.

non-steady state regime Regime which takes account of the observed fact that the size of the cone of depression increases as a function of pumping time. Also termed non- equilibrium regime, (cf. steady-state regime).

nutrient flux Inflow of dissolved minerals essential for bacterial proliferation. Increased nutrient flux is often linked to high entrance velocities across the screen.

observation well Syn: piezometer A generally narrow well or tube designed to measure the water-table level or hydraulic head at a particular point. *In theory, observation wells are set out along two rectangular axes centred on the control well and at increasing distances away from it, the distance of the last one being close to the estimated radius of influence; observation wells must have a very low response time.*

open area coefficient Syn: open area Open area coefficient: $C = f / f + 1$ where f is the dimension of the slot between two coils and l is the width of the encasing wire. Fundamental parameter controlling the inflow of water from the aquifer towards the well.

operation (of water wells) The techniques or actions used for extracting groundwaters from aquifers. Planning the extraction of groundwaters depends on the evaluation of water reserves and resources. More generally, the extraction of mineral resources or Earth materials from the surface or subsurface, (the term abstraction is used for groundwaters, and mining for mineral ores).

organic compounds Dissolved organic matter present in water (e.g. polyaromatic hydrocarbons, phenols, carbon tetrachloride, pesticides).

orthophosphoric acid Symbol: H_3PO_4 Water-soluble, transparent crystals, melting at 42°C; phosphoric acid is obtained commercially from phosphate rock.

overextraction, overdevelopment, groundwater overdraft Abstraction of groundwaters at a rate above the safe or sustainable yield, not to be confused with overpumping.

overpumping Simplest method of well development, which consists of pumping at a rate far above the estimated water extraction capacity. There are risks of irregular development as a result of vertical variations in the permeability of the terrain. This type of development can provoke a compaction of the fine sediments, which causes a reduction in permeability, while sand bridging formed by unidirectional flow may also lead to a lowering of yield, (cf. controlled overpumping).

oxygen corrosion Chemical attack of ferrous metals in aerated waters, due to a gas electrode process involving oxygen. Characterized by the formation of ferric hydroxide blisters which accumulate around the anode.

pasteurization A process involving the elevation of temperature for an appropriate period of time, for the purpose of either inactivating microorganisms, particularly pathogens, or decreasing their number to a specified level.

percussion drilling Method that consists of raising a heavy tool (churn drill bit) and letting it fall onto the terrain to be traversed, the height and frequency of the drop being varied according to the hardness of the formation. Also known as percussion boring, percussion system or percussive drilling.

permanent storage Fraction of the total reserve that is not replenished. In the case of unconfined groundwater, its upper surface corresponds to the mean minimum water table. For confined bodies, the permanent storage is very similar to the total groundwater storage, (cf. reserve).

pH The negative logarithm of hydrogen ion concentration, a measure of the acidity or the basicity of a solution ranging on a scale from 0 (acidic) to 7 (neutral) to 14 (basic). *pH measurements are of considerable interest since they reveal contamination by cement or by water from the aquifer, and also indicate the risks of flocculation of the drilling mud*, (cf. hydrogen potential)

photo-interpretation A rapid and cheap method for drawing up a structural or geological sketch map, thus representing a valuable adjunct to geological and soil surveys which provide the essential information for choosing the location of a borehole. Photographic interpretation may be applied to conventional aerial photogrammetry and satellite imagery, (cf. remote sensing).

physical treatment Pneumatic or hydraulic technique of well development/restoration used to break up mechanical clogging in the gravel pack, (e.g. controlled overpumping, air-lift pumping, compressed-air treatment).

piezometer A generally narrow observation well or tube designed to measure the water-table level or hydraulic head at a particular site.

piezometric surface Syn: potentiometric surface The level to which water in a confined aquifer will rise in observation wells, mapped by interpolation between piezometer measurements.

Pito(t) meter Syn: Pitot tube A measuring device consisting of a delivery main entering a rigid conduit with a diameter ensuring that flow takes place over the full cross section and along a minimum length. At the end of the conduit, there is a section of tube of diameter identical to that of the intake. The end of this tube is obstructed by a metal plate with a circular orifice at the centre. On the side of the tube, there is a transparent manometer tube for the direct reading of discharge rate.

pneumatic development A procedure based on the same principle as surging. Makes use of the forward and return flow of groundwaters around the screen brought about by the large volume of air introduced into the borehole. Two distinct methods can be adopted: the open hole method and the closed hole method.

polymer mud Product containing chemical compounds with a high molecular weight resulting from the association of several simple molecules having a low molecular weight. Can be used directly as a drilling mud or as an additive to bentonic mud, (cf. natural polymers).

polyphosphate treatment Addition of a solution of polyphosphate in order to break down the cohesiveness of clay. Used to disperse mud cake, infiltrated mud or the clay present in cutting samples, (cf. deflocculation).

post-production Phenomenon involving the displacement of fluids towards the well in order to equilibrate the pressure. Appears during the recovery in water-level after drawdown or cessation of pumping.

potentiometric surface, piezometric surface, water table The level to which water in a confined aquifer will rise in observation wells. It is also the top surface of a groundwater body defined by the set of piezometric levels measured at different points at a given date. Since this surface corresponds to the upper boundary of the aquifer, it is the hydrodynamic limit of the system, mapped by interpolation between piezometer measurements.

pressure head Syn: piezometric head Potential energy of a unit mass of water at any point compared with a pressure of one atmosphere at the same elevation. Note: do not confuse with static head.

protected perimeter e.g. attenuation zone, remedial action zone, well field management zone, (cf. sanitary zone of well protection).

pump discharge column Syn: rising main Tube used for extracting water from a well, connects the submersible pump with the distribution network.

pumped water discharge The rate at which water can be pumped out of a well.

pumping discharge Yield obtained from a well during pumping.

pumping test analysis Procedure for interpreting data from pumping tests, usually based on the graphical representation of drawdown vs. time curves and a comparison with tabulated well functions, (e.g. Berkaloff, Boulton, Hantush, Jacob, Theis and Walton methods). The experimental drawdown vs. time curve is superimposed on standard curves in order to assess departures from the ideal state, (cf. pumping test).

pumping tests, aquifer tests Procedure for measuring well efficiency, well performance or aquifer characteristics, (e.g. step drawdown test, constant discharge test, absorption test). Such tests make it possible to determine: the characteristics of an aquifer/catchment structure; the hydrodynamic parameters; the operating conditions of the well and the evolution of drawdown. Pumping tests performed before the start of water abstraction or after a regeneration phase allow the satisfactory completion of cleaning and development procedures, including natural development, (cf. pumping test analysis).

purifying capacity A measure of the ability of a given terrain to absorb a pollutant from the groundwater, equivalent to the path length required for complete removal of the pollutant from the liquid phase, (cf. effective transfer capacity).

quadratic head loss Syn: turbulent well loss. Symbol: CQ^2 Non-linear reductions in hydraulic head arising from turbulent flow within the well structure, screen and casings. *Quadratic head losses become all the more important as the drawdown curve becomes increasingly convex.*

quasi steady-state regime An eventual stabilization of the hydrological system in which the aquifer restores its water balance.

quaternary ammonium compounds Strongly alkaline substances that are derived from ammonium by replacing one or more of the four hydrogen atoms with organic radicals, (cf. ammonia).

radius of influence (of well pumping) Syn: area of influence Zone within which water-levels are influenced by pumping.

recharge area Syn: intake area Zone located between radius of influence of well pumping and upstream boundary of the hydrological system. Corresponds to undisturbed aquifer, (cf. radius of influence).

redox potential Symbol: Eh A scale of electric potential measured in volts indicating the ability of a substance or solution to cause reduction or oxidation reactions under a given set of conditions. The higher the Eh, the more oxidizing are the conditions and the greater is the capacity of the medium to accept electrons.

redox potential vs. pH diagram Plot of redox potential versus hydrogen potential used to display stability of minerals or solutions in natural systems.

reduction Reverse of oxidation, being equally important in aerated groundwaters as in surface environments.

regulating storage The volume of mobile water contained in the temporary zone of saturation, (cf. reserve).

remedial protection zone Immediate protected perimeter for pollution control around a water well.

remote sensing Technique based on data obtained from artificial satellites in Earth orbit. The periods of the shots and the repetitivity of the information makes it possible to select the images of most interest. Satellite imagery leads to a better integration of major fractures (on the scale of several km), but the lack of relief on these images hinders satisfactory correlation with the ground truth, and the user is then obliged to fall back on conventional methods. Where high groundwater discharges are sought, satellite imagery can prove to be a valuable guide to the hydrogeologist in the selection of borehole sites.

reserve Syn: storage Quantity of water contained in a hydrological system at a given date or stored over a period of time, expressed in terms of volume (hm^3 or km^3). Four types of storage are distinguished: total groundwater storage, regulating storage, permanent storage and extractable storage. The groundwater storage is evaluated using the volume of the slice of aquifer in question and either the effective porosity (in the case of an unconfined aquifer) or the storage coefficient (for a confined aquifer).

residence time The average time that formation waters remain in contact with a given volume of rock, directly linked to the rate of infiltration of water within the massif, (c.f. convection time, transfer time).

resistivity The coefficient r is proportional to the electrical resistance R of a homogeneous conductor of length l and cross-sectional area s; if the current density parallel to l is uniform and the temperature is constant, then R = r l/s. The units of r are ohm.metre. The measurement of resistivity allows an estimate to be made of the concentration of dissolved salts (or salinity) in a water sample at a given temperature. It is also possible to determine the amount of dissolved matter in a water/sediment mixture. Continuous on-line recordings are commonly carried out with a resistivity meter, often in combination with a nephelometer for turbidity measurement.

resource Quantity of water that can be extracted from a defined volume over a given period of time. Evaluation of the resource is based on the hydrodynamic and hydrochemical behaviour of the aquifer, the resource is measured in terms of mean discharge rate (m^3/s, hm^3/yr or km^3/yr).

reverse circulation System used for drilling large-diameter water wells in unconsolidated formations. In the cleaning of holes with large diameters, the mud can be injected into the annular space to bring up the cuttings inside the drill string.

Reynolds number A dimensionless parameter which represents the ratio between forces of inertia and forces of viscosity, i.e.: Re=v.d /V, where Re is the Reynolds number, v is the Darcy velocity, d is the characteristic length and Vr the kinematic viscosity. Serves to calculate the linear loss in head occurring in a conduit under an established steady-state regime, assuming that the conduit is rectilinear, the discharge rate is relatively stable and the diameter of the conduit is homogeneous. The transition between laminar and turbulent flow is a function of a large number of parameters, but it is generally accepted that the laminar flow regime ceases for Reynolds numbers greater than unity, (c.f. laminar flow, c.f. turbulent flow).

river bank effect Uptake of pollutants onto particulate matter as waters percolate through the first few metres of river bank deposits towards the aquifer. Associated with filtration and bacterial activity.

rotary drilling A tool (drill bit) is attached to the end of a drill pipe string, and the assembly is driven with a rotary movement at variable speeds and under vertical compression. The rotary movement is transmitted to the drill string and to the tool by a motor situated at the well-head. The drill pipes are hollow so that mud can be injected into the bottom of the borehole.

run off, runoff Part of the precipitation flowing towards drainage systems into free bodies of water.

Ryznar's index Chemical parameter commonly used in corrosion problems, equivalent to 2 pHS - pH.

salts Chemical compounds formed by the reaction between an acid and a base, (e.g. sulphates, chlorides). Ionic species present in water are derived in part from the dissolution of minerals such as calcite, dolomite, gypsum and halite, during their residence in the subsurface. Groundwaters take up a certain number of mineral substances into solution, including limestone, dolomite, gypsum, sodium chloride and potassium chloride, (cf. water chemistry).

sand bridging Clustering of grains in the gravel pack due to unidirectional flow, leading to reduced permeability.

sanding up Influx of sand into a well due to inappropriate design, screen corrosion or overextraction. Can be treated by air-lift pumping, well scraping, surging, etc.

sanitary zone of well protection, catchment protection zone cf. protected perimeter.

saturated aquifer formation An aquifer whose reserve has been fully recharged. A saturated aquifer formation will bring about dilution of the drilling fluid.

scale Syn: encrustation

scouring Syn: water flush Process by which well is cleaned by injection of water using compressed air, (cf. air-lift development).

scraping Syn: brushing Method of well development making use of wall scratchers (e.g. porcupine brush).

screen A slotted or perforated liner placed in the well to control sand influx. The screens should be placed adjacent to the points of greatest water inflow and, generally speaking, over the entire thickness of the tapped aquifer zone, (cf. bridge-slot screen, louvre-slot screen, Johnson-type screen).

sediment influx Movement of sediment particles into the well across the screen face, arises from screen corrosion, overextraction or poor well design, (cf. sanding-up).

seismic reflection Geophysical surveying method based on the recording of reflected seismic waves, whose travel-times depend on the nature and structure of the geological formations traversed. Although relatively little used in hydrogeology, small- scale seismic reflection surveys can provide a valuable adjunct to seismic refraction, covering a range of investigation from a few tens of metres to some hundred metres depth, (cf. seismic refraction).

seismic refraction Seismic surveying based on the study of travel-times of fully refracted waves, depending on the nature and structure of geological formations. Although seismic refraction suffers from some limitations, its use in hydrogeology is preferred to seismic reflection because better results are obtained in the depth range 0-200 m and the implementation is easier (cf. seismic reflection).

self-development A process in which fine-grained material is washed away from the immediate vicinity of the well. As a result, the flow velocity in the aquifer increases near the well, thus leading to increased drawdown and enhanced performance. If the 90-percentile is greater than 0.25 mm, then it is considered that the formation can be developed naturally. This phenomenon is also termed suffosion.

sequestering power Capacity of a given metal salt to combine with soluble species to form a complex with covalent properties.

solubility of gases Volume of a gas that can be absorbed by a volume of water depending on the pressure and concentration of gases in the medium. Waters enter into contact with gases especially in the zone of infiltration (e.g. only certain gases, such as carbon dioxide, ammonia or hydrogen sulphide, exhibit high solubilities). Being governed by Henry's law, gas solubility in water decreases with increasing temperature and increasing salt concentration, (cf. Henry's law).

solutizing power Capacity of certain substances to form soluble complex salts, thus incorporating insoluble species into their structure and promoting solvation. This property is useful in solubilizing carbonate scale.

solvation Process in which the hydrating power of water (which is a strongly bipolar molecule) leads to a partial or complete breaking of the various electrostatic bonds between the atoms and the molecules of the substance entering solution, thus forming new bonds and structures in the liquid state. Dissolution is said to occur when solvation is complete, (cf. dissolution).

sorting coefficient Abbr.: SC Quotient of the diameters of the 40- and 90-percentile classes on the cumulative grain-size curve.

specific capacity (of a well) Discharge rate obtained per unit height of drawdown; it has the dimensions of surface/time and is denoted as Q/s, (cf. specific drawdown curve).

specific drawdown Symbol: s/Q Drawdown level measured in the well divided by the pumped discharge; it is expressed in $m/m^3/h$, being obtained from the equation of C.E. Jacob.

specific drawdown curve Plot of specific drawdown versus discharge used to analyse step drawdown tests. Provides an estimate of aquifer and well losses, in addition to information on transmissivity and storage coefficients, (cf. specific capacity).

specific yield (of an aquifer) Syn: effective porosity In the case of recharge, the volume of water that fills the available void space of the porous medium per unit volume; in an unconfined aquifer, the volume of water drained under the effect of gravity from the porous medium per unit volume.

Spot Recent satellite that provides an improved image resolution at ground level (30, 20 or 10 m) and offers the possibility of obtaining stereoscopic views. As a result, such images can reveal just as must detail as aerial photographs at a scale of 1:50,000.

square kelly A square drill pipe.

stabilized foam Physical mixture of liquid and gaseous components, in which the bubbles are maintained in stable suspension. The foaming solution is sometimes supplemented with polymers or bentonite in order to increase its density and improve its viscosity qualities and also stabilize the borehole walls. The foam must maintain a certain consistency, comparable to that of shaving foam.

standard electrode potential, equilibrium potential The potential of a couple in which the left-hand half cell is the standard hydrogen electrode and the right-hand cell is the electrode system in question (using 1M solutions at 1 atmosphere H_2 pressure, e.g.: Zn/Zn^{++} (1M)//Cu^{++} (1M)/Cu). Oxidation takes place at the left-hand half cell, (cf. electrochemical series).

standing water-level, rest water-level, standing level The level of the free water surface in an unpumped well or the piezometric level of an unconfined aquifer. It corresponds to the supposedly stable upper surface of an undisturbed aquifer. In the case of an unconfined aquifer, the standing water-level is always found beneath ground level; in the case of a confined aquifer, the standing water-level remains virtual as long as a borehole or a piezometer has not attained the aquifer, lying always above the base of the overlying impermeable layer. It is also incorrectly termed static level.

static head, hydrostatic head Pressure exerted at any point in a liquid, represented by the height of the overlying water column. Equivalent to pressure head plus elevation potential energy, (cf. hydraulic head).

steady-state regime, equilibrium regime Regime in which, after a short pumping time, the geometry of the cone of depression remains constant. In strict terms, the steady-state regime does not exist except under exceptional conditions. In reality, the appearance of a quasi-steady state regime is accepted for the purposes of calculation, (cf. non-steady state regime).

step drawdown test Used to assess the efficiency and performance of a well, and to measure the response of performance to varying discharge rate, (cf. pumping test).

Stoke's law A sphere of radius r moving with velocity v through a large expanse of fluid of viscosity n will experience an opposing force, or viscous drag, F, such that $F = 6\pi\,r\,n$. The equation describes the rate of settling of solids particles in a suspension or dispersion, but is valid only for restricted conditions (laminar flow and low Reynolds number). In practice, corrections need to be made for boundary conditions.

storage coefficient Parameter which characterises an aquifer. Also known as storativity or bed storativity.

subartesian well A well where water rises from the confined aquifer but does not reach the ground surface without pumping, (cf. artesian well).

sulphate-reducing bacteria Strict anaerobes that obtain their source of energy by converting sulphate to sulphide (e.g. *Thiobacillus*). These bacteria are ubiquitous in groundwaters, soils and muds, but can only develop in the absence of oxygen and at very low Eh values, (cf. iron bacteria).

sulphate reduction Chemical reaction between an electron-acceptor and sulphate ions to form sulphide. A secondary phenomenon which can bring about modifications in the chemical composition of groundwaters and lead to biological clogging; it is associated with bacterial activity in de-oxygenated aquifers, (cf. sulphate- reducing bacteria).

surfactant A surface-active substance used for its detergent properties.

surge block Syn: desanding plunger.

suspended particulate matter Abbr.: SPM Sediment particles which do not settle out under normal conditions of flow.

swelling marls *The degree of swelling of marls increases with the alkalinity of the drilling mud.*

Theis method, Theis's bi-logarithmic method Technique for the analysis of pumping tests assuming a fully confined aquifer and no leakage from or into the aquifer. Used when the pumping time is short, when the distance between the borehole and the observation well is very large or in the case of pumping from an unconfined aquifer with a small drawdown; in such circumstances, the data are difficult to treat with the Jacob method, (cf. pumping test analysis).

thixotropy Capacity for a mixture of particles in suspension to pass from a solid state to a liquid state while being agitated and to return to its initial state when the agitation ceases.

total capacity Maximum flow rate physically obtainable from a water well, being a function of the local characteristics of the aquifer and the groundwater recharge. Also known as potential yield or physical yield limit, (cf. water-yield).

total groundwater storage Quantity of mobile water contained in a volume comprised between the substratum and the upper boundary of the aquifer. The mean total groundwater storage is bounded at its top by the mean annual piezometric surface, (cf. reserve).

total static head, hydrostatic head, pumping discharge head Pressure exerted at any point in a liquid, represented by the height of the overlying water column. It is equivalent to pressure head plus elevation potential energy. *The nature of the pumping equipment will vary according to the expected discharge and the total static head and remains dependent upon the diameter of the equipped borehole*, (cf. hydraulic head).

trace elements Chemical elements present in water whose concentration per unit volume does not exceed a few tens of parts per million (ca. 0.01%), e.g.: Al, Fe, Mn, Ni, Cr and Pb.

transfer time The time taken for a tracer to travel from an intake point to a given discharge in a hydrological system, (cf. convection time).

transmissivity A measure of the discharge of a water-bearing layer over its entire thickness per unit width, subject to a unit hydraulic gradient. The discharge capacity of a well in an aquifer depends on the hydraulic conductivity K and the thickness e of the aquifer, the discharge capacity of a well can also be estimated by means of a parameter T, denoting transmissivity, using the equation: $T = Ke$, where: T is expressed in m^2/s, K in m/s and e in m. Transmissivity should not be confused with hydraulic conductivity, which is calculated over unit thickness. Long pumping tests are used to calculate the transmissivity, based on measurements of drawdown and recovery both in the control well and also in observation wells.

turbidity Cloudiness or opacity of fluid media caused by dispersion or absorption of transmitted light, due to the presence of suspended particulate matter (clay, silt, organic matter, plankton) or coloured chemical pollutants. The turbidity of water becomes even greater with increasing amounts of colloid in suspension; in practice, turbidity is evaluated by means of a nephelometer which measures the intensity of light diffused laterally by the water sample. The unit of measurement is the IU (International Unit), which corresponds to 1 mg of formazine per litre.

turnover rate Mean annual recharge of the aquifer, IE, expressed in volume, divided by the mean total groundwater storage, WM.

turnover time Theoretical duration needed for the cumulative recharge volume of an aquifer to become equal to its mean total storage, WM, which is equivalent to the discharged volume of underground waters over the long term, QW; the turnover time is expressed in years: WM/IE=WM/QW.

unconfined aquifer A groundwater body having an upper boundary corresponding to a water table that is free to fluctuate in response to hydrodynamic factors, (e.g. Champigny Limestone, Beauce Limestone, Alsace valley alluvium, Lorraine karstic limestone). The water-bearing formation is not saturated throughout its entire thickness, (c.f. aquifer, confined aquifer, leaky aquifer).

uniform inflow distribution A regular pattern of flow into a well, usually achieved by installing a second screen with openings that ensure constant entrance velocity throughout the tapped aquifer zones, (cf. entrance velocity).

unsaturated zone Syn: vadose zone Soil-water zone situated between the ground surface and the water table. Water percolates rapidly downwards through this zone, and is not easily extractible due to capillary forces, (cf. zone of infiltration).

unstable terrain, running ground Uncohesive formations with quicksand-like behaviour. Soft or unconsolidated formations require feed pipes during drilling to prevent collapse.

void space volume Volume of air contained in pores, (cf. effective porosity).

Walton method A method proposed by B. Walton makes it possible to characterize the condition of a well in terms of its C value.

water balance Syn: water budget A measure of the difference between input and output flow rates in natural hydrological systems. The global water budget is the amount of water involved in the hydrological cycle each year. The calculation of a water balance provides a means of checking the consistency of data with respect to the recharge and flow behaviour within groundwater basins.

water chemistry Compositional specification of dissolved salts and gases present in water. The composition of a particular water results from dissolution and chemical reactions with substances present in the formations traversed by the water. The concentration of dissolved substances in water may be expressed in terms of the amount of gases, salts or ionic species per unit volume, (c.f. salts, dissolved gases).

water cycle Syn: hydrological cycle Natural cycle in which water evaporating at the Earth's surface - mostly from the oceans - passes into the atmosphere and falls back as precipitation. Vast circulation process occurring at the surface of the Earth, including the movement of water within the soil and subsoil; the *origin, storage and flow of groundwaters is controlled by the functioning of the hydrological cycle, which can be represented as an equation taking account of the water balance: P=E+R+I, where: E=evapotranspiration, P=precipitation, R=runoff and I=infiltration.*

water flush Syn: scouring Procedure for cleaning a well by water injection.

water hardness, hardness of water The quantity of calcium and magnesium salts contained in a water sample, expressed as the equivalent mass of calcium carbonate dissolved in 1 litre. *The hardness of water is due mainly to the presence of calcium and magnesium salts in the form of bicarbonates, sulphates and chlorides; a water hardness of 1° corresponds to 10 mg/l on the French scale, 0.7° on the British scale.*

water inflow Syn: influx Movement of water from tapped aquifers into a well.

water-level recorder, gauge e.g. water-level dippers, pressure transducer system.

water-producing zone Syn: tapped aquifer zone Section of an aquifer that can be tapped by a well.

water quality Standards set by regulating authorities for the control of physical, chemical and microbiological pollution in the drinking water supply (e.g. colour, turbidity, taste, physico-chemical properties, undesirable substances, toxic substances, bacteriology, pesticides). For each listed parameter, water operators must ensure that the maximum permissible dose or level is not exceeded.

water-retention agent Product used to control loss of circulation.

water-yield Syn: capacity Flow rate of a well. Various types are defined, including safe yield, potential yield and economic yield, (cf. total capacity).

well development Operation designed to repair damage to the well face and aquifer matrix caused by drilling. Mainly aimed at removing cake from well walls and improving the transmissivity of the near-field well environment in order to restore the well's performance, (cf. aquifer development).

well effect Changes in permeability due to the influence of drilling mud. Becomes apparent during phases of pumping. It is commonly observed in oil-wells, which are almost always drilled with a rotary system and show productive capacities that are relatively low. In the oil industry, it is customary to refer to this phenomenon as the "skin effect."

well field Group of wells designed for the abstraction of groundwaters from an aquifer.

well field management zone Far-field protected perimeter for pollution control around a well or group of wells

yield-depression curve Plot of well discharge versus drawdown obtained from pumping tests.

zone of infiltration Unsaturated zone containing some air in its void space volume. It is located between the water table and the ground surface, or up to the base of the overlying aquitard where this exists, (cf. unsaturated zone).

zone of water-table fluctuation Syn: temporary zone of saturation Belt of oscillation of the free upper surface of a groundwater body.

Index

A

abandonment 164, 206, 231, 281, 308, 309, 331
acid
    acetic 307
    Ascorbic 282
    citric 286
    hydrochloric 281, 282, 285, 286, 287, 301, 302
    hydrochlororic 91
    orthophosphoric 276
    phosphoric 276
    polyphosphoric 276
    pyrophosphate 66, 276, 278
    sulphamic 282, 285, 292
    sulphuric 302
    tripolyphosphoric 276
acidification 209
acidizing 81, 91, 160, 287, 288, 357
acoustic log 173
acrolein 292
ACTIF 320
active chlorine 292, 299, 300, 357
active management of groundwaters 312, 357
additive 60, 67, 83, 282, 286, 357
adjuvant 278, 357
aerobic 230, 357
AESN 260
ageing 115, 203, 206, 208, 210, 216, 217, 231, 235, 241, 246, 248, 250, 267, 269, 281, 306, 357
air
    binding 245
    drilling 71
air-lift 60, 90, 95, 97, 270, 273, 274, 357
alkalimetric titre 46, 357
alkalinity 46, 63, 66, 253, 357
alternating pumping 89, 357
ammeter 205, 207, 213
ammonium 48, 357
anaerobic 48, 230, 242, 263, 265, 357
analogue recorder 110, 357
anti-water-hammer valve 208
aquiclude 7, 357
aquifer
    Albian 12
    confined 10, 360

    heterogeneous 8
    homogeneous 8
    leaky 9
    semi-confined 11
    unconfined 9
aquifer development 236, 254, 358
aquifer losses 287, 358
aquifuge 7, 358
artesian water 54, 60, 79, 358
artesian well 61, 358
artificial recharge 235, 254, 255
atmospheric zone, 218
attenuation zone 194, 198, 199, 358
Atterberg limits 65, 66, 358
Aubergenville 138, 254, 255, 258, 261, 289

B

bacteria
    Gram-negative 230
    iron 228
    oligotrophic 242
    pedunculate 229
    sheathed 229
    siderocapsacean 228
    siderophilic 218
    sliding 229
    sulphate-reducing 218, 228
bacterial proliferation 67, 272, 294, 298, 358
bactericide 67, 282, 293, 298, 301, 358
BADGE© 320
bailer 54, 55, 89, 270, 273
barite 60, 62, 358
base exchange 44, 47, 358
Benoto process 55, 358
bentonite 65, 358
Bernouilli equation 358
bio-forte 275
biomass 229, 242, 291, 293, 294, 298, 299, 358
bonding 213
Bouguer anomaly 42
Boulton method 147, 150, 358
branch pipe 205, 214
buckling 69, 73

C

cake 57, 60, 63, 67, 87, 91, 100, 164, 170, 238
caking 60

capacity-related volume 300
carbon steel 218, 254
carbonate precipitation 268
casing 73
    flange 275
    rupture 268
cavitation 217
CD-ROM 328
cement slurry 84, 85, 86
cementation 83, 182, 359
centralizer 81, 178, 184, 235, 359
chelating power 277, 278, 359
chemical attack 44, 280, 281, 291, 303, 359
chemical clogging 216, 239, 240, 241, 359
chemical corrosion 221, 252, 302, 304, 359
chemical treatment 251, 269, 275, 276, 278, 282, 283, 292, 308, 359
chloride 43, 46, 47, 48, 219, 252, 281, 286, 359
chlorimetric titre 299, 359
chlorination 204, 292, 307
chlorine dioxide 292
CHRONO© 321
cleaning 88
clogging
    biological 240, 256
    carbonate 280
    chemical 238
    mechanical 235
Clonathrix 228
colloid 45, 237, 280, 359
colloidal deposits 291, 359
Compaq 315
complexation 228
complexing agent 277
conductance 45
conductivity 45, 359
cone of depression 105, 106, 130, 131, 133, 190, 192, 193, 210, 360
continentality 49, 360
controlled overpumping 273, 360
corrosion 43, 214
    acid 223
    bacterial 227
    biological 228, 358
    electrochemical 251
    oxygen 223
cracking corrosion 218
critical discharge 111, 114, 118, 183, 186, 360
critical radius 130
critical velocity 114, 360
crushing 73, 74, 76, 83
cyberspace 328

D

Darcy's Law 14, 15, 16, 50, 103, 113, 135, 360

DEC 315
declogging 274, 275, 281, 282, 283, 293, 295, 300, 360
deflocculation 277, 278, 279, 360
degasification 250, 257
denitrification 262, 263, 265, 360
density 60
detergent 275, 282
development 88
dewatering 148, 149, 205, 211, 214, 216, 244, 245
diffusivity 20
disinfection 186, 209, 291, 292
disodium orthophosphate 276
downhole logging 215
downtime 83, 164, 186, 293
drainage basin 4, 5, 24, 25, 361
drainage factor 150, 361
drill
    bit 54, 55, 58, 59, 135, 163, 166
    collar 55, 57
driller's log 165
driller's log 361
drilling
    fluids 59, 68
    mud 63
    rate 166, 168, 361
drilling by cutting 55
DTH 58, 59, 71, 164, 360
DWS 75, 170, 360

E

e-mail 328
earthing 33, 34, 361
ecology 330
effective porosity 17, 19, 22, 149, 150, 361
effective precipitation 1, 4, 5, 361
Eh 226
electric log 171, 361
electric panel array 39, 362
electric survey 33, 38, 362
electrochemical
    corrosion 362
electrochemical corrosion 219
electrochemical series 220, 227, 251, 362
electrofiltration 173, 362
electrolyte 173, 221, 252, 305
electron
    acceptor 228
    donor 221
electroosmosis 173, 362
enzyme 301
evapotranspiration 2, 3, 4, 5, 362
explosive 285
external zone 218

F

ferric hydroxide 224, 229, 230, 238, 240, 279
ferruginous deposits 235, 239, 290, 362
fictitious radius 156, 362
fishing 165, 362
float shoe method 85, 362
flocculation 60, 63, 279, 363
flow meter 227, 249, 363
flow superposition 104, 131, 137, 363
foaming agent 68, 70, 71
foragum 67
fracturing 17, 25, 26, 34, 36, 39, 40, 70, 90, 363

G

Gallionella 227
galvanization 304
gamma-gamma log 173, 187, 363
gases 43
geophysical log 79, 363
goethite 224, 226
grab bucket 55
GRAPHER® 319
gravel pack 53, 55, 67, 78, 81, 87, 88, 100, 134, 162, 165, 170, 180, 181, 235, 236, 237, 239, 273, 275, 304, 363
gravimetry 41, 363
groundwater
    basin 4, 24
    body XI, 4, 9, 54, 68, 141, 285, 363
    level 109, 129, 204, 212, 244, 288, 321, 363
guar gum 67

H

Hantush method 143, 364
Hantush-Berkaloff method 143
hardness 46
Henry's law 43, 364
hot-forming 304
hours indicator 205, 213
hydration 43, 364
hydraulic conductivity 16, 18, 50, 99, 106, 134, 192, 364
hydraulic gradient 1, 14, 16, 18, 198, 364
hydraulic head 13, 364
hydraulic vibrator 55, 364
hydrochloric acid 68, 279, 281, 282, 286, 301, 364
hydrofluoric 308
hydrogen sulphide 43, 48, 208, 302, 364
hydrographic basin 2, 364
HYDROLAB© 324
hydrological cycle 2, 364
hydrological system 4, 364
hydrolysis 43, 364
hydrostatic pressure 62, 364

hyphomicrobium 228

I

IBM 315
ICRP 50
impervious fault 25, 364
index
    fineness 81
    Langelier 45, 365
    plasticity 65
    Ryznar 45, 253
infiltration
    basin 255, 256, 365
    rate 256, 257, 365
inhibitor 237, 275, 281, 282, 292, 308
INRIA 318
IPSN 264
iron bacteria 228, 231, 365
iron precipitation 268
iron-manganese
    clogging 291, 365
    hydroxides 365
isotopes 49

J

Jacob method 121, 365
javel water 291
jet cleaning 308, 365
Johnson-type screen 75, 271, 365

K

karst 17, 28, 41, 88, 201, 307, 365

L

laminar flow 12, 14, 15, 114, 135, 365
Landsat 27, 365
latency time 263
leakage coefficient 142, 365
leakage factor 142, 365
Leptothrix 228
Leptothrix lopholea 228
Leptothrix ochracea 227
lining tube 71, 74, 178, 181, 252, 253
loading chamber 208
Lugeon 168, 366
Lyonnaise des Eaux 204, 255, 259, 300

M

magnetite 44, 225, 226
manganic incrustation 291
manometer 110, 205, 207, 214, 366
manometer tube 108, 366
mechanical
    clogging 231, 235, 272, 278, 366
    declogging 366
metallogenium 228
microgravimetry 41
mineralization 228
monosodium orthophosphate 276

mud filtrate 63, 366
mud loss 60, 62, 366
mud pit 60
natural polymers 67, 68, 366

N

Nernst equation 220
non-steady state 106, 366
NPSH 204
Nu-well® 282
nutrient flux 242, 294, 299, 366

O

observation well 19, 54, 106, 110, 122, 129,
131, 140, 143, 153, 154, 155, 261, 366
Odex 59, 71
OECD 314
organic compounds 367
overextraction 245, 367
overpumping 88, 367
oxidation 44
oxygen corrosion 223, 224, 367

P

pasteurization 305, 367
patrimonial 204
percussion drilling 54, 367
permeability 16
pH 45, 63, 221, 226, 262, 281
phosphoric acid 276, 277, 281
photo-interpretation 26, 51, 161, 367
pHS 45, 362
piezometer 106, 367
Pitot tube 108
pitted corrosion 218
plugging 95, 119, 235, 309
pneumatic development 89, 367
polyphosphate 283, 307, 367
porosity 17
post-production 134, 368
potassium permanganate 292
pressure detector 206
pumping test 105

Q

quadratic head loss 114, 115, 135, 368

R

radionuclide 50, 264
radius of influence 105, 368
re-lining 304
readjustment 78
recharge area 193, 197, 368
redox potential 262, 368
reduction 44
remote sensing 27, 369
residence time 49, 189, 275, 297, 369
Revert 67
Reynolds number 15, 369

river-bank effects 259
Rochelle salt 286
rotary drilling 55, 369

S

sacrificial anode 253
sampling tap 205, 214
sand bridging 87, 89, 370
sand content 60, 63
sand trap 214
sanding up 269, 370
screen 74, 370
screw pump 275
sealer 305
self-potential 173, 227
service life 206, 216, 250, 265, 303
Siderocapsa 228
Siderocystis 228
skin effect 134
sodium hexametaphosphate 277
sodium hypochlorite 299
solutizing 291
sorting coefficient 82, 370
specific capacity 72, 79, 90, 99, 182, 186,
210, 249, 370
specific drawdown 116, 370, 371
Sphaerotilus 228
SPM 256
Spot 27, 371
square kelly 55, 371
stainless steel 218, 252, 254, 281, 305
steady state 106
sterilization 205, 208, 215, 281, 291
Stoke's law 236, 371
storage coefficient 19, 371
submerged zone, 218
submersible pump 305
sulphates 48
SURFER® 319
surge effect 297
surging 89, 270
swelling marls 66, 372
swivel 274

T

telluric currents 302
tensile stress 73
thinner 68
thixotropy 65, 372
torque 168
total alkalimetric titre 46
total capacity 210, 211, 215, 244, 372
Toxothrix trichogenes 227
transducer 206
transfer time 197
transmissivity 18, 20, 103, 105, 119, 129,
130, 132, 138, 191, 288, 372
treatment protocol 293, 297, 301

trisodium orthophosphate 276
turbidity 45, 372
turnover
    rate 49, 372
    time 22, 372

U

UID 272
ultrasound treatment 305
ultraviolet irradiation 305
uniform corrosion 218, 223
USA 189
USGS 318

V

Villeneuve-la-Garenne 298
virtual communities 328
viscosifier 68, 71
viscosity 63
voltmeter 205

W

Walton method 142, 373
water balance 5, 373
water chemistry 42, 373

water cycle 4, 5, 373
water hammer 236
water injection 249, 274
water meter 205, 207, 208, 213
water quality 46, 50, 206, 215, 260, 373
water-retention agent 68
water-tight cementation 302
well development 87, 373
well field 112, 155, 194, 199, 204, 242, 260, 330, 373
well head 87
wetting agent 275, 282, 301
WHO 50

Y

yield-depression curve 103, 111, 112, 115, 119, 268, 373

Z

zone of infiltration 43, 373

Masson Éditeur
120, boulevard Saint-Germain
75280 Paris Cedex 06
Dépôt légal : juin 1997

Achevé d'imprimer sur les presses de la
SNEL S.A.
Rue Saint-Vincent 12 – B-4020 Liège
tél. 32(0)4 343 76 91 - fax 32(0)4 343 77 50
mai 1997